Facing The Future

Global Issues in the 21st Century

Ron Chasmer /
Pamela Perry-Globa

Toronto
Oxford University Press

Oxford University Press Canada
70 Wynford Drive Don Mills Ontario M3C 1J9

Oxford   New York
Athens Auckland Bangkok Bogotá Buenos Aires
Calcutta   Cape Town   Chennai   Dar es Salaam
Delhi Florence Hong Kong Istanbul Karachi
Kuala Lumpur   Madrid   Melbourne   Mexico City
Mumbai Nairobi Paris São Paulo Singapore
Taipei   Tokyo   Toronto   Warsaw

and associated companies in
Berlin Ibadan

Oxford is a trade mark of Oxford University Press.

Canadian Cataloguing in Publication Data

Chasmer, Ron
    Facing the future: global issues in the 21st century

Includes bibliographical references and index.
ISBN: 0-19-541136-6

1. Human geography. I. Perry-Globa, Pamela. II. Title.

GF41.C42 1998   304.2   C96-932372-7

Editor: Elaine Aboud
Designer: Brett Miller
Illustrations: Rob McPhail; VISUTronX
Cover photo: Gary John Norman/Tony Stone Images

Acknowledgements
The publisher and authors wish to thank the following people for their contribution in reviewing the
manuscript:

Dave Knox
Head, Department of Geography
Collingwood Collegiate
Collingwood, Ontario

Fraser Scott
Head, Department of Geography
Langstaff Secondary School
Richmond Hill, Ontario

Jim Cottrell
Head, Department of Social Studies
Crescent Heights High School
Calgary, Alberta

The authors wish to thank the following people for
their advice, support, and encouragement:
Pam Chasmer
Jim Tilsey

Mr. and Mrs. D. R. Perry
Don P. Globa

Printed and bound in Canada

This book is printed on permanent (acid-free) paper ∞.

3 4 5 – 01 00 99

Contents

CONTENTS

INTRODUCTION

Each day, millions of people engage in a continuous hand-to-mouth struggle to find sufficient food to survive. Millions of other people—those with safety nets of abundant food and adequate shelter and health care—have the privilege of pursuing a variety of leisure activities at the end of the workday. Some regions of our planet are home to ever-increasing human populations, straining the abilities of the natural environment and human resourcefulness to provide even the most basic needs. Other regions are sparsely inhabited and rich with natural resources, which are exploited to accommodate lives of plenty. In developing nations, the adoption of sustainable practices will improve each country's ability to support its population and manage its own political affairs. In developed nations, where overuse and overconsumption are depleting the earth's resources, the approach to resource use has to be altered.

All life on our planet is interconnected on a daily basis. People are bound to each other as members of humankind, providers for our families, traders and consumers of goods and services, dwellers in rural and urban areas of our countries, and citizens of the globe. Being a global citizen means deliberating a range of issues, and participating in vital decisions. It involves our creative and critical thinking skills, as well as our personal value systems, as we work towards immediate and long-term solutions. Global citizens recognize humankind's interdependence and shared experience. As a result, they are committed to preserving the planet and enhancing the quality of life for all citizens. Students of global citizenship need to envision the world as they would like it to be. This vision will provide direction as they learn about world issues and take part in decisions that will help to shape the earth's future.

Facing the Future looks at the issues—cultural, resource, economic, environmental, or political in basis—that confront us as we approach the millennium. While some of these issues may be pertinent to specific regions, they all have global ramifications for the future. By examining these issues, whether they be industrialization, sexual equality, civil war, etc., and working on the various activities in this textbook, you will develop your analytical abilities and apply your knowledge and skills in making informed decisions about world issues.

PART 1

GEOGRAPHIC APPROACHES TO WORLD ISSUES

CHAPTER 1

STUDYING ISSUES

The *Oxford English Dictionary* defines a problem as "a doubtful or difficult matter requiring a solution." Many of the simple problems we experience as children have a limited number of solutions, and so they are quite easy to solve. Later in life our problems become more complex and more difficult to solve. Nevertheless, we strive to find solutions.

Figure 1.1 Headlines in current newspapers illustrate a wide range of world issues.

Similarly, the world is faced with a variety of complex problems such as hunger, disease, and overpopulation. Just as individuals may solve their problems in a variety of ways, the world's diverse peoples and cultures also find solutions that are best suited to them. The discussion of these problems and of possible solutions is the basis of this course. By learning how to solve global problems, you can become a better decision maker.

Read any newspaper, or watch a television news broadcast, or just listen to people talking. It is likely that the topics they discuss relate to issues. An **issue** refers to a topic that has engendered considerable disagreement among people and has reached the stage of open debate. Most people have opinions about important issues and are eager to share them. Sometimes the variety of differing viewpoints on a single issue can be confusing. But if we dialogue openly, it helps us to understand the numerous perspectives on an issue and to realize that a single issue may be viewed differently by various people.

GLOBAL ISSUES VERSUS LOCAL ISSUES

Much of what we learn through the media seems to be a series of unconnected events that have little to do with one another. But is this perception really true? Sometimes it is necessary to look deeper into these events to discover underlying trends. For example, there may be a series of incidents involving students at local high schools. In one case, a boy is beaten by two other students in a plaza adjacent to the school. The following week a drive-by shooting wounds a student outside a suburban school. Later that same month, a teenage girl has to fight off a former boyfriend in the school parking lot following a dance. Each is an isolated incident in one particular city. Yet together they represent an important national issue: to what extent is violence increasing in Canadian schools?

This course deals with world issues. You might think that violence in schools is not a global issue—or is it? Is it possible that three isolated incidents in a Canadian city could be representative of a growing trend towards violence among young people around the world? If so, this is an important global issue. We need to analyse and discuss it in order to determine differing points of view and suggest possible solutions.

How do we determine if violence among young people is a growing trend? Further investigation may reveal other incidents of violence. Young skinheads in Europe attack people from diverse cultures. Two ten-year-old boys murder a toddler in Britain. Student riots break out in Thailand. The number of incidents around the world suggests that violence among young people is indeed a global problem. We must look beyond the surface of local or national events to discover if they have any global implications.

CONSOLIDATING AND EXTENDING IDEAS

1 Read the headlines in Figure 1.1 and suggest how each one may indicate a possible global issue.

2 Survey a large daily newspaper. Clip out a variety of articles and discuss with the class how each article may relate to a possible global trend.

KINDS OF ISSUES

Five types of issue confront us today: cultural, resource, economic, environmental, and political. **Cultural issues** relate to people and their values, attitudes, and institutions. Violence in Canadian schools, for example, is a cultural issue. **Resource issues** centre on the earth's natural resources. The depletion of the cod fishery off the coast of Newfoundland is a resource issue. **Economic issues** involve money and finance. Canada's national deficit is an important economic issue. **Environmental issues** are those problems that affect the earth's environment. Climatic change as a result of global warming is an environmental issue. **Political issues** involve the power structures within and between governments. Some Quebeckers' desire for sovereignty is a political issue.

It is essential to analyse an issue from a variety of viewpoints in order to understand the attitudes of the **stakeholders**—that is, those people directly involved. Consider the controversy over the logging of old-growth forests near Clayoquot Sound on the west coast of Vancouver Island in the early 1990s. Although this was a local issue, it could be interpreted as part of a much larger global issue—the destruction of the world's rainforests. The forest company viewed these old-growth trees as a valuable economic resource to be harvested. Environmentalists considered the logging of these forests as the destruction of an irreplaceable wilderness region. Economists argued that the clear-cutting of these forests would create jobs in an economically depressed region, provide profits for the forest company during a time of recession, and raise tax revenues for the provincial government. Native peoples were concerned that their ancestral lands were being destroyed, and along with them a vital part of their Native culture. Ultimately, politics became the primary focus of the issue as all sides lobbied the provincial government to support their points of view.

Figure 1.2 Protesters at Clayoquot Sound

CONSOLIDATING AND EXTENDING IDEAS

a) Working with a group, choose one issue from the following list: global warming, the North American Free Trade Agreement (NAFTA), species extinction, the rights of cultural groups, the conflict in Bosnia, disarmament, or any other current issue.
b) Determine whether the primary focus of the issue is cultural, resource, economic, environmental, or political. Explain your decision.
c) Using a copy of Figure 1.3, indicate the connection between the primary focus and the various stakeholders who have diverse perspectives on the issue.
d) Present your findings to the class. Defend your choice of primary focus.

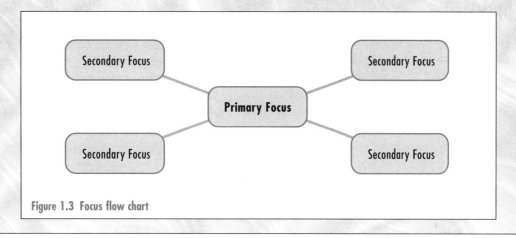

Figure 1.3 Focus flow chart

BIASES

The more we analyse media, the more we realize that all viewpoints and perspectives are biased. As an audience, it is our job to recognize these biases. Only then can we understand and evaluate an issue. In investigating any issue we come across a variety of information. In evaluating this information, we must first analyse it to determine how subjective or objective it is— whether it is reporting facts or expressing the writer's personal bias.

Factual articles are those with substantiated evidence. They provide us with objective reporting, which allows us to form our own conclusions. Editorials are based solely on the writer's opinion. As such, they are subjective, and so they have their own bias. When analysing information to determine its level of objectivity or subjectivity, we should follow these steps:

1 Read through the article once to get a general idea of the writer's point of view.
2 Reread the article and highlight all the *facts* in one colour.
3 Read the article a third time and highlight all the *opinions* in another colour.
4 Count up the number of facts and express this as a ratio to the number of opinions. This will give you a measure of how objective or subjective an article is. The greater the ratio of facts to opinions, the more objective an article is. If the opinions far outweigh the facts, the article is extremely biased.

When evaluating information, it is also useful to know the credentials of the writer. The work of noted authorities often carries a great deal of credibility, even if they express personal opinions. In addition, not all so-called experts in the popular media are true authorities on the issues they present to their audiences. We should consider our sources carefully before accepting their credibility and, therefore, the validity of their information.

PARADIGMS

We all have biases, even though we may not be aware of them. Where do our biases originate? In the late 1980s, noted American psychologist Joel Barker popularized the concept of **paradigms**. These are the rules and conditions we use to understand those things we perceive. According to Barker's theory, everything that happens to us is processed by our brain and related to our own life experiences. For example, imagine you are speeding down a winding country road. As you pass an approaching motorist, she yells "pig!" at you. What do you think she means? Your immediate reaction is probably that she is insulting you by calling you a pig. Once you round the next bend, you almost hit a pig standing in the middle of road. You suddenly realize that the motorist was warning you, not insulting you! Seeing a pig in the middle of the road is an unfamiliar situation. So you originally dismissed the warning as an insult because it is more familiar within your personal experience. (Joel Barker, *The Business of Paradigms* [Burnsville, MN: Chart House Learning, 1989], videocassette.)

Different people may view a situation in different ways, depending on their paradigms. For example, a hunter sees a forest as a good place to hunt game. A forester assesses a forest for its resource potential. A farmer considers the forest as an obstacle that must be cleared before the land can be used for field crops. A city dweller may appreciate the forest for its natural beauty. Thus each person has a different paradigm about forests.

Sometimes it is impossible for us to see other people's points of view because we do not understand their perspectives and experiences. A situation may become controversial as a result of people's different paradigms. Several years ago, hunting harp seal pups was banned on the ice floes off the coast of Labrador. Early each spring seal hunters headed onto the ice as the harp seals gave birth to their pups. The baby seal pups were clubbed to death and then skinned for their soft white pelts. The hunters considered this a *harvest*. Animal rights groups viewed it as a *slaughter*. Protesters clashed in demonstrations with seal hunters. Clearly, each side had a different paradigm about the seal hunt, and

Figure 1.4 The "pig paradigm"

neither side could understand the other's point of view. In response to international pressure, the federal government banned the hunt. Yet understanding the views of others is essential if we are to solve world issues. We must consider all points of view objectively and without emotion. Only then can we reach sound, workable solutions.

Often, society at large will change its paradigm on a certain issue in a relatively short period of time. In North America, before the Second World War, women were expected to get married, stay at home, and raise children. Today, the paradigm has changed. Women are no longer expected to remain at home. Instead, they are encouraged to seek a lifestyle of their own choosing, whether it be within the home or in the business world.

Figure 1.5 A seal with her pup

CONSOLIDATING AND EXTENDING IDEAS

1 Find several articles and/or videotape documentaries about a particular issue. Analyse each item to determine its objectivity—that is, its ratio of fact to opinion—using the method described on page 5.

2 In your own words, define the term *paradigm*.

3 Write down what each of the following words means to you: *girls, boys, freedom, race*.

Share your ideas with a group of students and determine the paradigm that each member has about each word.

4 Find out what paradigms once existed in Canada for each of the following situations, then explain how these paradigms have shifted today:
a) attitudes towards the Soviet Union/Russia;
b) attitudes towards recycling;
c) attitudes towards smoking.

CHAPTER 2

STATISTICAL MODELS AND GEOGRAPHY

Statistics are to the geographer what experiments are to the scientist. They allow information to be measured, or **quantified.** Only then can true relationships be established. The Appendices on pages 433 to 457 provide statistics that measure population, culture, economics, resources, and technology for major countries of the world. Studying these statistics reveals present conditions and future trends and helps us to formulate solutions to global issues.

GRADED SHADING MAPS

When statistics are mapped they reveal global patterns. The **graded shading map** is an important statistical tool. It is easy to create and simple to interpret. Different shades of one colour are used to represent each statistical range. The higher the range, the darker the colour. Figure 2.1 is a graded shading map showing the average annual rate of population growth. It reveals at a glance the regions with rapid population growth and those with slow growth.

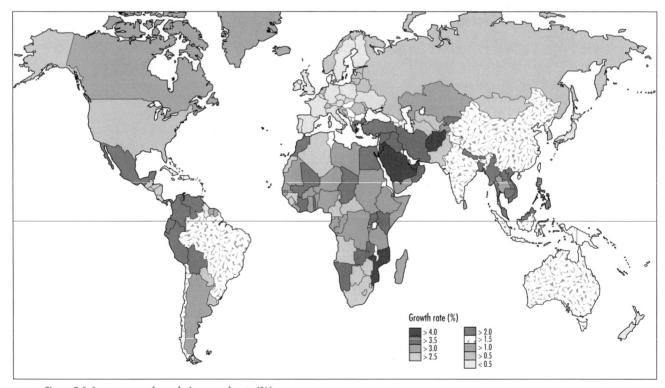

Growth rate (%)

> 4.0	> 2.0
> 3.5	> 1.5
> 3.0	> 1.0
> 2.5	> 0.5
	< 0.5

Figure 2.1 Average annual population growth rate (%)
Which regions are growing at less than 1.5 per cent a year? 0.5 per cent?
Source: Copyright 1992 Broderbund Software, Inc.

STATLAB

Working with a group, choose a category of statistic from Appendix A, B, or C and prepare a graded shading map for it. (**Note**: Creating graded shading maps involves a lot of work if you are preparing them manually. There are computer software programs available that can help you create these maps quickly and easily.)

DETERMINING THE RANGE

1 Find the largest value and the lowest value in your set of statistics. For example, if you are studying birth rates, the largest value is 52 (for Uganda) and the lowest value is 10 (for Greece).

2 Divide the difference between these values by the number of shades you plan to use. This should be no less than 4 and no more than 6. In this case we will use 5. So 55 - 10 = 45, divided by the number of shades (45 ÷ 5). The ranges would therefore be 10–19, 20–29, 30–39, 40–49, and 50–59.

3 Confirm that the range is suitable by checking the approximate number of countries for each shade. If, for example, 75 per cent of the countries have a birth rate under 30, the map would not show patterns well because most of the countries would be the same shade. Take a sample of every fifth country. List it in the appropriate category, then add the totals for each to determine if they are evenly distributed.

10–19	20–29	30–39	40–49	50–59
Cuba	Algeria	Gabon	Burundi	Côte d'Ivoire
Austria	Indonesia	Bangladesh	Kenya	Uganda
Finland	Lebanon	Honduras	Sudan	
Hungary	Morocco	Turkmenistan	Madagascar	
Norway	Sri Lanka	Jordan	Rwanda	
Sweden	Trinidad and Tobago	Vietnam	Oman	
Latvia	Chile			
New Zealand	Peru			
	Azerbaijan			
8	9	5	6	2

Figure 2.2 (a) An uneven distribution pattern

Source: Statistics from *World Resources,* 1996–97 (New York: Oxford University Press, 1996).

4 If the distribution is uneven, as in Figure 2.2 (a), revise the range. It is not always necessary to have the same numeric value for each range. In the example, the sample would be more evenly distributed if the values were changed to 10–15, 16–25, 26–32, 33–42, and >42.

10–15	16–25	26–32	33–42	>42
Austria	Cuba	Morocco	Sudan	Burundi
Finland	Sri Lanka	Lebanon	Bangladesh	Côte d'Ivoire
Hungary	Trinidad and Tobago	Vietnam	Jordan	Kenya
Norway	Chile	Peru	Gabon	Madagascar
Sweden	New Zealand	Azerbaijan	Honduras	Oman
Latvia	Indonesia	Algeria		Rwanda
		Turkmenistan		Uganda
6	6	7	5	7

Figure 2.2 (b) A more evenly distributed pattern
Source: Statistics from *World Resources, 1996–97.*

ORGANIZING THE DATA AND CREATING THE MAP

5 Organize your data in an organizer or a chart, such as the one shown in Figure 2.2 (b). List each of the countries under the appropriate category.

6 Choose several coloured pencils that are different shades of one colour. Use the darker shades to represent the higher values and the lighter shades to represent the lower values. Prepare a colour key to show the ranges. Countries for which there are no data should be shaded a neutral colour, such as grey.

7 On an outline map of the world, colour each country the appropriate shade and add a title. (Very small countries such as Vanuatu may not be large enough to show on the map.) The finished product should look similar to the map in Figure 2.1.

CONSOLIDATING AND EXTENDING IDEAS

1 Answer the following questions to analyse the patterns revealed by the map you created in the Statlab.
   a) Which geographic regions had high values?
   b) Which geographic regions had low values?
   c) Were there any **anomalies**—that is, countries that did not follow the pattern of their neighbours?
   d) Describe the general patterns your map reveals.
   e) How could your map be used in the study of world issues?

2 Based on your answers in activity 1, write a brief analysis about the statistics you mapped. Include it with your map to complete this statistical project.

CORRELATIONAL ANALYSIS

Often one set of statistics is dependent on another. For example, student marks increase as school attendance increases. Therefore, students who want to do well in school should attend classes regularly. The study of these **correlations** is instrumental in the decision-making process, whether it relates to personal or global issues.

When both sets of statistics in a sample increase, there is a **positive correlation** between the two sets of data. This means that as one set of statistics rises, so does the other set. Cigarette smoking and lung cancer could be an example of this type of correlation. As the number of cigarettes smoked increases, the incidence of lung cancer increases. **Negative correlations** occur when one set of data increases while the other set decreases. For example, the more cigarettes a person smokes, the shorter that person's life expectancy is likely to be. In this case, as one set of statistics increases, the other set decreases.

The absence of any correlation is also a powerful tool in analysing an issue. It can prove that there is no relationship between two sets of data and can be just as effective in a debate as either a positive or negative correlation. Suppose a study were conducted to compare intelligence and sex. If there were a correlation to show that one sex had more intelligence than the other, chauvinists would have an argument to support their claims of superiority. The fact that there is no correlation between intelligence and sex proves that no one sex has more intelligence than the other.

Once a correlation has been established, it is necessary to analyse the pattern. For example, there is a negative correlation between birth rates and literacy rates. As literacy rates rise, birth rates drop. One explanation might be that literate people are able to read about family planning. They are then able to make informed decisions about family size. People who cannot read, on the other hand, cannot benefit from family-planning literature. Instead of making an informed choice, they may follow traditional practices that produce large families. Of course, there may be other factors that influence both sets of data. In this case, government-sponsored pension funds may reduce the need for parents to rely on their offspring to support them in their later years. This could contribute to lower birth rates regardless of literacy rates. Therefore, we must be careful not to make simple assumptions about complex issues.

SHOWING CORRELATIONS

There are three methods of illustrating correlations between two sets of data: graphically, using a **scattergraph**; spatially, by comparing two graded shading maps; and mathematically, by calculating a **linear correlation coefficient**. Each method has its advantages and limitations.

THE SCATTERGRAPH

Scattergraphs show relationships very well and are easy to construct. Start by drawing a simple line graph with an x and a y axis. Using an appropriate scale, plot the location where the two sets of data, or **ordered pairs**, intersect on the grid. Once all ordered pairs have been plotted, draw a faint line connecting the points that mark the outer extremities of the data. (Outlying points far from the general pattern are anomalies that should be ignored at this point.) Using a ruler, find the long axis of the shape you drew on the graph. Draw a straight line through the middle of the space defined by the points. There should be an equal number of points above and below this **line of best fit**. If the line extends from the lower left to the upper right, there is a positive correlation. If the line extends from the upper left to the lower right, there is a negative correlation. The closer the points are to the line of best fit, the stronger the correlation. If the line you have drawn is more or less circular—in other words, if there is no long axis—there is no correlation. Figure 2.3 shows different types of scattergraphs.

Scattergraphs provide a visual image of a correlation, but they do not measure the degree of correlation, nor do they show spatial relationships. Other methods of analysis, described later in this chapter, expand on the visual impression of the scattergraph.

SAMPLE POPULATIONS

W	X	Y	Z
$w_1 = 8$	$x_1 = 7$	$y_1 = 3$	$z_1 = 20$
$w_2 = 2$	$x_2 = 9$	$y_2 = 4$	$z_2 = 15$
$w_3 = 12$	$x_3 = 11$	$y_3 = 5$	$z_3 = 7$
$w_4 = 0$	$x_4 = 15$	$y_4 = 6$	$z_4 = 3$

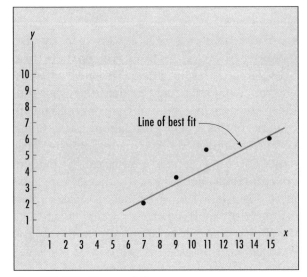

Figure 2.3 (a) Positive correlation

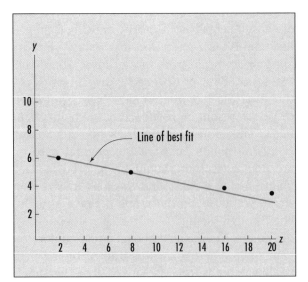

Figure 2.3 (b) Negative correlation

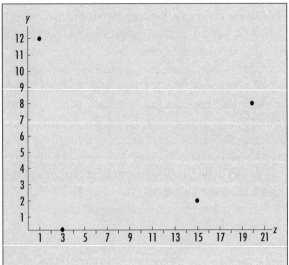

Figure 2.3 (c) No correlation
There is no line of best fit because the points are so scattered.

STATLAB

1 Study the statistics in Figure 2.4.

	BENIN	EGYPT	SOMALIA	BHUTAN	THAILAND	CUBA	CHILE	BULGARIA	SPAIN	FIJI
Life expectancy (years)	48	64	47	51	69	75	74	71	78	72
Infant mortality (000)	86	67	122	124	37	12	16	14	7	23

Figure 2.4 Life expectancy and infant mortality, selected countries (1990–95)
Source: Statistics from *World Resources, 1996–97*.

a) Prepare a scattergraph showing the relationship between life expectancy and the infant mortality rate. (The infant mortality rate is the number of infants per 1000 births who die before their first birthday.)
b) Draw the line of best fit. Determine if there is a negative or a positive correlation between the two sets of data, then write an explanation that could account for this.
c) What other factors might influence these two sets of statistics?

2 Study the statistics in Figure 2.5.

	BENIN	EGYPT	SOMALIA	BHUTAN	THAILAND	CUBA	CHILE	BULGARIA	SPAIN	FIJI
Life expectancy (years)	46	62	47	49	69	57	72	72	78	72
Gross domestic product/capita	$398	$611	n/a	$160	$1774	$399	$2538	$1414	$13 510	$2015

n/a: not available

Figure 2.5 Life expectancy and gross domestic product, selected countries (1993)
Source: Statistics from *World Resources, 1996–97*.

a) Prepare a scattergraph showing the relationship between life expectancy and gross domestic product (GDP) per capita. (GDP is the total value of goods and services per person.)

b) Draw the line of best fit. Determine if there is a negative or a positive correlation between the two sets of data, then write an explanation that could account for this.

c) What other factors might influence these two sets of statistics?

d) What anomalies did you notice?

3   Study the statistics in Figure 2.6.

	BENIN	EGYPT	SOMALIA	BHUTAN	THAILAND	CUBA	CHILE	BULGARIA	SPAIN	FIJI
Life expectancy (years)	48	64	47	51	69	75	74	71	78	72
Total land area (000 km²)	111	995	627	47	511	28	749	111	499	18

Figure 2.6 Life expectancy and land area, selected countries (1990–95)
Source: Statistics from *World Resources, 1996–97.*

a) Prepare a scattergraph showing the relationship between life expectancy and total land area.

b) Draw the line of best fit. Determine if there is a negative or a positive correlation between the two sets of data, then write an explanation that could account for this.

GRADED SHADING MAPS

When you created your graded shading map (page 10), you may have noticed that other students' maps looked similar to yours even though they displayed different statistics. Other maps looked opposite. Where your map showed dark shades, they showed light ones. Comparing statistics in graded shading maps is an important method of analysis. Maps that appear to be similar have a positive correlation between the two sets of data. Maps that appear to be opposite have a negative correlation. Where no distinct patterns appear to exist, the correlation is either weak or non-existent.

If you study Figure 2.7 (Infant mortality rates) and Figure 2.8 (Crude birth rates), you will notice that the two maps look similar. Africa,

southern Asia, and northern South America are shaded dark, while North America, western Europe, Russia, and China are shaded light. What type of correlation is indicated between these two sets of data? How could you explain the relationship between birth rates and infant mortality? What other factors could influence this data?

Figure 2.7 (Infant mortality rates) and Figure 2.9 (Female life expectancy) show a negative correlation. Notice that regions where infant mortality is high are also regions where people die at a relatively young age. What explanations could account for this pattern?

The *amount* of correlation cannot be measured on a graded shading map. However, this technique does show global patterns clearly. In addition, anomalies are easily identified.

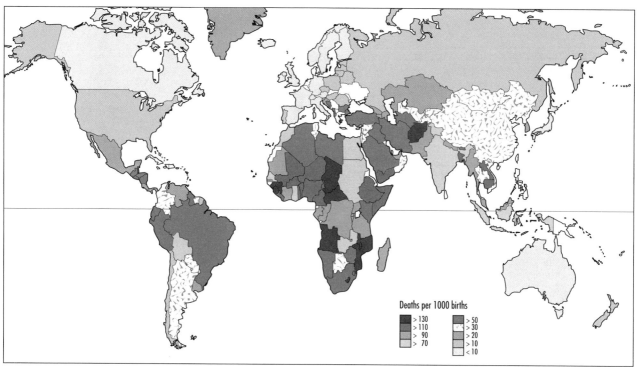

Figure 2.7 Infant mortality rates
Source: Copyright 1992 Broderbund Software, Inc.

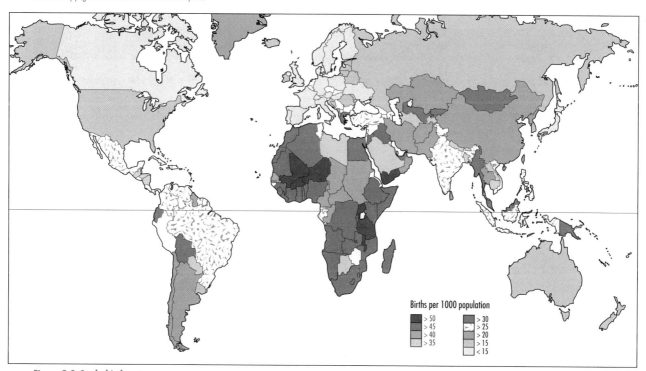

Figure 2.8 Crude birth rates
Source: Copyright 1992 Broderbund Software, Inc.

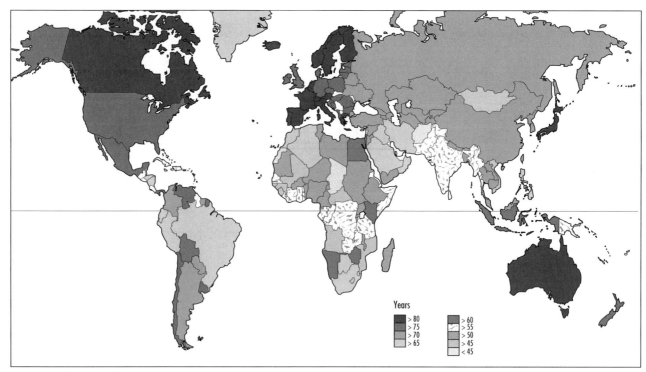

Figure 2.9 Female life expectancy
Source: Copyright 1992 Broderbund Software, Inc.

STATLAB

1 Pair up with another student and compare the maps you completed in the assignment on page 10.
   a) Determine the type of correlation that exists between the two sets of data.
   b) Describe the pattern in the two maps.
   c) Account for the similarities between the two sets of data.
   d) List any anomalies and give possible reasons for them.
   e) How could one set of data be manipulated to change another set of data?

2 Randomly select twenty to thirty countries from the tables in Appendix A, B, or C. Draw a scattergraph to confirm your conclusions in activity 1.

3 Display the maps and graphs, along with your written work, on a piece of Bristol board. Include an appropriate title.

LINEAR CORRELATION COEFFICIENT

Neither the scattergraph nor the graded shading map shows the degree of correlation between two sets of data. The linear correlation coefficient does. While the calculation appears complicated, it is actually quite simple. Try calculating the linear correlation coefficient a few times following the steps below to help you understand how it works. Once you have grasped the idea, a programmable calculator, a simple computer program, or a computer spreadsheet can do the "number crunching" for you. Use organizers or charts such as the ones in Figure 2.10 (a–d) to help you to set up the data.

Step 1: List samples of the population for the first set of data in the first column—that is, X = {7, 9, 11, 15} and Y = {3, 4, 5, 6}. Add up the numbers for each sample, then calculate the **mean,** or average, for each of X and Y.

Figure 2.10 (a) Linear correlation coefficient for X and Y						
X	**Y**	$x = X - \bar{X}$	$y = Y - \bar{Y}$	x^2	xy	y^2
7	**3**					
9	**4**					
11	**5**					
15	**6**					
$\Sigma X = 42$	$\Sigma Y = 18$					
$\bar{X} = 42 \neg 4 = 10.5$	$\bar{Y} = 18 \neg 4 = 4.5$					

Step 2: Calculate the **variance,** or difference, between each sample of the population and the mean. You can check your calculations by adding them up. If the sum is 0, the calculations are correct.

Figure 2.10 (b)						
X	Y	$x = X - \bar{X}$	$y = Y - \bar{Y}$	x^2	xy	y^2
7	3	7 - 10.5 = -3.5	3 - 4.5 = -1.5			
9	4	9 - 10.5 = -1.5	4 - 4.5 = -0.5			
11	5	11 - 10.5 = +0.5	5 - 4.5 = +0.5			
15	6	15 - 10.5 = +4.5	6 - 4.5 = +1.5			
$\Sigma X = 42$	$\Sigma Y = 18$					
$\bar{X} = 42 \neg 4 = 10.5$	$\bar{Y} = 18 \neg 4 = 4.5$					

Step 3: Find the **deviation** between each sample of the population and the mean by squaring the variance. This eliminates negative signs. The **total deviation** is the sum of the deviations for each member of the population—that is, $\Sigma X = 35.0$.

Figure 2.10 (c)						
X	Y	$x = X - \bar{X}$	$y = Y - \bar{Y}$	x^2	xy	y^2
7	3	7 - 10.5 = -3.5	3 - 4.5 = -1.5	$-3.5^2 = 12.25$		$-1.5^2 = 2.25$
9	4	9 - 10.5 = -1.5	4 - 4.5 = -0.5	$-1.5^2 = 2.25$		$-0.5^2 = 0.25$
11	5	11 - 10.5 = +0.5	5 - 4.5 = +0.5	$+0.5^2 = 0.25$		$+0.5^2 = 0.25$
15	6	15 - 10.5 = +4.5	6 - 4.5 = +1.5	$+4.5^2 = 20.25$		$+1.5^2 = 2.25$
$\Sigma X = 42$	$\Sigma Y = 18$			$\Sigma x^2 = 35.0$		$\Sigma y^2 = 5.0$
$\bar{X} = 42 \neg 4 = 10.5$	$\bar{Y} = 18 \neg 4 = 4.5$					

Step 4: Calculate the slope of xy. This vector is actually the line of best fit. Simply multiply the two variances of each ordered pair together—that is, -3.5 x - 1.5 = +5.25. Calculate the sum of each product. It is important to include either the plus (+) or minus (-) sign because it shows the correlation as either positive or negative.

Figure 2.10 (d)

X	Y	$x = X - \bar{X}$	$y = Y - \bar{Y}$	x^2	xy	y^2
7	3	7 - 10.5 = -3.5	3 - 4.5 = -1.5	$-3.5^2 = 12.25$	-3.5x - 1.5 = +5.25	$-1.5^2 = 2.25$
9	4	9 - 10.5 = -1.5	4 - 4.5 = -0.5	$-1.5^2 = 2.25$	-1.5 x 0.5 = +0.75	$-0.5^2 = 0.25$
11	5	11 - 10.5 = -0.5	5 - 4.5 = +0.5	$+0.5^2 = 0.25$	+0.5x + 0.5 = +0.25	$+0.5^2 = 0.25$
15	6	15 - 10.5 = +4.5	6 - 4.5 = +1.5	$+4.5^2 = 20.25$	+4.5x + 1.5 = +6.75	$+1.5^2 = 2.25$
$\Sigma X = 42$	$\Sigma Y = 18$			$\Sigma x^2 = 35.0$	$\Sigma xy = +13.0$	$\Sigma y^2 = 5.0$
$\bar{X} = 42 \neg 4 = 10.5$	$\bar{Y} = 18 \neg 4 = 4.5$					

Step 5: Once you have completed Figure 2.10 (a–d), the appropriate values are substituted in the following formula:

$$\frac{\Sigma xy}{\sqrt{(\Sigma x^2)(\Sigma y^2)}} =$$

In this example, $\dfrac{\Sigma xy}{\sqrt{(\Sigma x^2)(\Sigma y^2)}} = \dfrac{+13}{\sqrt{(+35)(+5)}} = \dfrac{+13}{\sqrt{175}} = \dfrac{+13}{13.2} = +.98$

The coefficient is always less than 1 and is expressed as an integer. The closer the value is to 1, the stronger the correlation. In the example, +.98 indicates a strong positive correlation. Any value under ±.50 indicates a weak correlation or no correlation at all.

STATLAB

Refer to Figure 2.3 (Sample populations).

1 For each of the following pairs of data, hypothesize the type of correlation you expect to find and calculate the linear correlation coefficient:
   a) X and Z   b) W and Z   c) Y and Z

2 a) Calculate the linear correlation coefficient for the map and scattergraph assignment you completed on page 16. Does it support the assumptions you made when you completed the graded shading map and the scattergraph?
   b) Present your findings to your group.
   c) How do correlations help the problem-solving process for both personal and global issues?

3 Devise a way to use a spreadsheet program to calculate linear correlation coefficients.

4 Develop a computer program to calculate linear correlation coefficients.

Chapter 3

Diversity and Disparity

As we attempt to confront global issues, we need to reflect on what we share as members of the global community—our global environment and future—and what we enjoy individually and collectively—geography, resources, economic and political systems, lifestyles, opportunities.

Diversity and disparity exist within our modern world. When we consider **diversity**, we recognize differences within and between nations of the world, diversity that may be geographic, cultural, political, and/or economic. Geographic diversity reflects differences between countries (Ecuador, Columbia, and Peru, for example) and between regions (Asia, Europe, North America) with respect to landforms, soils, climate, natural vegetation, water and mineral resources, and energy sources. Cultural diversity emphasizes the seemingly endless variety in human behaviour—language, rituals, ethnic derivations, **indigenous peoples**, religions, customs. Human behaviour is as diverse within nations as it is between them. Political systems around the globe range from one-party Communist states to federal parliamentary monarchies, from military Marxism to multi-party constitutional democracies. When we also take into consideration political alignments such as defence alliances and international relationships, we realize the extent of political diversity in our world today. Economic diversity is characterized by differences within and between nations. It is associated with economic development, participation in primary, secondary, tertiary and quaternary industries, competition for resources, and involvement in global trade networks.

Disparity also reflects differences within and between nations, but these differences identify inequalities between individuals, groups, areas, regions, alignments, and zones. Statistical indicators often are used to help us understand global disparity because they offer us a numerical means for comparison. Statistical data may describe infant mortality rates, life expectancy (male/female), persons per physician, GNP per capita, **literacy** rates, percentage of labour force in agriculture, and nonrenewable resource consumption. Each type of data helps to reveal the inequalities that exist among us. These disparities, in turn, are linked frequently with diversity because geography and cultural, political, and economic factors have an impact on national statistics.

When you completed the correlational analysis on page 18, you may have seen that not all countries have the same standard of living. Standard of living refers to the quantity and quality of food and other goods and services that people are able to purchase or otherwise attain. In the developed world, many people have a high standard of living. In other words, they have sufficient goods and services to accommodate their needs and wants. The opposite is often true for many people living in developing countries. They are frequently deprived of the goods and services that people in the developed world take for granted. Generally, the countries of Africa, South America, and most of southern Asia have fewer economic opportunities than those of western Europe, North America, and some Pacific Rim countries.

It is important not to confuse the terms

Figure 3.1 *top,* the skyline of Singapore; *middle,* a shopkeeper in Morocco; *bottom,* Chor Bazaar in Bombay, India.

Identify these places on a world map. How do the above photos illustrate the impreciseness of the terms developed *and* developing?

standard of living and **quality of life**. Quality of life is affected by material standards of living, but it also includes environmental, social, and political factors. A clean environment, human rights, political freedom, and freedom of opportunity are only a few examples of non-material elements of life that most people desire.

In broad terms, wealthy countries are referred to as the **developed world** while poorer countries are called the **developing world**. However, it is important to use these terms in their most general sense. Only the truly wealthy and the truly poor countries can be simply categorized. The majority of countries fall somewhere in between. There may even be different levels of development within countries. Cities such as Nairobi in Kenya, São Paulo in Brazil, and Bombay in India are as modern and vigorous as any city in developed countries such as Canada and Japan. Suburban shopping centres, skyscrapers in the central business district, and busy highways make these cities atypical of developing countries. Similarly, there are economically depressed regions within developed nations that may share many of the characteristics of developing countries. In Canada, for example, some communities in the North are economically deprived. In the United States, especially along the border with Mexico, there are rural slums with conditions very similar to those in slums south of the border in Mexico.

The terms *developed* and *developing* also raise another question: When does a country stop being a developing country and become a developed one? Is Brazil a developed country? It has more in common with the United States than it does with a country such as Mali in North Africa. Yet many people would still consider Brazil to be developing. And what about the oil-rich countries of the Middle East? Their incredible wealth puts them in a category apart from even the most developed nations, but they also have many of the demographic, social, and economic characteristics of developing countries. Finally, the term *developing* is inappropriate for those countries that are economically stagnant. The problems of these nations are so immense, it may be unlikely that they will ever become developing nations.

PREDOMINANT ECONOMIC ACTIVITIES

In order to describe the regions of the world more accurately, four classifications have been established based on predominant economic activity: **diversified economies, emerging industrial economies, resource-based economies**, and **subsistence economies**.

Diversified economies are those in which there is a balance among **primary industries**, such as mining, lumbering, and agriculture; **secondary industries,** such as auto-making, paper products, and telecommunications equipment; **tertiary industries,** such as banking, education, and health care; and **quaternary industries**, such as medical research, computer software development, and publishing. Because secondary, tertiary, and quaternary industries provide a greater income per unit of work than primary industries, nations with diversified economies usually have higher living standards, more health and educational facilities, and greater monetary wealth than countries that rely solely on primary industries. Canada, the United States, Japan, Australia, and the countries of western Europe all have diversified economies.

Emerging industrial economies are found in those countries that are in the process of industrializing. These countries once depended on two or three primary industries for most of their income. Today, they have developed a manufacturing sector that makes up the major part of their economic base. Many emerging industrial nations are found in the Pacific Rim. Countries such as South Korea, Thailand, Malaysia, and China are becoming world leaders in the manu-

Figure 3.2 Calgary—a modern city in a diversified economy

Figure 3.3 Bangkok—a bustling city in an emerging industrial economy

Figure 3.4 Al Kuwayt—Kuwait's resource-based economy provides extensive social services to its citizens.

facture of a variety of products, from inexpensive consumer goods to sophisticated electronic equipment.

Resource-based economies are found in those nations that rely on the export of one or two natural resources. The nations of the Middle East have amassed great wealth because of their vast oil resources. Kuwait, Qatar, Dubai, Saudi Arabia, and other countries in the region enjoy a high standard of living. Many social programs are provided at virtually no cost to the people. Yet these nations also share some of the characteristics of subsistence economies. Birth rates are usually high, literacy rates are low, and infant mortality rates are often much closer to those of subsistence economies than diversified economies.

Historically, countries move through growth stages from one level of development to another. Canada once had a subsistence economy. Pioneers struggled to live off the land. In time, Canada developed a resource-based economy. Wheat, timber, furs, and fish were exported to Britain, France, the United States, and other countries. In the twentieth century Canada started to emerge as an industrial nation. Today, we have a diversified economy based not only on primary resources, but also on manufacturing and service industries.

Subsistence economies are found in the world's poorest nations. Generally, these countries have barely enough food and other basic resources to sustain their people. Poverty is often compounded by overpopulation, environmental degradation, and reliance on few natural resources. These countries depend on **aid** from developed countries to help them improve living conditions. Many countries in sub-Saharan Africa and southern Asia have subsistence economies.

ECONOMIC DISPARITY

Because we live in a diversified economy, we might think of it as the norm. But, in fact, it is our lifestyle that is unusual. The number of people in the world who live in wealthy countries such as Canada is extremely small. The vast majority of people live in emerging, resource-based, or subsistence economies.

How is the standard of living in diversified economies different from the other three economic regions? Consider health care as an example. Our excellent health-care system and abundant food supply allow Canadians to live long lives. People in other regions have lower life expectancies as a result of poor health care, inadequate diets, and environmental deterioration. Canadians succumb to diseases of advanced age, such as Alzheimer's disease, heart disease, and cancer. These illnesses are relatively rare in developing countries because the people do not usually live to advanced ages. In subsistence economies, they are more likely to have diseases of malnutrition, such as kwashiorkor and beriberi, or environmental diseases, such as dysentery, malaria, or schistosomiasis. Individuals of all ages may be afflicted with these diseases, but the toll is usually highest among children. Nevertheless, people in subsistence economies are often more physically fit than Canadians because of their rigorous lifestyles. The living standards in emerging industrial economies and resource-based economies are usually somewhere along the continuum between diversified economies and subsistence economies.

Most people living in diversified economies have greater financial wealth than people living in developing countries. Gross domestic product per capita may be ten to twenty times higher than in nations with subsistence economies. However, the cost of living is also much higher. It is far less expensive to live in a developing country than in a country with a diversified economy. The demand is so much lower for goods and services that they are relatively inexpensive in these countries. Conversely, it is more costly to live in an economy where people have a great deal of disposable income. Moreover, people living in Western economies often have higher expectations. For example, bicycles and buses are more common forms of transportation than automobiles in many emerging industrial economies. Yet most Canadian families "need" a car, or perhaps even two or three cars, to get around.

Technology also contributes to make diversified economies different from other economic regions. We tend to rely on high-tech solutions to our problems. Computers, robots, satellites, and other sophisticated tools monitor our environment, build our consumer goods, and process enormous amounts of information. Most of the world uses a more direct approach. Simple tools, often made from local materials, help some people live much as they have for hundreds, even thousands, of years. Fuel consumption is low; renewable resources such as wood and animal dung meet most of the fuel needs. In our society, enormous power plants burn fossil fuels to power our energy-intensive lifestyle. We utilize many synthetic fibres made from nonrenewable oil stocks, while many other societies usually rely on natural fibres grown from crops or livestock.

While the people of industrialized economies have many material things, to say we are rich and people in developing countries are poor is not necessarily true. It all comes down to one's perspective. The article "Development Discovers Poverty" on page 24 provides better insight into what the terms *rich* and *poor* really mean.

> A traveller returning to Bangkok, Thailand, after a ten-year absence was shocked at the economic changes. Until recently, Thailand had a subsistence economy. Today, it is an emerging industrial nation. The traveller noted the changes in the cost of living: "In 1982, I could buy a delicious snack of noodles, wrapped in a newspaper, for about 25 cents. Today, I have to go into a restaurant to eat, and the same snack costs 5 dollars."

DEVELOPMENT DISCOVERS POVERTY

I could have kicked myself afterwards. Yet my remark had seemed the most natural thing on earth at the time. It was six months after Mexico City's catastrophic earthquake in 1985 and I had spent the whole day walking around Tepito, a dilapidated quarter inhabited by ordinary people but threatened by land speculators. I had expected ruins and resignation, decay and squalor, but the visit had made me think again: there was a proud neighbourly spirit, vigorous building activity, and a flourishing shadow economy.

But at the end of the day the remark slipped out: "It's all very well but, when it comes down to it, these people are still terribly poor." Promptly, one of my companions stiffened: *"No somos pobres, somos Tepitanos!"* (We are not poor people, we are Tepitans.) What a reprimand! I had to admit to myself in embarrassment that, quite involuntarily, the clichés of development philosophy had triggered my reaction.

"Poverty" on a global scale was discovered after the Second World War. Whenever "poverty" was mentioned at all in the documents of the 1940s and 1950s, it took the form of a statistical measurement of per capita income whose significance rested on the fact that it lay ridiculously far below the US standard. When size of income is thought to indicate social perfection, as it does in the economic model of society, one is inclined to interpret any other society which does not follow that model as "low income." This way, the perception of poverty on a global scale was nothing more than the result of a comparative statistical operation ... "Poverty" was used to define whole peoples, not according to what they are and want to be, but according to what they lack and are expected to become. Economic disdain had thus taken the place of colonial contempt.

Moreover, this conceptual operation provided a justification for intervention: wherever low income is the problem the only answer can be "economic development." There was no mention of the idea that poverty might also result from oppression and thus demand liberation. Or that a culture of sufficiency might be essential for long-term survival. Or even less that a culture might direct its energies towards spheres other than economic.

Towards the end of the 1960s, when it was no longer possible to close one's eyes to the fact that "economic development" was patently failing to help most people achieve a higher standard of living, a new conception of "poverty" was required. Whoever lived below an externally defined minimum standard was declared "absolutely poor." ...

The stereotyped talk of "poverty" has disfigured the different, indeed contrasting, forms of poverty beyond recognition. It fails to distinguish, for example, between frugality, destitution, and scarcity.

Frugality is a mark of cultures free from the frenzy of accumulation. In these, the necessities of everyday life are mostly won from subsistence production with only the smaller part being purchased on the market. To our eyes, people have rather meagre possessions: maybe the hut and some pots and the Sunday costume, with money playing only a marginal role. Instead, everyone usually has access to fields, rivers, and woods while kinships and community duties guarantee services which elsewhere must be paid for in hard cash. Nobody goes hungry. In a traditional Mexican village, for example, the private accumulation of wealth results in social ostracism—prestige is gained precisely by spending even small profits on good deeds for the community. Such a lifestyle only turns into demeaning "poverty" when pressurized by an accumulating society.

Destitution, on the other hand, becomes rampant as soon as frugality is deprived of its foundation—community ties, land, forest, and water. *Scarcity* derives from modernized poverty. It affects mostly urban groups caught up in the money economy as workers and consumers whose spending power is so low that they fall by the wayside. Their capacity to achieve through their own efforts gradually fades, while at the same time their desires, fuelled by glimpses of high society, spiral towards infinity...

Up until the present day, development politicians have viewed "poverty" as the problem and "growth" as the solution. They have not yet admitted that they have been largely working with a concept of poverty fashioned by the experience of commodity-based need in the Northern Hemisphere... They have encouraged growth, and often produced destitution, by bringing multifarious cultures of frugality to ruin. For the culture of growth can only be erected on the ruins of frugality; and so destitution and dependence on commodities are its price.

The core of the development idea was that the essential reality of a society consists in nothing else than its functional achieve-ment: the rest is just folklore or private affairs. From this viewpoint the economy overshadows every other reality: the laws of economy dominate society and not the rules of society the economy. This is why, whenever development strategists set their sights on a country, they do not see a society that *has* an economy, but a society that *is* an economy.

In societies that are not built on the compulsion to amass material wealth, economic activity is also not geared to slick zippy output. Rather, economic activities like choosing an occupation, cultivating the land, or exchanging goods are understood as ways of enacting that particular social drama in which the members of the community see themselves as the actors. The economy is closely bound with life but it does not stamp its rule and rhythms on the rest of society. Only in the West does the economy dictate the drama and everyone's role in it.

It seems my friend from Tepito knew of this when he refused to be labelled as "poor." His honour was at stake, his pride too; he clung to his Tepito form of sufficiency, perhaps sensing that without it there loomed only destitution or never-ending scarcity of money.

Wolfgang Sachs. From *Edges*, October 1992 (Essen, Germany: The Institute of Cultural Studies), 34–36.

CONSOLIDATING AND EXTENDING IDEAS

1 Write a detailed paragraph about the limitations of the terms *developed* and *developing* in describing the countries of the world.

2 Using the Appendices on pages 433 to 457, prepare an organizer, a chart, or a database comparing fifteen nations from five continents. Show at least eight sets of statistics from the following categories: Population and Culture; Economic and Resource Factors; Technological Development.
   a) List the countries with similar characteristics in four separate groups. Label each group as a diversified economy, an emerging industrial economy, a resource-based economy, or a subsistence economy.
   b) Describe each set of statistics for the four groups. Make a general statement about the statistics for each group of countries.
   c) On an outline map of the world, indicate the fifteen countries by economic category. Use a different colour for each type of economy and include a legend.

3 Define these terms in your notebook: *diversified economy, emerging industrial economy, resource-based economy,* and *subsistence economy.* Write a paragraph explaining how these terms might lead to misunderstandings about economic disparity within countries.

4 Using the Appendices, prepare an organizer or a chart comparing countries with diversified economies, such as Germany, Canada, and the US, with developing countries, such as Mali, India, Afghanistan, and Albania. Include such criteria as health, wealth, lifestyle, technology, population, and location.

5 Read the article "Development Discovers Poverty" on page 24. In a written summary of the article, answer the following:
   a) What is Sachs's thesis? What evidence does he give to support his point of view?
   b) How might the attitudes of people in the developed world be offensive to people in other regions?
   c) Do you agree or disagree with Sachs's ideas? Explain your answer.

CHAPTER 4

THE INQUIRY MODEL

Figure 4.1 *Top,* Manitoba—stockpiling wheat after a bumper crop; *bottom,* Mozambique — Canadian grain being distributed by CIDA workers

Conducting research is essential to intellectual growth. When you started doing research in elementary school, you probably summarized all the information you found in point-form notes, whether or not it was relevant to the topic. Then you rewrote the information "in your own words." This was an essential first step. In senior grades, however, you need sophisticated research techniques. It is not enough to restate someone else's ideas. You must be able to express your own thoughts.

Each subject you take relies on some sort of model to help you write essays. Often these models enable you to develop a hypothesis and support it with facts or logical arguments. These models are essential if you are to develop the writing skills you will need in post-secondary education. Models help you to improve your writing by providing focus.

USING THE INQUIRY MODEL TO STUDY ISSUES

World issues is unlike many other courses. Instead of the paramount goal being the acquisition of knowledge, the focus of a world issues course is on problem solving. The inquiry model used in this course goes beyond those in many other subjects. Consider the model shown below. Which features make it different from other models you have used?

At first, this inquiry model may seem much like others you have used. You analyse information from secondary resources or statistics. But once you reach step 6, the ideas you present are your own. You do not rely on someone else to provide the answers; instead, you figure them out for yourself. Whether you are suggesting solutions to a major environmental issue or solving a personal problem, the strategy of analysis, synthesis, application, and evaluation will help make you a better problem solver. Of course, knowing the facts is essential to formulating good solutions!

THE INQUIRY MODEL

1 **DEFINE THE ISSUE:** In simple terms, what is the problem?

2 **IDENTIFY HOW PATTERNS OF PHYSICAL GEOGRAPHY AFFECT THE ISSUE:** How have patterns of climate, natural vegetation, location, landforms, soil, and so on influenced the issue?

3 **IDENTIFY HOW PATTERNS OF HUMAN GEOGRAPHY AFFECT THE ISSUE:** How have patterns of population, land use, transportation, political boundaries, natural resources, and so on influenced the issue?

4 **DESCRIBE CAUSE-AND-EFFECT RELATIONSHIPS THAT AFFECT THE WHOLE PLANET:** How have issues at the local or national level had an impact on people around the world?

5 **ANALYSE THE EFFECTS OF THE ISSUE ON THE ENVIRONMENT AND ON PEOPLE:** How has the issue had a major influence on the environment and on people locally and around the world?

6 **SOLUTIONS TO THE ISSUE:** What are the possible solutions to the problem over the short term? Over the long term? Which solution do you think is best?

7 **COST-BENEFIT ANALYSIS:** How would you fund your solution? Is your solution feasible in light of the cost?

8 **POSSIBLE CONSEQUENCES OF SUGGESTED SOLUTIONS:** How will people be affected by each of the possible solutions? Have enough compromises been made so that all people will be satisfied? *Will your solution work?*

CONSOLIDATING AND EXTENDING IDEAS

1 a) Read the article "The Hunger Remains."
  b) Conduct your own research to find out more about this issue. Add your own ideas to your notes.
  c) Using the format in the inquiry model, write an essay describing this issue and the possible solutions. You might write a separate paragraph for each category.

2 Find an article in a newspaper or magazine that deals with a world issue. Summarize the information using the inquiry model.

THE HUNGER REMAINS

Despite the heroic efforts of thousands,
hunger in the Third World is still
a daily reality for half a billion people.

Benjamin Franklin once said there is nothing certain except death and taxes. If he were still alive, he might be tempted to add Third World hunger to his list. More than half a billion people are undernourished every day, and their ranks are swelling every year.

The problem goes far beyond how much food is produced. Since the 1970s, enough food has been grown to feed everyone and then some. Although most of the surplus comes from North America and Europe, dozens of poor countries are also food exporters. Yet Third World hunger persists, both in the "self-sufficient " nations and elsewhere.

To end hunger we need to sever its social, political, and economic roots. For hunger is a product not only of bad harvests, but of wars, of injustice, and of getting priorities in the wrong order. Let's examine these causes in more detail, starting with the hungriest place on earth.

The last decade was in many respects a lost one for Africa. Per capita grain production fell by more than 10 per cent. Meanwhile, population levels increased by 3 per cent a year. The net result is that one in three Africans go to bed hungry.

Against this backdrop of daily misery, famine is a frequent visitor. In 1985, 2 million Africans perished. Hundreds of thousands have died every year since then. And, this spring [1991], aid agencies warned that more than 29 million lives in Sudan and other countries were hanging by a thread.

African famines typically follow droughts. Despite this, many aid experts believe they can be prevented. If it weren't for military campaigns and related factors, the impact of uncooperative weather would be small. Consider Ethiopia, which bore the brunt of the 1985 famine. During the eleven preceding years, civil war raged between rebels in Eritrea, a northern province, and government-led forces. Endless fighting took an incredible toll on agriculture, as farmers were uprooted and crops destroyed. Meanwhile the government poured money into huge state-run farms, which fed few peasants but kept soldiers' stomachs full.

When the rains failed, it was the last straw. Poor people died by the hundreds of thousands. The toll would have been even higher if it were not for an outpouring of generosity from around the world. Rock stars alone raised $300 million in relief supplies. But getting the food to the needy proved difficult. About 70 per cent of the famine victims lived behind rebel lines. On several occasions, the government prevented grain from reaching them.

This year's crisis [1991] features the same plot, but a different cast. Sudanese troops are fighting a rebel army. The country has run out of food. And, both sides in the dispute are seizing food relief shipments. To add insult to injury, less than half of Sudan's farmland is currently under cultivation.

War brought hunger to several other African nations during the 1980s, including Mozambique, Angola, and Uganda.... But even if peace breaks out tomorrow . . . it will take years to get food production back on track. One reason is debt. During the 1970s, the Third World borrowed heavily from Western lending agencies to finance development. Then, in the early 1980s, a global recession sent interest rates soaring. Cash-strapped nations found themselves owing more and more to foreign lenders. Between 1980 and 1987, Africa's total debt ballooned from $120 billion to $250 billion.

This ground further development to a halt. Many nations slashed social programs to the bone. And, to earn the foreign exchange necessary to pay down their debts, they became more export oriented. Africa increased the amount of land devoted to "cash crops" such as coffee, cotton, and peanuts. This meant that there was less land available for growing the food Africans actually eat.

Higher prices enticed farmers to make the switch. But, it was a change many would later regret. To ensure decent harvests, fields planted with cash crops need to be watered, fertilized, and sprayed with pesticides. Many poor farmers cannot afford these "agrichemicals," yet try to grow the plants anyway. The problem is that when denied fertilizer, cash crops quickly rob soil of nutrients. Land becomes useless, even for growing food.

Farmers who can afford agrichemicals, or get loans to buy them, face different headaches. Over the past few decades, the cost of pesticides and fertilizers has risen faster than the price farmers can get for their crops . . . To maintain the same standard of living, cash-croppers are forced to cultivate more land by swallowing up neighbouring farms. . . .

What can be done to reverse this trend? The most radical idea is to cancel the continent's debts. If the world pretended they never existed, Africa would feel less pressured to export crops. Not surprisingly, this solution goes too far for most bankers. But, they have shown support for "debt relief" schemes such as the Brady Plan. Invented by, and named after, the US Treasury Secretary, the Brady Plan encourages banks to write off some of their Third World debts. In return, they receive guarantees that interest will be paid on the balance.

Another remedy is to remove existing trade barriers between the Third World and its customers. North America, Europe, and Japan currently levy taxes on imported foods such as roasted coffee and cocoa powder. But, in their unprocessed form, coffee and cocoa can freely cross borders. This system ensures that Africa loses out on profits. A lot more money is made by food processors in wealthy countries than by the farmers who supply them. If trade barriers came down, developing nations could put more effort into canning coffee and less into growing it.

In the meantime, African agricultural policies could stand some fixing. In most nations, farmers are unable to earn a decent living growing food. As a result, there's rarely a surplus. One of the few exceptions is Zimbabwe. Since 1980, its government has raised the price of maize, a staple food, several times. It has also offered credit to poor farmers. Together, these initiatives increased maize yields from 80 000 to 800 000 t a year. And, although one in five people is still hungry, living standards are improving. A 1987 health survey in Zimbabwe showed that 6 per cent of newborn children were underweight, down from 15 per cent in 1982. . . .

For their part, African leaders accuse the rich Western nations of making the situation worse. We are too generous with food aid, they argue. Free food is welcome during a crisis. But at other times, it hinders development efforts. African farmers can't compete with overseas shipments, so their fields lie fallow.

This does not mean we should ignore Africa between its famines. Peasants may resent us sending them our leftovers, but they do appreciate other forms of assistance. Basically, any tools that help them produce more food with less effort, cost little, and don't harm the environment are high on their wish lists . . .

John Eberlee. From "The Hunger Remains," *Canada and the World*, October 1991, 14–18.

PART 2

CULTURAL
ISSUES

CHAPTER 5

IDENTIFYING CULTURAL ISSUES

THE IMPACT OF CHANGE

Change is the one constant in today's world. Environments are changing because of the impact of a growing human population. The types of resources we use change as natural resources become depleted and new synthetic materials are developed. Economic and political systems change and evolve as people adjust to the stresses of living in an unpredictable world. It is impossible for most of us to resist change. Yet the rapid nature of change today makes it difficult for us to adjust.

This is especially true of cultural issues. At one time, different cultures were separated by physical geography. Landforms such as sheltered mountain valleys and remote plateaus led to isolated pockets of people who developed unique languages, religions, attitudes, and values. This isolation may have caused them to be suspicious or mistrustful of outsiders. It was sometimes difficult for them to make trade deals or political alliances because of their mistrust. This dislike of strangers is known as **xenophobia**.

The way in which a group of people communicates and the language it uses affect its interaction with other peoples. When languages are translated, certain nuances often take on different meanings in other cultures. For example, the Chinese expression "may you live in interesting times" might be considered as a blessing by many North Americans who enjoy the challenges of change. However, the expression is intended as a curse! In traditional Chinese culture, harmony is valued above all else,

so to "live in interesting times" is unfortunate because "interesting" implies a time where the natural order is upset.

Certain traditions may be acceptable in some cultures but insulting in others. Many European cultures considered it polite to look into the eyes of the person to whom you are speaking. Other cultures may consider this practice to be impertinent and prefer that you look away from the eyes of the person you are addressing. Symbols of respect differ in various religions. In the Islamic faith, women and men remove their shoes before entering a mosque. In the Jewish religion, men wear a *yarmulke*, or skullcap, in synagogues. The practice in the Christian religion is for men to remove their hats when they enter a church. When outsiders visit a foreign country, especially one that evolved in isolation, they should make sure they understand cultural taboos and attitudes so that they do not offend the people in the host country.

Because of these subtle differences, when two different cultural groups first encounter each other, there may be an atmosphere of suspicion and distrust. As they become more familiar with one another, both groups usually benefit from their association. The introduction of new technologies, attitudes, values, and approaches usually creates a vibrant new culture that is distinct from any other.

Often, the superior technology of one culture may enable it to dominate another. Military technology may include better weapons, a different attitude to warfare, and advanced military organization. Better weapons allow the invader to inflict more damage than the defender. Imag-

Figure 5.1 Children from diverse cultures on the islands of Guadeloupe

ine a battle where muskets are used over spears. The spear throwers have little chance of defending themselves because the range of their weapons is so much shorter than the range of the muskets. When the invader has the attitude that the enemy must be crushed and no prisoners will be taken, this can have a devastating effect on a defender who views war as a ceremonial event where captives are used for religious sacrifice or are enslaved. A well-organized army also has a distinct advantage over a force that is poorly led. If an invading army has any combination of these three advantages, it will most likely be successful in an armed conflict with a weaker foe.

Such a situation occurred in Mexico in the sixteenth century. Spanish *conquistadors*, led by Hernando Cortés, invaded and colonized

> The Caribbean nation of Trinidad and Tobago has developed a rich and unique cultural blend as Spanish, English, African, and Indian settlers have mixed with the indigenous peoples. A British model of government has combined with customs, traditions, religions, and languages from many parts of the world to form a unique society.

Mexico. The indigenous peoples had a sophisticated culture during the Aztec Empire. They had the most advanced calendar in the world. Their capital city of Tenochtitlán was very grand. The rich, descriptive language, complex religion, and well-developed social structure established the Aztec Empire as one of the most prosperous of its time. The Spanish invaders, however, had steel weapons, horses, and gunpowder. They were able to capture the Aztec leader and gain political dominance over Mexico. After a time of persecution, destruction, and reconstruction, a uniquely Mexican culture evolved. Spanish values, religion, and customs blended with the culture of the Aztecs and the other indigenous peoples.

Interestingly, isolated cultures within the new Mexican nation did not develop the same national character. Although the Mayan Empire had been in decline for centuries before the arrival of Cortés, the Mayans, cut off from the rest of Mexico by the tropical rainforests of the Yucatán Peninsula, continued in their age-old traditions. Today, Mexicans of Mayan heritage in southern Mexico are seeking cultural, economic, and political equality so that they too can share in the development of modern Mexico.

DIVERSE CULTURES

Today, people of different cultures are interacting more than at any other time in history. Print and broadcast media, the Internet, and other forms of modern technology give us insight into people and places around the world. We are travelling more and over longer distances than ever before. Airplanes can take travellers anywhere in the world, often within hours. Canadian tourists are visiting distant regions such as Nepal in the Himalaya Mountains and the island of Bali in the South Pacific.

Imagine how Nepalese farmers would feel if Canadian tourists entered their country and were inadvertently disrespectful of religious and cultural practices. Western business people who make deals in China, eastern Europe, and other emerging industrial nations must understand the cultures and religions of their hosts. If they fail in this regard, conflict is bound to develop.

Immigrants to Canada sometimes have customs unfamiliar to many Canadians. At first, we may find it difficult to accept some of the attitudes and opinions. But as we get to know people from various cultures, we come to understand and respect other values and traditions. For example, many Canadian schools

> The Canadian government establishes quotas each year for the number of immigrants and refugees it allows into the country. As a rule of thumb, the quota is usually about 1 per cent of the total population. In 1996, the quota was set at 250 000 people.

have "safe school" policies that do not allow students to bring knives into the classroom unless they are being used as part of the school program. This helps to ensure a safe learning environment for students. However, this policy collides with a tradition in the Sikh religion, in which young men carry a *kirpan*—a small ceremonial dagger—in their traditional clothing. It would be inappropriate for students to be suspended from school because they practised a cultural tradition. At the Peel Board of Education in Ontario, the solution involved compromise between the Sikh community and school officials. The young men now carry replicas of their knives. In this way they can continue to practise the tradition and still obey the rules of the school board. Another example, in 1989, involved three RCMP officers who petitioned the Alberta Court to prohibit the wearing of turbans by Sikh Mounties when they were on duty. They argued that the policy of the force, which allowed Sikh officers to wear turbans—a symbol of their religion—was unconstitutional because it violated the RCMP's obligation to be secular or neutral in religious matters. The RCMP, backed by the Canadian Sikh community, maintained that the banning of turbans would effectively bar religious Sikhs from joining the force. The petitioners are now awaiting a decision from the Supreme Court of Canada as to whether or not it will hear their appeal against the Alberta Court's decision.

All over the world, diverse cultures are demanding political, cultural, and economic equality with the people who represent the majority in their nations. Studies by Thomas Homer-Dixon, a Canadian expert on hostility and government policy, show a relationship between government policy, resource availability, and rebellion among various cultures (see Chapter 22, page 372). Diverse cultural groups within nations may become resentful if the essential resources they need are unavailable, and the government does not respond by helping the people. In the southern Mexican state of Chiapas, regional hostilities came to a head in 1995 after years of population pressures and unsustainable farming practices reduced available resources to the point where this Mayan enclave could no longer live decently. The central government in Mexico City was not prepared to help with relief and development projects, but seemed more concerned with developing the heartland of the nation in response to the recent NAFTA agreement. The Mayan people, after years of being ignored, felt they had no recourse but to rebel in order to show the world the gravity of the situation. Similar internal conflicts between cultural groups and governments are evident in South Africa, as the nation attempts to restructure after the demise of apartheid; in Iraq, Iran, and Turkey with the Kurdish people, who seek autonomy in local issues; and in the mountains of the Philippines where the many indigenous groups live, including the Tagalog in Luzon and the Cenuano in Mindanao.

Political instability in many parts of the world makes nations such as Canada seem peaceful by contrast. When people move from one place to another, it is a combination of **push** and **pull factors** that causes them to move. The forces "pushing" them out of a region may be unemployment, environmental change, or political instability. The "pull" factors determine to which country they will move. Immigrants come to countries such as ours for many reasons. These might include job opportunities, environmental factors, political stability, democratic elections, the protection of individual rights and freedoms as enshrined in the Canadian Charter, and the Canadian government's multicultural policy. In many parts of the world the basic rights and freedoms that we take for granted are non-existent.

Canada's multicultural policy encourages people of different races and religions to settle here. All Canadians are guaranteed "freedom of conscience and religion; freedom of thought, belief, opinion, and expression" in the Charter of Rights and Freedoms. **Multiculturalism** is evident in many of our urban centres. People from many different cultures live and work in communities throughout these centres. Temples, churches, mosques, synagogues, and numerous other religious buildings are found throughout the cities. Cultural events, such as the Hispanic Festival at Toronto's Harbourfront, Oktoberfest—a German celebration of the harvest—and Caribana—a Caribbean folk festival held in Toronto in mid-summer—highlight Ontario's cultural diversity. In Alberta, Canadian multiculturalism is celebrated each August at Dreamspeakers—a Native arts and film festival in Edmonton, at the Nikka Yuko Japanese Garden in Lethbridge, and at the Ukrainian Canadian Archives and Museum, and the Ukrainian Cultural Heritage Village in the Edmonton area.

Ethnic newspapers and radio and television programming, cultural groups, ethnic food stores, and various organizations also contribute to the cultural mosaic. The federal government supports multiculturalism through funding for the arts and other cultural events. All levels of government, as well as businesses and community organizations, sponsor a variety of cultural celebrations. While Canada's policy of multiculturalism encourages all Canadians to be tolerant and accepting of one another, this is not always the policy in other countries.

Diverse cultures may be treated with suspicion and intolerance. They sometimes face discrimination simply because they are not members of the dominant cultural group. In France, North Africans from former French colonies comprise about 2 million of the population. These French citizens often face discrimination in their daily lives and have difficulty finding work. In Moscow, Chechens and other cultural groups from the Caucasian Republics are looked upon with suspicion. Turkish workers in Germany have been subjected to acts of violence. In the southwestern United States, Mexican immigrants are not readily accepted into the dominant society. Ethiopian Jews who have immigrated to Israel are not welcomed by some citizens because they are not of European or Middle Eastern descent.

Throughout history, racism has been part of government policies in many countries. In Germany prior to and during the Second World War, the Nazis systematically set out to exterminate the Jewish population of Europe. In the United States, the segregation of African Americans was officially sanctioned policy from the days of slavery in the eighteenth century into the 1950s and 1960s.

While Canada today accepts cultural groups from around the world, we have not always offered the same understanding and tolerance to the aboriginal peoples of Canada and some immigrants. In the past, Native peoples were forced to **assimilate**, or become part of, the dominant society. Many Native groups lost their ancestral languages, attitudes, values, and traditions. Forced to abandon the traditional lifestyles that had sustained them for centuries, many aboriginal peoples have found themselves living between two worlds.

In the nineteenth century, many Chinese workers came to Canada to help build the railway system. In 1872, people of Chinese descent were not allowed to vote and, in 1885, the Canadian government imposed a "head tax" on Chinese workers in an attempt to persuade them to leave the country. During the Second World War, because their ancestors had come from Japan, Japanese Canadians were forced into camps where they were interred until the war ended. Also during the war, a ship carrying Jewish refugees who were fleeing the Nazis was refused entry at Halifax, even though the refugees faced certain death if they returned to Germany. Today, the Canadian government is attempting to redress the injustices that were imposed on Native peoples and certain immigrant groups in the past.

Racism still pervades societies all around the world. In order to maintain our reputation as a country that strives to be fair and open-minded, Canadians must ensure that we continue to treat people of all origins with respect and understanding.

> Racism can also be practised by a dominant minority group. In South Africa, a policy of **apartheid**, in which black South Africans were segregated from the ruling white minority, was officially adopted in 1948 and not abandoned until 1994.

REFUGEES

Political, economic, and environmental factors cause many people to seek refuge in foreign lands. People usually flee their homes because they are politically opposed to their government. Some fear imprisonment, torture, and even death if they stay in their country of origin. Often, refugees are willing to risk their lives to find sanctuary in a new country.

The Office of the United Nations High Commissioner for Refugees (UNHCR) was established in 1951. Its mandate was to help

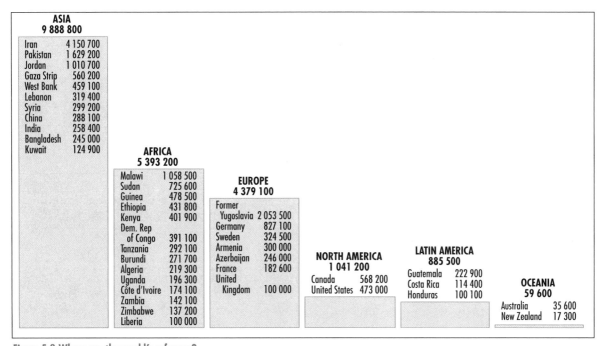

Figure 5.2 Where are the world's refugees?

the millions of refugees displaced after the Second World War. In 1959, the number of refugees worldwide was estimated at 1.2 million people. By 1995, there were 27.4 million people seeking temporary refuge from life-threatening situations in their home countries (see Figure 5.2). The number of refugees more than doubled in the ten years from 1985 to 1995. Part of the reason for this increase is related to how the UNHCR defines a refugee. In the past, it counted only those people who fled their country and moved to another one. Because of the current situation in the former Yugoslavia, it now includes people who move from one place to another within their country. The hope of this UN agency is that once displaced people are identified as refugees, they will be given assistance without actually having to leave their own country. This is a good strategy because it should encourage international aid and thus alleviate the tensions that often

develop between refugees and residents of host countries.

The main reason for the increase in refugees is war—not the kind of world wars that plagued the earlier half of the twentieth century, but civil wars and local disputes between neighbouring countries. In the past twenty years, most of the major wars have occurred in developing countries. The conflicts that erupted in Southeast Asia in the 1960s and 1970s devastated the lives of millions of Vietnamese, Cambodians, and Laotians. Refugees from these countries fled to neighbouring countries such as Thailand, Hong Kong, and Malaysia. Dubbed "boat people" because they often escaped by boat, many eventually immigrated to countries such as Canada and Australia where they started new lives. Some Cuban and Haitian refugees have used all manner of vessels to travel the sometimes treacherous waters of the Caribbean to reach the coast of southern Florida. Political instability or fear of persecution in their home countries made the dangers of the trip seem insignificant to them by comparison.

In the 1980s, 3 million people escaped from Afghanistan to Pakistan and Iran as the Soviet government sought to take over Afghanistan. While some returned home when the war ended, 2.6 million Afghans continued to live in exile. Sometimes civil wars can lead to **genocide**, the systematic extermination of a national, cultural, or religious group. Members of the persecuted group are forced to flee their homeland simply to survive. In 1994, 1.8 million Tutsi fled from Rwanda to neighbouring Burundi, Tanzania, Uganda, and Zaïre because ruling Hutu extemists launched a campaign of genocide against them. Three other troubled areas in Africa have led to large numbers of refugees. There are currently 785 000 Liberian, 526 000 Somali, and 525 000 Sudanese refugees living in neighbouring countries until the political situation calms down in their own countries. In Europe, Bosnia and Hercegovina top the list of internally displaced people with over 2.5 million refugees.

Displaced during the Arab-Israeli War of 1947 to 1948, the Palestinians of Israel have been housed in "temporary" refugee camps for fifty years. For years, Israeli and Palestinian politicians have looked for ways to solve this situation. The Palestinians are gaining political autonomy but it is a slow process with frequent setbacks. In 1993, a peace agreement was signed between the Palestine Liberation Organization and Israel. In May 1994, most Israeli soldiers left the Palestinian territories and new Palestinian police forces took over. The Palestinians were given the political power to govern themselves. Despite these new freedoms, the Palestinian territories face incredible obstacles. Not only do they have to contend with the political instability of the region, but they must also build a nation out of densely populated land

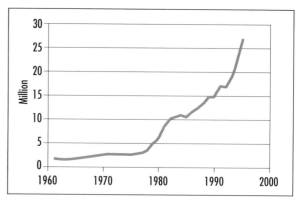

Figure 5.3 (a) Refugees receiving UN assistance, 1961–1995

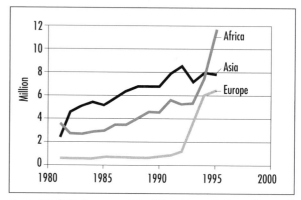

Figure 5.3 (b) Refugees receiving UN assistance in Asia, Africa, and Europe, 1981–1995

Source: *Vital Signs 1996: The Trends That Are Shaping Our Future* by L. R. Brown, C. Flavin, and H. Kane. Copyright © 1996 by Worldwatch Institute. Reprinted by permission of W. W. Norton & Company, Inc.

where no development has occurred for years; a land where unemployment hovers at around 50 per cent and over half the people are under fourteen years of age. As this book is being written, hostilities in the region have heightened because of the construction of an archaeological tunnel under land in Jerusalem that Muslims consider to be sacred. It remains to be seen if a lasting peace can take hold in the Middle East.

Unlike immigrants, refugees move because they are pushed out of a region. Pull factors have little to do with where they settle. Refugees usually leave their countries with little more than the clothes on their backs. They move to the closest available country that will accept them. It is not so much that they want to be in that country, but they have no other options. Often there is tension between the refugees and the citizens of the host country. Residents worry that the refugees will put undue pressure on the country's infrastructure. They are concerned about diseases that may be brought into their country, and about increased demand for food, clothing, and construction materials that may drive up prices. What will be the impact on the local environment of so many additional people? All these new residents need proper sanitation, which may take time to implement. In the meantime local water supplies could be polluted. These factors may lead to disapproval and rejection by permanent residents. Sometimes they force the government to reject the refugees. Unable to escape, the refugees have no choice but to move from place to place looking for a country that will shelter them. This situation occurred in Hong Kong when Vietnamese boat people sought sanctuary there during the Vietnamese War. Hong Kong, the most densely populated place on earth, residents argued, did not have the resources to house the refugees. They were pushed out to sea where they remained for some time until international pressure forced the city-state to take them in temporarily.

Western nations take in refugees for humanitarian reasons, but it is often only the most affluent refugees who are able to take advantage of the opportunity. Only those who can afford the airfare and have family, business, or academic ties are able to enter countries such as Canada or Great Britain. Many affluent countries are wary of people claiming to be refugees. They want to help people who are genuinely in fear for their lives.

It is not only governments that provide aid to refugees, but also service clubs and religious groups. These organizations provide medical and dental care and essential clothing, and help refugees find housing, language training, and suitable jobs. Most groups hope to have the refugee family self-supporting within two years. Unfortunately, these charitable groups do not have the resources to help very many displaced people. Most of the world's refugees never get the opportunity to make a fresh start in a country of their choice.

Figure 5.4 Rwandan Hutu refugees fleeing camp in northern Burundi

WESTERNIZATION

In many parts of the world today, traditional values and customs are being eroded by Western influences. American movies and television programs are viewed in countries around the world. Sometimes, local programming is taken off the air because American sitcoms have become more popular. Canadians often prefer television series such as *Friends*, *Baywatch*, and *Party of Five* to Canadian programming. In Japan, American adventure films are very popular, and in Lebanon during the 1980s the *Rambo* movies were watched more than any other movie. Some countries such as Saudi Arabia have banned Western television and movies in order to encourage people to value their own cultural traditions.

Western-style clothing is worn in world capitals and remote villages alike. Often, the local attire makes a lot more sense than that of the popular culture. In tropical countries, loose-fitting, light-coloured cotton robes are cooler than blue denim jeans or business suits. Yet the clothes of the popular culture are often adopted over local fashions. Some enterprising fashion designers are working to re-establish traditional fashions not only at home but also on the international scene. Designers, models, and stylists from various cultural groups are gaining popularity for the freshness of their ideas and appearance.

American fast-food chains have spread to every populated continent on the globe. As with clothing, sometimes foods from other countries are adopted into the popular culture, often in an Americanized version. The results can be seen in pizza that is unlike anything found in Italy; danish pastries that would never be baked in Denmark; tacos vastly different from what you would find in Mexico; and Chinese food mass-produced to North American tastes. The Americanization of international tastes results, some may argue, in rich and unique cultural identities being threatened. However, over the last twenty years, many restaurants have opened in North America that serve ethnic foods prepared in the traditional manner.

The consumer society depicted in Western television promotes a wasteful lifestyle to traditional cultures that may have had more conservationist roots. Many young people in developing countries would rather act like North Americans than carry on in the ways of their ancestors.

There is a new group of consumers evolving in developing countries such as Chile, Malaysia, and Mexico. They are learning to consume natural resources in the same way that North Americans do. Statistics on television ownership in developing countries give us some understanding about the size of this growing consumer group. In 1985, 570 million people in the developing world had televisions in their households. By 1991 the figure had almost doubled. This growth rate is six times greater than the population growth rate. In a world where some resources are finite and environmental degradation is a serious problem, this dramatic increase in the number of consumers does not bode well for the planet in the twenty-first century and beyond. Not only are resources being

> Sometimes governments take steps to preserve a country's culture. In 1995, the Canadian Broadcasting Corporation decided to eliminate over the next few years most foreign content in its television broadcasts. In this way the CBC will become much more of a national force in shaping and reflecting the culture of Canada.

used up, but global warming is intensifying, wildlife habitats are disappearing, soil erosion is increasing, and the state of the environment in general is deteriorating, all as a result of mass consumption. Imagine the consequences of a huge country such as China learning to consume resources in the way North Americans do. Could the planet survive if all people became mass consumers?

Consider just two aspects of consumption— energy and kilojoules. In 1995, the developed nations of the United States, Canada, Japan, and Europe consumed 3.75 times more energy than all developing countries combined. Projections into the years 2000 and 2010 suggest that while the requirements of developed nations will tend to stabilize at current rates, the demands for energy by developing countries will increase, almost doubling in only fifteen years. As far as kilojoule consumption is concerned, in 1994, the average intake for individuals in developed countries was 13 812 kJ per person (3976 kJ higher than the United Nations' recommended daily intake of 9386 kJ), while average intake for individuals in developing countries was 9208 kJ (628 kJ less than recommended consumption). We can only speculate on the increased pressure that will undoubtedly be placed on developed nations in the next decades to share the earth's bounty.

THE MYTH OF WESTERNIZATION

In recent years, Westerners have reassured themselves and irritated others by expounding the notion that the culture of the West is and ought to be the culture of the world.

This conceit takes two forms.

One is the Coca-colonization thesis. Its proponents claim that Western and, more specifically, US popular culture is enveloping the world. US food, clothing, pop music, movies, and consumer goods are more and more enthusiastically embraced by people on every continent.

The other has to do with modernization. It claims not only that the West has led the world to modern society, but that as people in other civilizations modernize, they also Westernize, abandoning their traditional values, institutions, and customs, and adopting those that prevail in the West.

Both these project the image of an emerging, homogeneous, universally Western world—and both are to varying degrees misguided, arrogant, false, and dangerous.

Advocates of the Coca-colonization thesis identify culture with the consumption of material goods. Yet the heart of a culture involves language, religion, values, traditions, and customs. Drinking Coca-Cola does not make Russians think like Americans any more than eating sushi makes Americans think like Japanese.

Throughout human history, fads and material goods have spread from one society to another without significantly altering the basic structure of the recipient society. Enthusiasms for various items of Chinese, Hindu, and other cultures have periodically swept the Western world with no discernible lasting spillover.

The argument that the spread of pop culture and consumer goods around the world represents the triumph of Western civilization depreciates the strength of other cultures while trivializing Western culture by identifying it with fatty foods, faded pants, and fizzy drinks. The essence of Western culture is the

Magna Carta, not the Magna Mac.

The modernization argument is intellectually more serious than the Coca-colonization thesis, but equally flawed.

The tremendous expansion of scientific and engineering knowledge that occurred in the nineteenth century allowed humans to control and shape the environment in unprecedented ways. Modernization involves industrialization; urbanization; increasing levels of literacy, education, wealth and social mobilization; and more diverse and complex occupational structures. . . .

As the first civilization to modernize, the West is the first to have fully acquired the culture of modernity. As other societies take on similar patterns of education, work, wealth, and class structure, the modernization argument runs, this Western culture will become the universal culture of the world. . . .

Only a few hundred years ago all societies were traditional. Was the world any less homogeneous than a future world of universal modernity is likely to be? Probably not. . . . To modernize, must non-Western societies abandon their own culture and adopt the core elements of Western culture?

From time to time leaders of great societies have thought it necessary. Peter the Great of Russia and Mustafa Kemal Ataturk, the founder of modern Turkey, were determined to modernize their countries and convinced that doing so meant adopting Western culture, even to the point of replacing traditional headgear with its Western equivalent. In the process they created "torn" countries, uncertain of their cultural identity. Nor did Western cultural imports significantly help them in their pursuit of modernization.

More often, leaders of non-Western societies have pursued modernization and rejected Westernization. Their goal is summed up in comments such as that of Saudi Arabia's Prince Bandar bin Sultan in 1994 that "'foreign imports' are nice as shiny or high-tech 'things.' But intangible social and political institutions imported from elsewhere can be deadly . . . Islam is for us not just a religion but a way of life. We Saudis want to modernize but not necessarily Westernize."

Japan, Singapore, Taiwan, Saudi Arabia, and, to a lesser extent, Iran have become modern societies without becoming Western societies. China is clearly modernizing but certainly not Westernizing. . . .

At the individual level, the movement of people into unfamiliar cities, social settings, and occupations breaks their traditional local bonds, generates feelings of alienation and anomie, and creates crises of identity to which religion frequently provides an answer. At the societal level, modernization enhances the economic wealth and military power of the country as a whole and encourages people to have confidence in their heritage and to become culturally assertive. As a result, many non-Western societies have seen a return to indigenous cultures. It often takes a religious form, and the global revival of religion is a direct consequence of modernization. In non-Western societies this revival almost necessarily assumes an anti-Western cast, in some cases rejecting Western culture because it is Christian and subversive, in others because it is secular and degenerate. The return to the indigenous is most marked in Muslim and Asian societies. . . .

East Asian societies have gone through a parallel rediscovery of indigenous values and have increasingly drawn unflattering comparisons between their culture and Western culture. For several centuries, they, along with other non-Western peoples, envied the economic prosperity, technological sophistication, military power, and political cohesion of Western societies. They sought the secret of this success in Western practices and customs, and when they identified what they thought might be the key they attempted to apply it to their own societies. . .

Now, however, a fundamental change has occurred. Today, East Asians attribute their dramatic economic development not to their importation of Western culture but to their adherence to their own cultures. They have succeeded, they argue, not because they have become like the West, but because they have remained different from the West. . . .

As Western power recedes, so too does the appeal of Western values and culture, and the West faces the need to accommodate itself to its declining ability to impose its values on non-Western societies. In fundamental ways, much of the world is becoming more modern and less Western. . . . The result is popular mobilization against Western-oriented élites and the West in general.

The powerful currents of indigenization at work in the world make a mockery of Western expectations that Western culture will become the world's culture. As indigenization spreads and the appeal of Western culture fades, the central problem between the West and the rest is the gap between the West's efforts to promote Western culture as the universal culture and its declining ability to do so. . . .

The West—and especially the US, which has always been a missionary nation—believes that the non-Western peoples should commit themselves to the Western values of democracy, free markets, limited government, separation of church and state, human rights, individualism, and the rule of law, and should embody these values in their institutions.

Minorities in other civilizations embrace and promote these values, but the dominant attitudes toward them in non-Western cultures range from skepticism to intense opposition. What is universalism to the West is imperialism to the rest.

The belief that non-Western peoples should adopt Western values, institutions, and culture, if taken seriously, is immoral in its implications.

The almost universal reach of European power in the nineteenth century and the global dominance of the United States in the latter half of the twentieth century spread many aspects of Western civilization across the world. But European globalism is no more, and US hegemony is receding, if only because it is no longer needed to protect the United States against a Cold War threat. . . . Furthermore, as a maturing civilization, the West no longer has the economic or demographic dynamism required to impose its will on other societies. Any effort to do so also runs contrary to Western values of self-determinism and democracy.

As Asian and Muslim civilizations begin to assert the universal relevance of their cultures, Westerners will come to appreciate the connection between universalism and imperialism and to see the virtues of a pluralistic world of many different civilizations.

From Samuel P. Huntington, *The Clash of Civilizations and the Remaking of World Order* (New York: Simon and Schuster, 1996).

SEXUAL EQUALITY

Since the 1970s, a **sexual revolution** has changed the economic, social, and cultural fabric of many developed nations. In countries such as Canada and the United States, women have sought equality with men in the workplace and at home. A 1992 UN study showed that Canadian women earned $41 for every $100 earned by Canadian men. This compares favourably to many developing countries, but is far below some European countries such as Sweden where the ratio is $71 for every $100 earned by a man. Still, the economic picture for Canadian women continues to improve. Thirty-two percent of all Canadian doctors are now women, nearly twice as many as there were twenty years ago. Ten per cent of the seats on the boards of publicly owned companies are now held by women. While this percentage is very small, twenty years ago women were virtually absent from corporate boardrooms. There is still a long way to go until women achieve pay equity with men and have the same employment opportunities.

In recent years the issue of violence against women has finally received the attention that it warrants. As stated in the Human Development Report of the UN Development Programme (1995), "violence stalks women throughout their lives from cradle to grave."

Violence against women undermines social and development programs in developed as well as developing countries. This largely preventable problem accounts for billions of dollars in health care costs worldwide. In addition, the time lost because of ill-health and the loss of women's contribution to growth and economic development are impossible to measure. Of course, the worst consequence of the issue is the pain, suffering, and debasement that many women experience each year in every nation of the world.

Women's groups have successfully lobbied to have issues relating to sexual equality included in the list of international human rights. The Vienna Declaration in 1993 was the first international recognition by the United Nations or any other body that violence against women was a violation of human rights. It is hard to believe how recently this recognition was finally made.

In much of the developing world, women are the primary providers for their families. Yet they have little political power. In some cultures,

PRACTICE	ESTIMATED OCCURRENCE
Population	• Sixty to 100 million women worldwide are considered missing, the result of abortion of female foetuses, infanticide, neglect, or murder.
Preference for sons	• In India, at least 300 000 more girls than boys die each year. • In 1987, 500 000 female infants disappeared.
Female genital mutilation	• Eighty-five to 114 million females worldwide have been mutilated. • More than 2 million girls per year are mutilated.
Wife beating	• Seventeen to 33 per cent of women have been assaulted by their spouses. • In Papua New Guinea, 18 per cent of married women receive medical treatment from injuries inflicted by their husbands.
Murder	• In 1987, 62 per cent of women murdered in Canada were killed by their spouses. • In India, one-quarter of all deaths of women sixteen to twenty-four years of age is due to "accidental" burns.
Rape	• An estimated one woman in five will be a rape victim in the United States. • More than 20 000 Muslim women were raped in Bosnia in 1992.

Figure 5.5 Selected estimates of violence against women

Source: Adapted from *Vital Signs 1996: The Trends That Are Shaping Our Future* by L. R. Brown, C. Flavin, and H. Kane. Copyright © 1996 by Worldwatch Institute. Reprinted by permission of W. W. Norton & Company, Inc.

Figure 5.6 Young girls in developing countries helping to support their households

In some societies, girls are isolated from men who are not members of their family. In Yemen, where 90 per cent of the primary teachers are men, and where girls live a secluded life, female education is almost impossible for cultural reasons. In other societies, it is not safe for girls to go to school because of the distances that must be travelled and the potential for violence against girls travelling on their own. Child marriage and teenage pregnancy are other factors that work against the education of teenage girls. Girls are unable to go to school because they are expected to stay home to look after their families. Pregnant students in many countries are routinely expelled.

females may be valued more for the money they can earn for the family than as individuals. In some parts of the world, women are deprived of an education and therefore have difficulty moving ahead. Girls in developing countries have less chance of getting an education than their brothers. Many young girls share the responsibility of maintaining the household with their mothers. They simply don't have time for school. They look after infants, fetch water and firewood, tend the animals, and take care of the crops. Many families cannot afford to send all their children to school. The cost of educating a child in Kenya, for example, amounts to about one-quarter of a family's average income. It is usually the boys who are sent to school. The reasoning is that they are more likely to find a paying job and leave the family home.

The type of education that young women receive is also different from what their male peers experience. Many are educated for specific trades. In Denmark and Ghana, two very different countries, most young women entering technical schools specialize in textile design, the clothing trade, and catering. In many countries, admittance to professions such as medicine, engineering, and finance is largely reserved for young men.

The advantages of literacy for women are numerous. The most obvious is increased employment income. It has been estimated that each year a girl stays in school results in a 26 per cent increase in earnings when she leaves school. Another advantage relates to fertility. Educated females are more likely to have smaller, healthier families. Recent surveys indicate that fertility can be reduced by three to

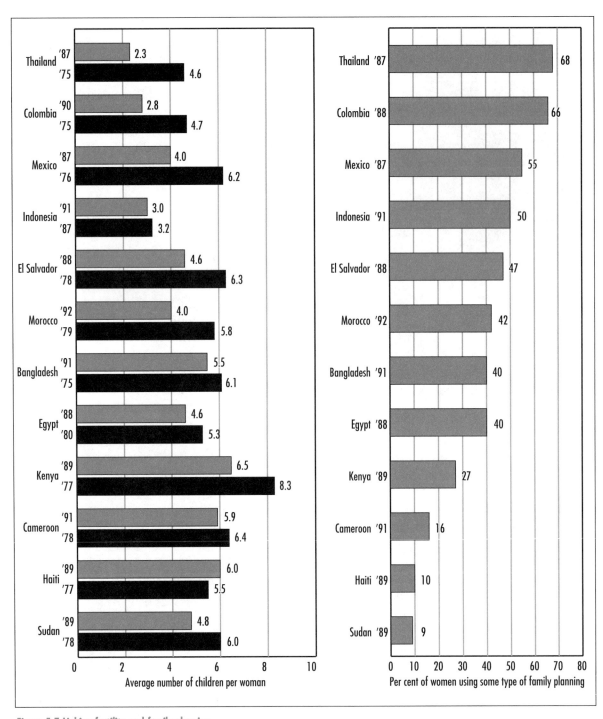

Figure 5.7 Linking fertility and family planning

The survey was based on 300 000 women in 44 developing countries.

four children for girls who are educated up to grades 5 or 6. Results are even better for girls with secondary and post-secondary education. If population growth is to stabilize, it is essential that girls be given the same educational opportunities as boys. Increased education also empowers women to demand better treatment from their spouses and from society at large.

The Report of the Independent Commission on Population and the Quality of Life (*Caring for the Future*, 1996) recommends the following policies to improve female education: establishing child-care centres so that girls can go to school instead of looking after their younger siblings; encouraging girls to follow technical and scientific subjects in order to increase their earning potential; reducing gender bias through teacher education; establishing segregated schools for girls in cultures where they are not allowed contact with males; expanding the teaching of sex education; and allowing pregnant girls to continue their education.

Now that there is international recognition of the problems that women face, perhaps international efforts will centre on the solutions to the problems. Young boys need to be brought up so that they value the roles their mothers and other women play in society. If the United Nations and other international organizations work to censure those countries that tolerate violence against women, it is likely the situation will start to improve.

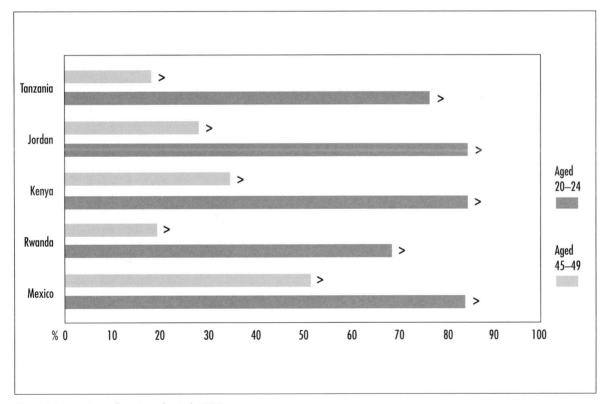

Figure 5.8 Percentage of women educated, 1994

CONSOLIDATING AND EXTENDING IDEAS

1 a) List the cultural issues described in this chapter. For each one, consider the contrasting opinions people may have. Explain why they may have these opinions. Summarize your work in an organizer or a chart like the one in Figure 5.9.

ISSUE			
Opinion 1			
Rationale			
Opinion 2			
Rationale			

Figure 5.9 Cultural issues

   b) Express your own opinion of each issue and discuss it with a group of students.

2 Research current government policies regarding aboriginal peoples for one of the following countries: Canada, United States, Brazil, New Zealand, Australia.
   a) Use electronic retrieval systems to conduct a search. Try searches of key words such as *Canada, aboriginal, policy.*
   b) Summarize the information in point form.
   c) Assess the policy to determine if you think that Native peoples are treated fairly.
   d) Summarize your findings in a brief presentation to the class or your group.

3 The rate at which developed countries consume resources is vastly greater than that of developing nations. Examine World Energy Consumption on page 144 of the *Canadian Oxford School Atlas*, 6th ed. Write generalized statements about (i) energy consumption in the Northern Hemisphere as compared with the Southern Hemisphere; (ii) energy consumption by continent.

4 Read the excerpt "The Myth of Westernization" on page 43.
   a) Use a concept web or mind map to illustrate the main ideas in this excerpt. You will need to identify key words/phrases within each paragraph before you begin to create your web/map. You might choose to use the word *modernization* as the core of your work, and relate other concepts to this central idea.
   b) In a well-reasoned paragraph, explain what the attempts of Peter the Great and Mustafa Kemal Ataturk have to tell us about the relationship between Westernization and modernization. How is it possible for a nation to modernize without Westernizing? You may quote directly from the excerpt to explain your thinking.
   c) If "economic prosperity, technological sophistication, and military power" reflect modernization, what information would you need to determine how modern Saudi Arabia, Japan, Singapore, and Taiwan have become? Using point-form notes, indicate how you would assess economic prosperity, technological sophistication, and military power.

d) Should developed countries set the standard for quality of life? In groups of three to five students, debate this issue.

5 Statistical data may suggest trends or patterns among social behaviours, groups of people, or nations. In this chapter, you learned that increased education of women in developing countries is affecting population growth. Study the bar graphs in Figure 5.7, which suggest a link between fertility (as measured by the average number of children per woman) and family planning (rhythm method, contraception, etc.). By interpreting the graphs, describe in paragraph form
a) the differing rates in numbers of children born per woman;
b) the relationship between the use of family-planning techniques and fertility;
c) the possible outcomes of continued family planning by the year 2000.

6 Working with a group, brainstorm a list of cultural issues that have *not* been discussed here. (Articles in newspapers and magazines may help stimulate your thinking.) If an issue is a local one, determine if it has any relevance in a global context. For example, while the Quebec government's language policy may seem uniquely Canadian, other cultural groups, such as the people in the Basque region of Spain, are also trying to preserve their languages.

Try using electronic data searches (computer databases) to establish global trends. In the case of language issues, key word searches may include *language, culture, minority, nationalism,* and *separatism.*

7 Select a cultural issue, either from this book or one of your own choice.
a) Collect articles from a variety of periodicals and newspapers. (Consider using electronic information retrieval systems available in some school libraries and most large reference libraries.)
b) Conduct an inquiry using the approach described in Chapter 4, page 28.

8 Choose *one* of the following nations, all of which have a large number of refugees:

Afghanistan	Angola	Cambodia	Chile
Ethiopia	Guatemala	Haiti	Iraq
Mozambique	Rwanda	Salvador	Somalia
Sri Lanka	Sudan	Tibet	Vietnam

With a partner, investigate and gather data to create a one-page "Refugee Report" that describes
a) the group(s) that were forced to leave their country;
b) numbers of people involved;
c) "push" and "pull" forces that caused them to flee;
d) places where they have they found asylum;

e) developments in their homeland since their departure;

f) appropriate maps and photos.

Consult sources that contain recent reports and information, such as periodical indexes for magazine and newspaper articles, vertical files, and electronic encyclopaedias.

9 During the twentieth century, there has been an overall trend of increased participation of women in government. Women have become heads of government in both developing and developed countries.

   Research and develop a biography on one of the women in the following list. Make an oral presentation of your biography to the class.

Isabel Perón	former president of Argentina
Begum Khaleda Zia	former prime minister of Bangladesh
Lidia Gueiler Tejada	former president of Bolivia
Kim Campbell	former prime minister of Canada
Eugenia Charles	former prime minister of Dominica
Edith Cresson	former prime minister of France
Ertha Pascal-Trouillot	former president of Haiti
Vigdis Finnbogadóttir	president of Iceland
Indira Gandhi	former prime minister of India
Mary Robinson	president of Ireland
Golda Meir	former prime minister of Israel
Maria Liberia-Peters	former prime minister of Netherlands Antilles
Violeta Barrios de Chamorro	president of Nicaragua
Gro Harlem Brundtland	prime minister of Norway
Benazir Bhutto	former prime minister of Pakistan
Corazon Aquino	former president of the Philippines
Hanna Suchocka	former prime minister of Poland
Maria de Lourdes Pintasilgo	former prime minister of Portugal
Siramavo Bandaranaike	prime minister of Sri Lanka
Chandrika Kumaratunge	president of Sri Lanka
Tansu Çiller	former prime minister of Turkey
Margaret Thatcher	former prime minister of United Kingdom

COUNTRY PROFILE: BELARUS

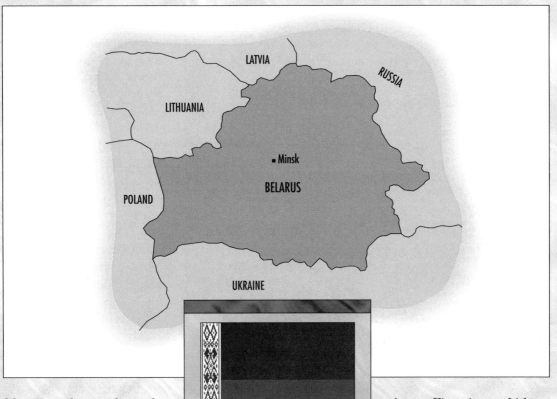

BELARUS—FACTS AND FIGURES

GNP per capita	$2 870
Population	10 141 000
Birth rate*	12
Death rate*	12
Infant mortality**	16
Female literacy	99%
Defence expenditures	n/a

* per 1000 population
** per 1000 births

Figure 5.10
Source: Statistics (1990–95) from *World Resources, 1996–97.*

Many people may be unfamiliar with Belarus since the nation was created quite recently. Some may remember it as a state in the former Soviet Union. At that time it was known as Belorussia or White Russia. It is bordered by Russia on the east, Ukraine on the south, and Poland, Lithuania, and Latvia on the west. Because of its location between rival countries, this region has been governed by foreign powers throughout its entire history up to independence. First, it was Lithuania that controlled the region starting in the thirteenth century. By the sixteenth century, Poland controlled Lithuania and also took over Belarus. As Russia gained power, Belarus eventually fell under its domination in 1795. Conquered by Germany in the First World War, the nation was absorbed into the Soviet Union after the war. It remained under the Soviet umbrella until its declaration of independence in 1991.

Figure 5.11 Terminal Square in Minsk, Belarus

At present, the country is going through very difficult times. The Belorussians have to build an infrastructure, which has been neglected for centuries. Many of the economic ties of the past seventy years with the former Soviet Union no longer exist. Energy supplies, raw materials, and investment capital are all scarce. Belarus also needs to develop new trading partners now that the Soviet Empire has collapsed. The country is looking to Poland, Lithuania, and other former Soviet satellite states as trading partners.

The biggest problem, however, is environmental. Belarus has to deal with the fallout from the 1986 Chernobyl explosion in neighbouring Ukraine. One-quarter of the nation's budget is devoted to cleaning up the damage from the accident. Twenty per cent of the population has been affected by the fallout. Many farmers have been given new land outside of the affected area in exchange for their old land, which is useless because of radioactivity. But these initiatives cost money—already in short supply.

Contrary to what one might expect, the culture of Belarus remained so strong throughout the centuries that conquering nations often embraced the language, religion, and other cultural manifestations. The Lithuanians adopted the Belorussian language and religion when they ruled Belarus. When Poland gained control in the Middle Ages, Lithuanians and Belorussians were often appointed as rulers in their own regions. During its rule, Russia adopted the Belorussian written language as the official Russian script. Today, the language and culture of Belarus are having a resurgence as the people finally have a country to call their own.

No longer a Communist state, Belarus's early experiments with democracy have been disappointing. Aleksandr Lukashenko, the democratically elected president, has vowed to cut out the corruption that developed after the fall of Communism.

CHAPTER 6

OVERPOPULATION: A CULTURAL ISSUE

POPULATION

Our world today is more interconnected and interdependent than ever before. While we acknowledge the variations and extremes of global diversity and disparity, within our global village we share concerns about the world's most prevalent issues:

- the impacts of colonialism
- hunger and the development and distribution of resources
- the complexities of international trade, debt, and foreign aid
- the use and overuse of the environment
- human health and rights, and employment and unemployment
- quality of life and standard of living
- political alignments and militarism
- the dynamics of change, which confound our continued understanding of each of the previous issues

Perhaps the single most important dynamic of change that interacts dramatically with all the others is that of population. Population increases and decreases have resounding impacts on all aspects of life on the planet, as indicated in the flow chart in Figure 6.1.

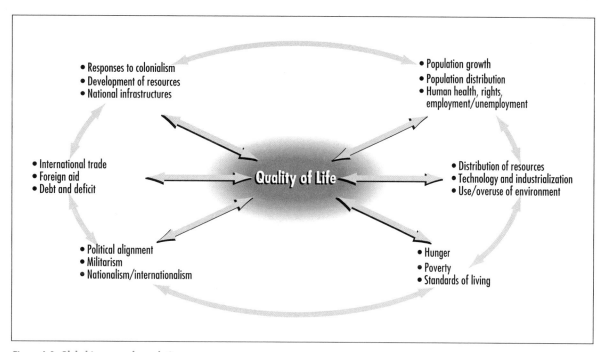

Figure 6.1 Global issues and population

As the populations of developing countries increase, and those of developed nations stabilize or decrease, social, economic, and political systems struggle to adjust. Population growth linked to poverty puts an ever-increasing strain on ever-diminishing local and even global environments. Economic activity based on heedless consumption in already wealthy nations exacerbates the cost of any population increase in these nations.

Overpopulation affects the distribution of global resources. The more people there are, the more natural resources are needed to support them. In many overpopulated regions of the world, there are shortages of the most basic resources of food, water, and shelter. The intensive extraction of minerals from the ground and the farming of fields and forests to provide for the growing population lead to all kinds of pollution. Streams are clogged with silt and hazardous chemicals. The air is filled with fumes and the land is strewn with slag, overburden, and industrial waste. If the population were lower, there would probably be plenty of natural resources and the local environment would be less polluted. On a global scale, the reduction in resource use would result in lower increases of greenhouse gases and a slower rate of global warming. Overpopulation also contributes to ozone depletion, species extinction, deforestation, desertification, and soil erosion.

All of these result because there are too many people living in a limited space on a planet that can only support a finite number of people.

Although people living in countries with diversified economies represent less than 20 per cent of the world's total population, they use over 80 per cent of the planet's resources. People who live in developing countries—80 per cent of the world's population—are faced daily with the limitations of their environments in their struggle for survival. While they are better conservationists than peoples of developed countries, increasing populations within the developing countries only intensify the problems of environmental overuse.

In the past fifteen years there has been a rapid increase in the clearing of the world's forests to develop agricultural land for growing crops, and to cut wood for use in building and manufacturing, and as fuel. Slash-and-burn agriculture and the misuse of forest products have caused acute shortages of fuel wood, fodder, timber, and a loss in biological diversity. The land that populations in the developing world depend on is degrading—erosion of topsoil and extensive flooding have made the land less able to sustain life. Local and global environmental pollution have also increased as developing countries repeat the patterns of industrialization of developed nations. In tropical, temperate, and boreal regions, forests are in crisis, as are the multiply-

> "It is not the poor who are responsible for the deterioration of the environment. When people have fewer resources, they husband them more carefully, because that is all that they have to fall back upon. Small farmers are a lot more efficient. They care for their land a lot better. It is the greed of the affluent which is using more of the resources than the poor do. [Mahatma] Gandhi said: 'In this land [India] there is enough for everyone's need but there is not enough for some people's greed.'"
>
> Vijay Vyas, India—South Asia Public Hearing. From *Caring for the Future* (New York: Oxford University Press, 1996), 29.

ing populations that rely on ever-diminishing resources. The recent Report of the Independent Commission on Population and Quality of Life (*Caring for the Future*, 1996) points out that developed and developing nations are currently being challenged by a period of ecological transition. We must attempt to harmonize patterns of human settlement with patterns of nature.

Economic and political systems are continually being tested to adjust to the dynamics of population change. Following the end of the Second World War (1945), we experienced improved standards of living globally. The development of national and international infrastructures provided a safety net of social services such as free education, "universal" health-care systems, family-planning assistance, and subsidies for food and housing. Nations of the developing world won their independence and sovereignty from former colonial empires, and borrowed to create social services crucial to improving the quality of life and economic growth. Increases in populations were regarded as both appropriate and advantageous, as resources seemed plentiful and readily available.

Within thirty years (1945 to 1973), however, this period of renewal and increased industrialization gave way to realizations of faltering growth, inflation, trade imbalances, debt crises, and government deficits within both developed and developing countries. Nations and international organizations that had previously had "deep pockets" to assist developing nations began to limit foreign aid. After the Second World War, the Cold War between the superpowers of the USSR and USA guaranteed economic assistance to many developing nations in return for political alignments. With the "thawing" of the Cold War in the 1980s and its official demise in November 1990, both powers reduced or halted financial investment in many

instances. Domestic economic problems—in a far less affluent America and in a Soviet Union witnessing the collapse of state ownership and central planning—rapidly superseded the needs of the developing world. Preoccupation with national indebtedness and deficit reduction at home took centre stage.

During the 1990s, developed nations around the globe have indeed become far more nationalistic and less internationalistic. In order to reduce their deficits, governments in many developed nations have adopted policies of reduced government spending and cut-backs in social services to their own populations. At a time when the need for greater internationalism—in the form of food, development, and military aid—has reached epic proportions, and emergencies of overpopulation demand both immediate and long-term solutions, nations of the developed world have reduced their commitments to support the populations of the developing world.

Furthermore, development agencies such as the IMF and the World Bank have insisted on the introduction of **structural-adjustment policies** in debtor countries: these programs require deep cuts in public spending to facilitate repayment of foreign loans. In regions most in need of improved social and economic conditions, such as health care, freshwater, education, sustainable food sources, adequate housing, the quality of life is declining year after year. Lower life expectancy, high infant mortality rates, malnutrition, reduced primary school enrolment, illiteracy, unemployment, and increased poverty levels have all resulted as the current economic trends fail to address the needs of developing nations.

These current economic strategies do not address the issue of population growth. In 1970, the United Nations estimated that 944 million people lived in absolute poverty. In 1997 this number increased to 1.5 billion people.

A recent international trend towards multi-national economic unions is the response to the aggressive economic climate of the 1990s. Expanding global markets have encouraged groups of nations, often geographically adjacent, to form economic unions, which can better compete and take advantage of the opportunities that these markets offer. The European Union—which includes Germany, Italy, France, Luxembourg, Ireland, Finland, Denmark, the United Kingdom, Spain, Austria, Sweden, Portugal, Greece, Belgium, and the Netherlands, and NAFTA, which comprises Canada, the United States, and Mexico, are the two most recent examples of multinational economic unions.

Developing countries have provided multinational corporations with opportunities for tremendous profits in the past decades. Manufacturing enterprises have moved out of the diversified economies of developed nations to take advantage of abundant low-paid labour, lax environmental protection legislation, and government policies that favour foreign investors and overlook the exploitation of their own workers (low wages, hazardous working conditions). The question is whether or not the impacts of multinational economic unions such as NAFTA will reverse economic patterns that encourage the rich to get richer and result in the poor becoming still poorer. How will Canada and the

> The debt of developing countries rose from $658 billion in 1980 to $1.9 trillion in 1994. Eight of the ten countries with the highest debt burden (1992) are located in the developing world. In order of descending amounts of debt, in millions of US dollars, these ten countries are:
> Brazil ($120 000), Mexico ($110 500), Indonesia ($85 000), the Russian Federation ($79 000), India ($785 000), China ($69 000), Argentina ($68 000), Turkey ($56 000), Poland ($49 000), and South Korea ($42 000).

United States balance their interests in employment and economic growth in Mexico with their own problems of unemployment and national/international production?

Just as multinational corporations want to increase profits year after year, the governments of most countries seek to increase their gross national product. The view that the combined value of goods and services should increase each year is flawed. It emphasizes production and consumption but does not look at the effect the consumer economy is having on the environment or the well-being of people. According to the Independent Commission on Population and the Quality of Life, "a world economy that thrives on relentless exploitation of natural resources, depending perilously on fossil fuels, causing limitless waste, and remaining oblivious to the precepts of equality and equity among different societies, is neither sustainable, nor tolerable. It is headed for disaster. The situation demands fundamental economic reorientation and restructuring—a transition that will require domestication of market mechanisms in terms of environmental and social objectives. Every human being of the present and the future, regardless of where he or she lives, must have equal and inalienable opportunities to profit from the Earth's natural resources." (*Caring for the Future*, page 50).

If there were fewer people in the world, markets would be smaller and less profit would be made. On the other hand, there would be less competition and greater local autonomy. Canada decided to unite economically with the United States and Mexico in the North American Free Trade Association (NAFTA) to ensure that Canada would be included in international markets.

POPULATION GROWTH RATES

The rate of population growth over the past 10 000 years has been incredible. While 10 000 years may seem like a long time, it is only a small fraction of the earth's 4.6 billion years. In 6000 BCE, the world's population was 5 million. People were primarily hunters and gatherers. The heaviest concentration of population was in western Africa, with scattered pockets of people in China and Europe. Four thousand years later, in 2000 BCE, the population had escalated to almost 90 million. The development of agriculture had improved the quality of life and had enabled a greater number of people to survive and flourish. By 1 CE over 250 million people inhabited the planet. Today, 2000 years later, the population of the earth is approaching 6 billion—over 1 million times more than it was just 10 000 years ago!

The rate at which the world's population is growing is also increasing. Figure 6.2 shows that the growth rate has increased steadily in each twenty-year period up to 1980. This **geometric growth** suggests that population pressures will increase even more in the twenty-first century. Is there some hope that this trend can be slowed down? If you look at the figures for the year 2000 in Figure 6.2, you will see that population growth is indeed slowing down. Demographers estimate that population growth should stabilize by the end of the twenty-second century. That may seem like a

long way off, but in terms of human life on the planet—about 1 million years—it is not. In fact, over the long term (1000 plus years) the human species may experience a situation where world population starts to decline. This may seem farfetched in view of the current world population figures, but death rates could actually exceed birth rates for both social and environmental reasons. In western Europe the birth rate equals the death rate in several countries, including Germany, Italy, and Denmark. The decline in the number of traditional family units and the choice of many young couples not to have children could result in reduced populations. In some regions of Canada, only 30 per cent of property owners have school-age children as members of their families. Thus, it is conceivable that as populations age, the proportion of the population of child-bearing age may decline to the point that deaths actually outnumber births.

The human species could have a reduced ability to reproduce because of environmen-

DATE	ESTIMATED POPULATION	PERCENTAGE GROWTH
1900	1.6 billion	
1920	1.8 billion	12.5
1940	2.2 billion	22.2
1960	3.0 billion	36.4
1980	4.4 billion	46.7
2000	6.1 billion	38.6

Figure 6.2 World population and percentage growth, 1900–2000
We can determine the rate of population increase by dividing the population figure for a given year by the previous population figure, then subtracting 1.
Source: Statistics from *World Resources, 1996–97.*

tal pollutants. Studies show that men in industrialized cities are starting to have lower sperm counts than in the past because of airborne pollutants. If the trend continues whereby the planet is polluted to the point that it is no longer habitable, human population will decline.

The next century will likely be the turning point for the human species. Will the human race will be able to maintain social and environmental structures so that the species can be preserved?

CONSOLIDATING AND EXTENDING IDEAS

1 Define the following terms using a dictionary of geography: *population, area, population density, population distribution, dense, sparse.*

2 To what extent do you believe that population growth is the single most serious global issue facing the world today? Write an explanation of your view.

3 a) Using the data on page 59, construct a line graph to illustrate world population figures from 6000 BCE to present day.
  b) Why has population accelerated at an ever-increasing pace?
  c) How do you account for the reduction in rate of increase from 1980 to 2000? What implications does this have for the future?

4 Copy Figure 6.3 into your notebook, then complete it as shown in the example for Canada.

REGION	POPULATION (000 000)	AREA (000 000 km^2)	DENSITY (people/km^2)	DENSE/SPARSE
Canada	29.4	10.0	2.94	sparse
China	1 221.5	9.6		
United States	263.3	9.4		
Hong Kong	5.7	0.001		
Australia	18.1	7.7		
Brazil	161.8	8.5		

Figure 6.3 Population patterns, selected sites, 1995
Source: Statistics from *World Resources, 1996–97.*

5 a) Explain how the following factors influence population distribution: rivers, coastlines, mountains, and other landforms; temperature, precipitation patterns, and other climatic factors; soils, minerals, and other natural resources; historic, economic, and political factors.

b) Present your information in an organizer or a chart, then discuss your findings with the class.

c) Based on the collective findings of the class, what conclusions can you make about population distribution?

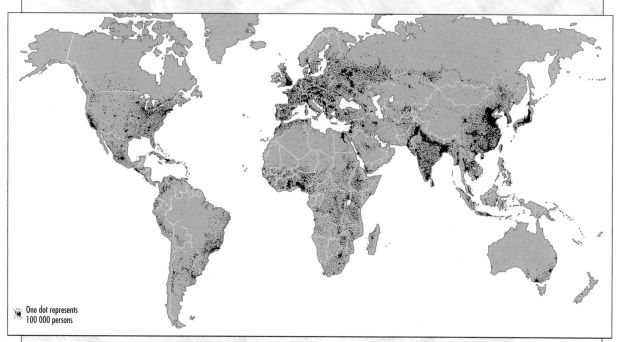

One dot represents
100 000 persons

Figure 6.4 Population distribution map
Source: G. Matthews and R. Morrow, *Canada and the World: An Atlas Resource,* 2d ed. (Toronto: Prentice-Hall, 1995). Reproduced with the permission of Prentice Hall Canada Inc. Copyright © 1985.

PEOPLE AND RESOURCES

The population distribution map in Figure 6.4 shows that some places support human populations better than others. Factors of physical geography such as climate, landforms, natural resources, and location influence where people live. It's a fact, for example, that more people live in temperate regions than in polar zones. This is especially true in Canada. It is no accident that 80 per cent of Canadians live within 200 km of our southern border, where temperatures are moderate.

It is interesting to examine the distribution of land area by continent to determine, in a broad sense, land resources available for settlement and use by the diverse populations (plant, animal, and human) that inhabit them. Of the earth's total land mass, Africa comprises 20.2 per cent; Oceania, 5.9 per cent; Antarctica, 9.7 per cent; Europe, 3.3 per cent; Asia, 33.7 per cent (including Russia and Turkey); South America, 12 per cent; and North America, 15.3 per cent (including Central America and the Caribbean). Of course, not all of these land areas are truly habitable. Flat coastal plains and river valleys generally have larger population densities than mountainous regions. Consider

Figure 6.5 *Top,* rice terraces in Java; *bottom,* farmers in the Nile Valley; *right,* a kibbutz in Israel

central India as an example. Millions of people live in the broad fertile valleys of the Ganges River while the Himalayan Mountains farther north are almost unoccupied. In China, abundant water supplies provided by the Yangtze, Huang He, and Xi Jiang Rivers in the fertile lands of the Great Basin have encouraged settlement of immense populations of more than fifty persons per square kilometre. Immediately to the west, the plateau of Tibet has a population density of less than one person per square kilometre.

Natural resources are also significant in understanding population density. If resources are abundant, people will thrive and the population will grow. The resources might be industrial minerals such as nickel and copper, or they might be agricultural. The rich volcanic soils of Java in Indonesia have resulted in one of the most densely populated rural areas in the world. Water is also an important resource. Without it, the ancient Egyptian civilization that once flourished along the banks of the Nile would never have existed.

Location is another factor that may determine why some places develop. Hong Kong, for example, is a cluster of barren, mountainous islands off the coast of China, with a hot monsoon climate. Yet it is one of the most densely populated places in the world. Its strategic location on the major trade route between China and the West ensured its development. In contrast, regions that are far from trade routes seldom develop such a dense population

Historical factors also play an important role in population distribution. The northeastern United States is the most densely populated region of the country mainly because it was the first area of European settlement. History played a role in today's dense population in Israel. Following the Second World War, millions of Jews fled Europe in the aftermath of the **Holocaust** for the sanctuary of the new Jewish homeland of Israel.

> The Holocaust was a black period in the history of the world. Over the six-year course of the Second World War, the Nazi government in Germany systematically annihilated an estimated 6 million Jews in Germany and throughout eastern Europe.

(3.6 billion) in Asia; 8.5 per cent (486 million) in Latin America; and 5.1 per cent (295 million) in North America. Looking at the sheer weight of numbers, we might be convinced that Africa and Asia—75 per cent of the world's total population live on these two continents—are overpopulated. Likewise, we might suggest that Oceania and North America are underpopulated (less than 6 per cent of the world's population live here). When we examine population densities in Figure 6.6, however, it appears that Europe is highly overpopulated and Africa is underpopulated!

Figure 6.7 compares the land mass of each continent with its proportion of the world's population. In examining this figure, we might be led to believe that there is still room for increases in population—particularly in North and South America, Oceania, and Africa.

There is certainly more to understanding overpopulation and underpopulation than merely comparing and contrasting statistics about

OVERPOPULATION AND UNDERPOPULATION

The concepts of **overpopulation** and **underpopulation** are somewhat more complex than we might think at first glance. In 1996 the world's population totalled 5.77 billion people. These people were located disproportionately in six regions: 12.7 per cent (732 million) lived in Africa; 0.5 per cent (29 million) in Oceania; 10.1 per cent (580 million) in Europe; 63.2 per cent

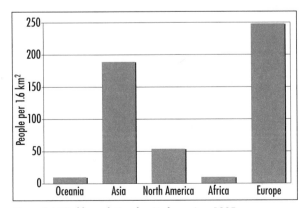

Figure 6.6 World population density by region, 1995

CONTINENT	% OF EARTH'S LAND MASS	% OF EARTH'S POPULATION
Africa	20.2	12.7
Oceania	5.9	0.5
Europe	3.3	10.1
Asia	63.0	33.7
North and South America	25.3	12.5

Figure 6.7 Land mass and population

population numbers, densities, and settlement patterns around the globe. The concepts of resource availability, development, and consumption are critical to understanding population issues. Let's begin by examining the extent to which an area is able to support a population, that is, a region's **resource base**.

Few people live in the immense island nation of Greenland, and for good reason. Greenland has a relatively small resource base. Much of the country is covered by a gigantic ice-cap. Its cold climate means that the land is unsuitable for most commercial agriculture. Most of Greenland's 57 000 people live along the southwest coast, where they derive a living from

Overpopulation as a result of a diminished resource base has occurred in Newfoundland and Labrador. The people have depended on fishing for generations. In the 1990s, this resource declined to the point where commercial fishing was no longer viable. Faced with long-term unemployment, many Newfoundlanders moved to other provinces. A solution may lie in developing new industries, such as **aquaculture** and expanding others, such as shipbuilding and tourism.

mining, fishing, and hunting sea mammals. These limited resources can sustain the population. But if the resource base became smaller, this sparsely populated country could become overpopulated.

Sometimes a region becomes overpopulated because the number of people has grown disproportionately to its resource base. This has happened in many parts of Africa, southern Asia, and South America. A prime example is the Caribbean country of Haiti. Unlike other island nations in the region, this former French colony has not developed its tourism industry but has relied on plantation agriculture and subsistence farming. Stripped of much of its natural vegetation, and

Figure 6.8 Expansion of Newfoundland's shipbuilding industry could help the province's economy.

with most of its rich topsoil eroded, this country is no longer able to sustain its people. As a consequence, Haitians escape by the thousands across the border into the Dominican Republic or to other islands where the resources have not been destroyed.

In regions of overpopulation too many people live on marginal lands with inadequate resources. The result is that many live in poverty, without adequate food, water, and shelter. As in Newfoundland, there are two solutions: migrate elsewhere or increase the resource base. Unfortunately, implementing these solutions is much more complex than identifying them. (The solutions to overpopulation are discussed further in Chapter 8.)

Underpopulation occurs when there are too few people to develop fully the resources of a region. Many people believe that Canada is underpopulated. Vast areas are virtually uninhabited, while many other regions are sparsely settled. With population projections indicating that Canada's population will actually decline in the twenty-first century, it is natural that our government continues to encourage immigrants to settle here. What other regions of the world do you think are underpopulated? Consider your opinions carefully.

As with all issues, of course, there is an opposing viewpoint that Canada already has enough people. Preservationists argue that Canada should maintain or reduce its population if environments are to be protected. It all depends on your point of view.

OPTIMUM POPULATION

In an ideal state, renewable resources are fully developed, there is full employment, and prosperity is universal. This is called the **optimum population**. Many western European countries have close to optimum populations. Denmark has a **static population**—one that is neither

Figure 6.9 *top,* Overcrowded living conditions in Jakarta, Indonesia; *bottom,* Irian Jaya, a sparsely settled province in eastern Indonesia

growing nor declining. It has a good agricultural base and derives many products from the sea. There are just enough people to farm the land and harvest the sea. Cities such as Copenhagen have become international trading centres, exporting local products and importing raw materials from other countries for **labour-intensive industries**—industries that require a great deal of labour but few natural resources.

Optimum populations most often occur in countries where the resource base has remained unchanged over a long period, and the population has had sufficient time to evolve. In the case of Denmark, the country could become overpopulated if the resource base were reduced. If, for example, pollution and overfishing reduced the annual fish catch, this could result in fewer resources for the population. The standard of living would decline unless the population decreased in keeping with the reduced resource base. Alternatively, if the resource base were to expand as a result of the discovery of offshore oil, then the country would theoretically be underpopulated. There would be more resources available for the population.

Countries try to reach optimum populations through a number of measures. In underpopulated countries such as Canada, immigration is often encouraged. Overpopulated countries often try to reduce population growth or to extend their resource base. China, the world's most populous country, is attempting to reduce fertility rates through strict laws. Japan, on the other hand, is expanding its resource base with a policy of economic imperialism. Japanese companies are developing vast amounts of raw materials throughout the Pacific Rim and are using these resources to expand the economic capabilities of their workforce. Even though Japan has a high population and few natural resources domestically, Japanese industrialists have compensated by expanding into foreign markets.

CASE STUDY

The Role of Technology:
The Development of the American West

The development of the American Great Plains in the eighteenth and nineteenth centuries illustrates the role of technology in economic development and population growth. Before the arrival of European settlers, Native Americans* had an optimum population. The natural resources of the region supported a small, relatively stable population. The main resource used by the indigenous peoples was the bison, more commonly known as the buffalo. These giant herbivores provided everything the people needed. Meat was dried and ground into pemmican, which could be stored indefinitely. Hides were used for tents, clothing, rugs, and blankets. Bones served as weapons, tools, and even building materials.

The stomachs were used as containers for carrying water. Hunting the great animals required team work and skill. They were often chased on foot and herded towards cliffs. Once the animals stampeded, nothing stopped them except the bottom of the cliff, where scores of buffalo died. One or two hunts a year were often all that was needed to sustain a tribe for the full year. Of course, smaller game and wild plants provided variety to the diet through the warmer months. Everything was in balance. Bison died each year but an equal number of calves were born. Similarly, human populations remained more or less static.

When horses and rifles were introduced by the European settlers in the seventeenth century, the economic potential of the bison increased. Hunting became much more efficient. It was not necessary to wait for the great annual migrations to direct a stampeding herd over a cliff. A hunter could simply go to the plain and shoot as many animals as were needed. The desire for new and expensive European goods, traded for buffalo robes, encouraged hunters to kill more animals than they needed for meat. At first, the increased availability of abundant buffalo meat and other by-products resulted in a population explosion. However, the growth was short-lived. The once abundant herds of bison became so depleted that Native peoples could no longer rely on the herds. This was further complicated by the mass extermination of bison by Europeans and Americans, who shot them from passing trains. Without food, thousands of Native peoples starved, while many more migrated north to Canada where there were still abundant game stocks. During this century Native peoples were assimilated into American mainstream society.

The settlers used new technology to develop other resources once the bison had been hunted to the point of extinction. The rich prairie soils proved to be valuable farmland, while other parts of the West were well suited for grazing. Three inventions allowed the region to prosper. With the self-scouring plough, farmers were able to cut through the tough grasses and roots to cultivate the land for planting. Barbed wire allowed cattle ranchers to contain their domesticated herds. Wells drilled into **aquifers** deep beneath the plains provided water for both cattle and farmers. With this new economic potential, the population soared.

New settlers arrived from the eastern United States and from Europe in the late nineteenth and early twentieth centuries. Most of the good farmland had been claimed in the eastern United States. In large families the eldest son usually inherited the family farm. The younger children sometimes moved west to homestead new land. Europeans migrated to the western United States for similar reasons. In addition, famine and hardship were common in the European countryside. The cities were squalid, polluted places with few opportunities and too many people. The frequent wars that swept across Europe, especially in the first half of the twentieth century, drove many people to escape to America.

Overpopulation may occur if the resource base becomes depleted or if the population grows too fast. This was the case in the Great Depression in the West during the 1930s, when several years of **drought** hampered the ability of the land to produce high yields. Years of ploughing deep furrows exposed the rich topsoil to winds. Without rainfall to hold together the soil and grow the vegetative cover that protects soil, huge dust storms developed. In a matter of days, soil that may have taken a thousand years to develop naturally was blown away. During the Second World War, food was needed to feed the troops. The price of grain rose accordingly and large-scale farming became profitable once again. But the farming practices were now different. Regions where rainfall was unreliable reverted to range country for pasturing livestock. Vegetation was not removed so that wind erosion could not destroy the fragile topsoil. Grain was grown in regions with regular rainfall, using new, sustainable practices.

The discovery of better agricultural practices has allowed the region to flourish once again. Deep ploughing is a thing of the past in some dry areas. Stubble from harvested crops is sometimes left in fields to reduce the effects of wind erosion. Wind-rows have been planted and irrigation techniques have improved so much that the dependence on rainfall has decreased. After many years of population fluctuations as the resource base shrank or expanded, it would appear the region has once again reached an optimum population. The population is appropriate for the resource base.

Today the American West maintains a balance between its economic potential and its population. The population is dropping as farms became larger and mechanization reduces the need for big farm families to perform the labour that was once required. Crop yields (the resource base) remain high because of chemical additives, irrigation, and hardy hybrid crops. Some people argue that the overuse of water from aquifers and rivers could create a desert in the next century. If these valuable water reserves are used up, the resource base will shrink and the region may experience consequences similar to those of the 1930s. The region would once again be overpopulated. Other conservationists are concerned about the overuse of

chemical additives. Today's pesticides are more potent than they once were, so lesser amounts are needed. They also biodegrade much more quickly (often within two weeks) than earlier chemical pesticides, which could remain in food chains for generations. Nevertheless, there is concern that beneficial insects and other organisms are being destroyed with the pests. Many of the older pesticides that have been proven to be dangerous have been banned. However, the World Health Organization estimates that 20 000 human deaths occur worldwide each year from pesticide use. Some pesticides, now banned in the United States, cause reproductive abnormalities, breast and testicular cancer, and a decline in sperm counts.

This has led to a small but growing movement in agriculture that seeks to grow crops organically, by using natural systems. Organic farming needs a larger population because more manual labour is required. More important, sustained yields can be maintained over an indefinite period because the preservation of natural systems is considered a priority. These pioneers are concerned about the ability of their farms to be productive, not for the next ten years or for their lifetimes, but forever.

CONSOLIDATING AND EXTENDING IDEAS

1 a) Write an explanation of the difference between the terms (i) *sparse* and *underpopulated*; (ii) *dense* and *overpopulated*.

b) Write an explanation of the term *optimum population*.

2 Study the photographs on page 65. Describe them in your notebook using terms from this chapter such as *sparse*, *resource base*, and *population distribution*.

CASE STUDY

China

China is an intriguing country. With over 1 billion people, an isolationist political system, and a vast land area, it remains an enigma to most Westerners. You know that China has a huge population. But you might not know that most of western China is as sparsely settled as the Canadian prairies! Most of the people live in fertile valleys and along the coastal plains of eastern China. Why is the population not evenly distributed across the country? The answer lies in the economic potential of each region.

Traditionally, eastern China has supported a rural peasant population. The people work the rich soil using hand tools and machinery supplied by government-run **cooperatives**. The abundant summer rainfall, warm temperatures, and long growing season support crops of wheat (in the north), rice (in the south), market vegetables, and hogs, ducks, and chickens.

With its location on the flood plains of the Huang He, the Yangtze, and the Xi Jiang, water is readily available for flooding rice paddies and irrigating other crops. In the past, with the Himalayan snow-melt of spring and the heavy monsoon rains of summer, these rivers flooded regularly. The floods carried sediments, which acted as natural fertilizers for the soils. Today, dams regulate seasonal flows and eliminate flooding. Fertilizer is now purchased from cooperatives. Greenhouses are used to start rice crops early in the spring. The seedlings are transplanted when the rains come and the paddies can be flooded. The introduction of hardy hybrid wheat and rice strains resistant to fungus has resulted in greater harvests.

Despite such a solid agricultural resource base, the east is still overpopulated. The population has flourished for so long that the land cannot support any more people. In 1979, the government instituted a "one-child" policy in the hope that the population will

Figure 6.10 *left*, Eastern China's burgeoning population; *right*, sparsely populated western China

come into line with the resource base. In urban areas, where most of the population lives, there are signs that the policy's goals will be met. However, enforcement of the policy is difficult in rural areas.

In addition to a reduction in population, it is expected that the resource base will be expanded through economic development. Manufacturing is starting to boom in the cities of the southeast. Raw materials from within China and from all over Southeast Asia are brought into the region. Heavy industry has grown rapidly since the Cultural Revolution. Coal-generating plants are being hastily built to meet the needs of growing domestic and export markets. The huge labour force produces inexpensive, durable goods such as textiles, shoes, and toys, and, increasingly, electronic and more expensive value-added goods.

The main resource of the region is becoming its *human* resource. Farm labourers who cannot find work in the agricultural regions come to the cities and work in the new factories and workshops that are springing up everywhere. While the east is overpopulated, the expansion of the resource base and the introduction of new technology in the factories are providing more opportunities for people. Combined with the burgeoning population, the economic potential of the region is enormous. Of course, economic success has its price. The use of lowgrade coal causes all kinds of respiratory diseases, fouls ecosystems, and increases the amount of carbon dioxide—a greenhouse gas—in the atmosphere.

Western China is quite different. Far away from the moderating influence of the South China Sea, the climate is cold and dry. Classified as cold desert, or semi-desert, the land is raw and lifeless. Soils range from sand to rocky desert pavement. Only coarse grasses and scrub grow here. Field crops are not generally grown because of the severity of the region. A few people are able to exist as **nomadic herders**. They migrate from highland regions, where higher rainfall provides some pasture in spring and summer, to protected valleys for the long, cold winters. Theirs is a subsistence economy. Virtually everything they need comes from the animals they herd. The herders trade animals or animal by-products for essentials that they do not produce, such as tea and sugar.

The government in Beijing is encouraging the nomads of western China to resettle in **sedentary** communities by offering them better health care, state education, and a reliable income. They will stay on large communes and raise their animals there for resale. However, their old ways are more sustainable. **Migratory** herding makes the best use of marginal lands because the animals do not overgraze one particular area. If the land is fenced, the animals are not free to move with the seasons. Hay that must be grown for the wintertime is harvested using tractors. If the land is too dry, wells pump water out of aquifers deep beneath the surface. Eventually, there will be no oil to run the tractors and the aquifers will dry up as these resources become overused.

The herders are reluctant to abandon the tradition of their ancestors and they do not want to be controlled by a government thousands of kilometres away. So while the population may be sparse in these western lands, the region may in fact be overpopulated, just as it is in the east, because there are too many people for the resource potential of the land.

CONSOLIDATING AND EXTENDING IDEAS

1 Study the **graded shading map** in Figure 6.11. Describe the population pattern it reveals.

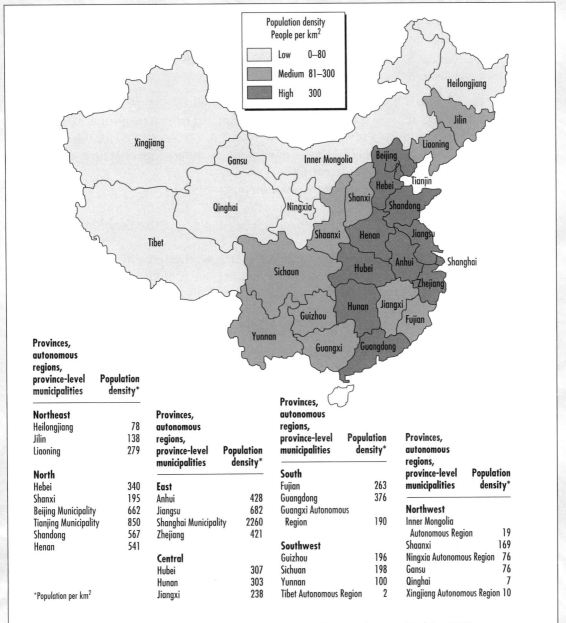

Figure 6.11 Population density by provinces, autonomous regions, and province-level municipalities, 1993

Source: Statistics from State Statistical Bureau, Beijing, China, 31 December 1993.

Population density
People per km²

	Low	0–80
	Medium	81–300
	High	300

Provinces, autonomous regions, province-level municipalities	Population density*
Northeast	
Heilongjiang	78
Jilin	138
Liaoning	279
North	
Hebei	340
Shanxi	195
Beijing Municipality	662
Tianjing Municipality	850
Shandong	567
Henan	541

*Population per km²

Provinces, autonomous regions, province-level municipalities	Population density*
East	
Anhui	428
Jiangsu	682
Shanghai Municipality	2260
Zhejiang	421
Central	
Hubei	307
Hunan	303
Jiangxi	238

Provinces, autonomous regions, province-level municipalities	Population density*
South	
Fujian	263
Guangdong	376
Guangxi Autonomous Region	190
Southwest	
Guizhou	196
Sichuan	198
Yunnan	100
Tibet Autonomous Region	2

Provinces, autonomous regions, province-level municipalities	Population density*
Northwest	
Inner Mongolia Autonomous Region	19
Shaanxi	169
Ningxia Autonomous Region	76
Gansu	76
Qinghai	7
Xingjiang Autonomous Region	10

2 Prepare an organizer or a chart comparing population, technology, and resources in western and eastern China.

3 Which of the following strategies do you think Chinese planners should consider? Why?
- moving people from the densely populated eastern half of the country to the sparsely populated western half
- enforcing the one-child policy (see "The Fertility Revolution in China," page 73)
- developing manufacturing to increase the resource base
- your own suggestions

4 Today, the earth has 5.77 billion inhabitants. By the year 2000, this figure will rise to about 6.2 billion and will continue to increase at a rate of about 100 million people per year. Some individuals believe that the world's population increase must be controlled for the betterment of humankind. Others believe that increases in world population should not be controlled. **Should world population be controlled?**
In this position paper you must defend a point of view.
Your final paper should include:
a) an introductory paragraph that identifies the issue and alternative positions (points of view).
b) a paragraph that presents one position on the issue as indicated above.
c) a paragraph that presents the other position on the issue as indicated above. Each paragraph (b and c) should argue the advantages of the position, and support the point of view with specific facts and examples.
d) a concluding paragraph that defends *your* position on the issue.

5 Examine the pie graphs in Figures 6.12 and 6.13.

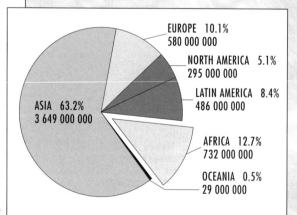

Figure 6.12 Distribution of world population, 1996
Source: Copyright © World Eagle 1996.

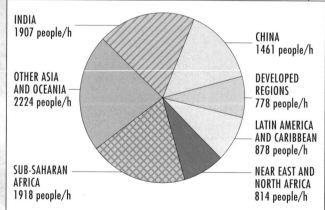

Figure 6.13 Population added each hour by region, 1994
Ten thousand people are added to the world population every hour.
Source: Copyright © World Eagle 1994. Reprinted with permission from World Eagle, 111 King St., Littleton, MA 01460 USA. 1-800-854-8273. All rights reserved.

Extrapolate from these graphs to produce your own pie graph that predicts distribution of world population in 2001.

6 Read the article "The Fertility Revolution in China" below.
   a) In an organizer or a chart, list each of the government's population policies in China and assess them according to their positive and negative effects.
   b) In a group, discuss your assessment of these policies' effects. Based on this analysis, do you think the experiment in China will work to reduce population? Explain your reasoning. Why could a country such as Canada never have the same type of controls?

7 In order to understand more immediately the complexities of world population and the indicators of standard of living and quality of life, consider the following:

> If there are thirty students in your class, twenty would live in Asia; three in Africa; seven in Europe; two in South America; two in Russia; one in North America.

What would their lives be like? Refer to the *Canadian Oxford School Atlas*, 6th ed., pages 132 to 137, and 140. For each student's home continent, determine
- population density;
- life expectancy at birth;
- infant mortality rate;
- population per physician;
- adult literacy rate;
- average nutritional levels;
- percentage of labour force in agriculture;
- GDP per capita.

Use a concept web to describe your findings.

THE FERTILITY REVOLUTION IN CHINA

In the early 1970s, China had one of the highest population growth rates in the world. Its population, dangerously nearing the billion mark, made up one-fifth of the total world population. Concerned, the Chinese government set out to introduce a series of family-planning policies.

What has happened, as many demographers have observed, represents one of the most rapid "fertility transitions" in human history.

But the real social effects of this remarkable turnaround are only now beginning to emerge. The results of the second phase of the In-Depth Fertility Surveys were released in China earlier this year [1990]. . . .

China's State Statistical Bureau conducted the surveys with the help of the International Statistical Institute in the Hague and the governments of Norway, Finland, and Denmark. The study involved interviews with married women aged fifteen to forty-nine in various parts of China. It documented some drastic changes.

At the time the Chinese government announced its first family-planning program, women in China had an average of nearly six children. The "later, longer, fewer" campaign of 1972, which emphasized later ages for marriage and childbearing, longer intervals between births, and smaller families, showed positive signs of getting the birth control message across. By the time the one-child-per-family policy was announced in 1979, fertility had fallen to 2.7 children per woman.

The one-child policy has led to even further reductions. Urban areas such as Beijing and Shanghai have essentially conformed to the policy, while rural areas have reached an average of about two children per family. . . .

Because China has not attained fertility levels impressively lower than most other developing countries, researchers are beginning to look at how Chinese society has adapted to the family-planning programs. Not all the developments have been positive.

Dr. Zeng Yi, Deputy Director of the Institute of Population Research at the Beijing University, says the discrepancy in statistics between rural and urban settings could be a major problem. He predicts that, in the future, only 8 per cent of women will comply with the one-child policy, with the average settling at 2.5 children per woman. But 61 per cent of urban women will only have one child. This, in the long run, could mean a shortage of young people in cities to care for their elderly parents. "Serious labour shortages and a significant health care burden imposed by an aging urban population will plague Chinese cities in the next century unless policies are introduced to mediate these effects," Zeng Yi says.

These proper policies, he says, should include the encouragement of rural youths to migrate to towns and cities and greater consistency in family planning in rural and urban areas. . . .

The birth control programs have also been unable to change the ingrained preference for a son among married couples. Despite government efforts, increased education, and modernization, more than 50 per cent of respondents in fertility surveys wanted a son as their first child, compared to only 5 per cent who wanted a daughter. The others did not express a preference. Often, those couples whose first child was a male signed the one-child certificate, whereas those whose first offspring was female went on to have other children in hope of having a boy.

Despite the potential problems of urban underpopulation and the inherent favouritism of son preference, the Chinese have managed to fuse successfully the modern notion of family planning with tradition. The results of the In-depth Fertility Surveys showed this fusion of attitudes and behaviour relating to fertility and family planning.

One of the main topics the study concentrated on was the age of Chinese couples at marriage and co-residence. An important component of China's birth control program has been the encouragement of late marriages and the postponement of childbearing. The legal age of marriage was raised from eighteen to twenty in 1980 but the officially recommended age has been higher—twenty-three for rural and twenty-five for urban women.

Since 1982, marriage age has averaged well over twenty years old in all of China's provinces. Indeed, the average marriage ages have been similar to those of developed countries for nearly two decades.

Dr. Carol Vlassoff and Dr. Iqbal Shah (World Health Organization). From *IDRC Reports* (October 1990). Reproduced with the kind permission of the International Development Research Centre, Ottawa, Canada.

CHAPTER 7

Theories of Population Growth

Most demographic experts today believe that population growth must be controlled. Yet this is a relatively new attitude. In the past, the majority of experts thought that steady population growth was beneficial. When the world had ample space and resources, population growth was a positive development.

ANCIENT PHILOSOPHERS

The Chinese philosopher Confucius believed that the number of people the land could support was limited. Nevertheless, he favoured large families, contending that productivity, defence, and tax revenues would all increase as the population grew. If too many people became concentrated in one region, Confucius believed that migration, either voluntary or enforced, would prevent the natural population checks of war and famine.

Two philosophers of ancient Greece, Plato and Aristotle, supported the view that a large population provided a military advantage. Since the Greek city-states were frequently at war with each other and with outsiders, a large population meant security. Both cautioned that each city-state should be self-sufficient and should not have to rely on trade for its essential needs. If the population grew so large that the city-state could not support it, the state would be vulnerable during times of siege. Politics, rather than economics, was the cornerstone of the philosophy of these founders of democracy. For democracy to work, a population had to be small enough so that citizens could vote on the affairs of state on a regular basis. Later in the Greek period (800 BCE to 146 CE), when populations increased, new city-states were created in neighbouring lands rather than allowing the existing city-states to become overpopulated.

The Romans, with their vast empire, were not concerned with population control. In fact, the larger the population, the more soldiers there were to fight for "the glory of Rome." When the empire became too large for the Romans themselves to defend, they relied on foreign mercenaries to guard their distant borders. Unlike the Roman soldiers, these foreign soldiers did not have a sense of loyalty to the empire, and so the Romans found themselves vulnerable to invasion in the third century. Thus the lack of an adequate Roman population to defend its territory contributed to the decline and fall of the Roman Empire.

Around the same time, Hebrew scholars also supported the notion of large populations. In the third century BCE, Kautilya equated a large population with both economic and military power. People were encouraged to have large families in order to offset the high death rates caused by famine, pestilence, and war in the Middle East.

Ibn Khaldūn, an Arab scholar of the fourteenth century, asserted that high concentrations of population were economically beneficial because they increased the division of labour. Some people could be farmers, while others could be traders, weavers, or money lenders. He also observed that population growth rates correlated to economic fluctuations. When the economy is good, there is political stability and population growth. The tax base increases and the quality of life improves. However, in time, political instability returns, and the economy declines along with the population. And so the cycle continues.

While some of these beliefs may seem simplistic or incorrect in the modern era, it is likely that they made sense for the time when these scholars lived. For example, in ancient times a large army was necessary to defend city-states. Today, the size of an army is no longer as important as the extent and sophistication of the military arsenal. Similarly, current ideas that may seem appropriate now may need to be rethought as times change and the population continues to grow.

MALTHUS

One of the first people to warn against the dangers of population growth was the English economist, minister, and philosopher Thomas Malthus. Malthus observed the clustering of populations in the rapidly growing cities of England during the Industrial Revolution and the dire conditions that resulted. Overcrowding led to inadequate housing, a lack of sanitation, widespread crime, and a variety of other social ills. Malthus concluded that, left unchecked, rapid population growth would destroy the quality of life.

He argued that population has the potential to double every generation. Food production, on the other hand, increases at a much slower rate. While he applied his theory only to England (see Figure 7.2), it had grave implications for the world at large.

Figure 7.1 An urban area in England during the Industrial Revolution

GENERATION (25 YEARS)	POPULATION	POPULATION THE FOOD SUPPLY CAN SUPPORT
1	11 000 000	11 000 000
2	22 000 000	22 000 000
3	44 000 000	33 000 000
4	88 000 000	44 000 000
5	176 000 000	55 000 000

Figure 7.2 Population growth rate and food supply in England (Malthus)

According to Malthus's calculations, human population would grow geometrically (1, 2, 4, 8, 16 . . .) while food production would grow arithmetically (1, 2, 3, 4, 5 . . .). Thus, by the third generation, population would exceed the food supply and the gap between the two would grow wider with each generation. The result would be poverty and starvation because the food supply would be unable to meet the demand.

Malthus's predictions for England did not materialize. This is because he was unable to foresee the consequences of the Industrial Revolution. New technological developments enabled the food supply to increase geometrically along with the population. Hybrid plants and animals and the cultivation of new territory in the Americas and Australia added to the food supply. Moreover, the population of England did not double every twenty-five years as Malthus had predicted. Average family size declined. England continued to prosper. Eventually its population stabilized. Today its growth rate is a slow 0.1 per cent a year. Still, in today's world of approximately 6 billion people, Malthus's theory is relevant. Many modern economists and scientists who support Malthus's theories believe that the time will come when natural systems will no longer be able to support the world's population.

As we become increasingly aware of the impact people have on natural systems, it might be wise to look again at the teachings of

> In 1845, Ireland was devastated by a potato famine. The Irish population had grown rapidly. Potatoes were the favoured crop because they provided the highest yield and food value per unit of land. When a blight destroyed most of the potato crop, there was widespread famine. Over one million people died; even more escaped the famine by immigrating to North America. Was Malthus right? Had the population exceeded the food supply? Or could the disaster have been avoided had there been a greater diversity of crops?

Malthus. As environments become overused and unable to deliver the natural resources needed by certain regions to sustain their populations, mass starvation may develop in specific regions. This is happening in many parts of Equatorial Africa and south Asia. The 1986 famine in Ethiopia occurred when the resource base was diminished as a result of reduced rainfall. Seasonal rains did not fall, pastures did not grow back, thousands of domesticated animals died, and the region became vastly overpopulated. (This topic is explored further in Chapter 13.) But what of the whole world? Is it conceivable that people will so upset natural systems that essential resources may no longer be sufficient for the world population? Of course, we cannot answer this question. But the planet is changing because of the impact human activities are having on natural systems. It is quite conceivable that one day the world will be uninhabitable because of what people are doing to the planet today! The natural checks of war, famine, and pestilence mentioned by Confucius, which we have been able to hold at bay for so long, may claim more and more human victims in the coming centuries. **Pandemics** of diseases such as Aids and influenza are becoming more frequent. They are spreading as a result of our increased mobility. Internal conflicts in developing nations from eastern Europe to central Africa are escalating, not only because of the lack of adequate resources, but also because of the inability of political forces to deal with problems related to resource distribution.

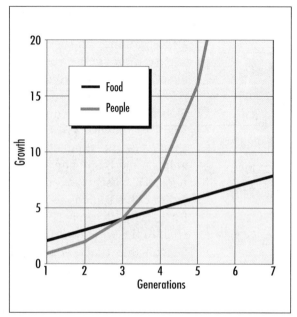

Figure 7.3 Food production

The graph shows food production as an arithmetic function, while population growth is geometric.

CATTON

In 1982 William Catton wrote *Overshoot: The Ecological Basis of Revolutionary Change*. This modern theory of **demographics** (the study of population growth) examined current population trends. Some people consider Catton to be a neo-Malthusian because he is reviving Malthus's theory that the world will run out of resources, resulting in mass starvation. After you read this brief summary of his ideas, decide whether you agree with this evaluation.

Catton defines **carrying capacity** as the number of people a given environment can support indefinitely. The environment is not harmed to the extent that it can no longer support the population, and the population does not

increase to the point that it places undue strain on the environment. In a sustainable economic system, the needs of the population are balanced against the health of the environment.

This idea is a relatively new one. In the sixteenth and seventeenth centuries, European expansion into underpopulated regions of the Western and Southern Hemispheres began to increase the earth's carrying capacity. There was a general belief that the earth's resources were limitless. Spain was exploiting the gold and silver deposits of Peru and Mexico. Britain and France were capitalizing on Canada's furs and timber. The Netherlands and Portugal were bringing spices, valuable woods, and precious gems from tropical islands. The earth's resources were being exploited at an unprecedented rate. According to Catton, this expansion laid the foundation for today's population crisis.

The process continued during the Industrial Revolution. New agricultural areas developed as farm technology improved and marginal lands could be made more productive. The development of the steam engine and the internal combustion engine, along with the discovery of electricity, revolutionized the way in which work was done. Human and animal muscle were replaced with the energy of fossil fuels. Nonrenewable resources, such as coal, oil, and natural gas, allowed Western economies to expand rapidly. The earth's carrying capacity continued to grow, and the world's population exploded.

Once growth and prosperity were commonplace, people expected the upward spiral to continue forever. This led to the **drawing down** of nonrenewable resources, that is, the stealing of resources from future generations. Today, we have come to realize the impact of industrial expansion on natural systems. Yet people in the diversified economies, such as Canada and the United States, continue to demand increasingly higher standards of living. At the same time, the populations in subsistence economies and emerging industrial economies are growing at unprecedented rates. These two phenomena cannot continue. Once the nonrenewable resources are gone, the earth's carrying capacity will drop dramatically. Then what will happen to the world's population? Since the Second World War, Western society has relied increasingly on technology to solve our problems. But this may be a dangerous reliance. There are no guarantees that technology can save the planet. The time may come when the earth's natural systems have been so abused by the human population that life as we know it may be seriously threatened.

Figure 7.4

Populations in diversified economies continue to demand a greater variety of goods and services.

CONSOLIDATING AND EXTENDING IDEAS

1 Summarize the views of the following in an organizational chart: Confucius, Aristotle and Plato, ancient Romans, early Christians, Kautilya, Ibn Khaldūn, Malthus, Catton.

2 a) Prepare a double-line graph comparing population growth and food production over five generations as Malthus predicted them.
   b) Why didn't Malthus's predictions for England materialize in the nineteenth century?
   c) Do you think that Malthus's theory will eventually prove to be an accurate one? Explain your answer.

3 How is Catton's view on population growth similar to Malthus's philosophy? How is it different? Do you consider Catton to be a neo-Malthusian? Explain your answer.

4 Assume that you are a population philosopher. Explain your views about population growth in a brief paragraph. Share them with the class in an oral presentation.

THE DEMOGRAPHIC TRANSITION MODEL

The world's population keeps growing, while resources are disappearing at an alarming rate. Surely the time will come when the earth's carrying capacity will be unable to support its population. Or will it?

Malthus's predictions for England never materialized and, so far, Catton's theory has not been proven correct either. One reason may lie in the **demographic transition model**. Human populations go through stages of growth that are dependent on technological and cultural changes. At various stages in the model, a population remains stable, increases rapidly, and even declines. When Malthus was writing, England was experiencing unprecedented population growth. He expected this growth would

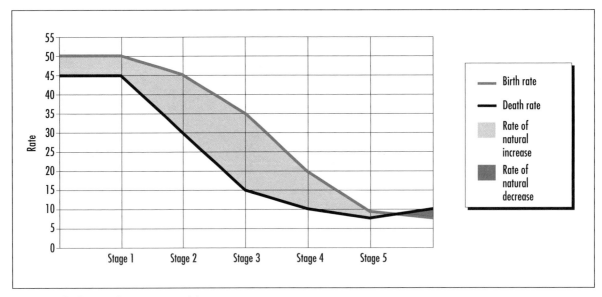

Figure 7.5 The demographic transition model

continue unchecked until food supplies ran out. Then there would be a period of population decline. In fact, what happened was a decline in England's population growth rate as technology became more advanced and social welfare improved. In the mid- to late nineteenth century enormous social reforms evolved in Britain. The establishment of labour standards regulated working conditions, restricted child labour, and made education compulsory. These three reforms, in particular, effectively eliminated the advantage of having a large family. Since child labour was no longer tolerated, there was no point in having large families because children could no longer work to augment the family income. Improved technology resulted in better health care and diet. The result was that more children lived to maturity. People didn't need to have as many children because survival rates had improved.

This pattern was repeated throughout western Europe and later in North America and other regions with diversified economic systems. We are beginning to see similar patterns evolving in regions with subsistence economies.

The demographic transition model is based on the statistical interpretation of birth rates and death rates. The **birth rate** of a country, also known as **crude birth rate**, is the average number of births during a year per 1000 population, at mid-year. **The death rate** of a country, or **crude death rate**, is the average number of deaths during a year per 1000 population, at mid-year. The **rate of natural increase** is determined by subtracting the death rate from the birth rate. When both the death rate and the birth rate are high, the rate of natural increase is low. Similarly, when both the birth rate and the death rate are low, the rate of natural increase is also low. Population growth occurs when the birth rate is much higher than the death rate. Conversely, if the death rate is much higher than the birth rate, the population declines.

Statistical analysis provides evidence to support this argument. While Malthus was a visionary whose ideas changed the way people thought about population growth, his ideas were unsubstantiated by statistical evidence.

STATLAB

1 a) Create a multiple-line graph for one of the countries in Figure 7.6. (Consider using a computer spreadsheet program with graphing capabilities.)

   b) Shade in the rate of natural increase as the area between the birth rate and the death rate. Indicate the period when population growth rates were increasing rapidly.

   c) Divide into groups. Compare your graph with those of other group members. Find periods when population patterns seemed similar for the countries in this study.

   d) Hypothesize as to what type of factors have caused changes in birth and death rates in these countries.

   e) Account for the following patterns:
      i) the increased birth rates in 1950 or 1960 for each country
      ii) the slightly increased death rates for Canada and Sweden in 1990

SWEDEN

DATE	BIRTH RATE	DEATH RATE
1740	36	34
1840	30	19
1870	27	17
1890	29	15
1910	24	14
1930	15	12
1950	16	10
1970	14	10
1990	13	11
1995	14	11

CANADA

DATE	BIRTH RATE	DEATH RATE
1740	n/a	n/a
1840	45	22
1870	37	19
1890	30	16
1910	29	12
1930	21	10
1950	28	8
1970	21	7
1990	14	8
1995	15	8

MEXICO

DATE	BIRTH RATE	DEATH RATE
1840	n/a	n/a
1870	n/a	n/a
1900	47	34
1920	43	38
1940	44	22
1960	45	11
1980	37	6
1990	30	6
1995	28	5

UNITED STATES

DATE	BIRTH RATE	DEATH RATE
1840	52	32
1870	42	19
1900	32	17
1920	28	13
1940	19	11
1960	24	10
1980	16	9
1990	16	9
1995	16	9

Figure 7.6 Birth and death rates in Sweden, Canada, Mexico, and the United States

Source: Statistics from *World Resources, 1996-97*.

2 Study the poem "Statistics."
a) Briefly restate the poem in your own words.
b) Outline how statistics misrepresented the true state of the nation described.
c) In what other ways can statistics misrepresent reality?

Statistics
Statistically
It was a rich island
income per capita
one million per annum

Naturally
it was a shock to hear
half the population
had been carried off
by starvation
Statistically
It was a rich island

A UN Delegation
(hurriedly dispatched)
discovered however
a smallish island
with a total population
of—2
Both inhabitants
regrettably

not each a millionaire
as we'd presumed.
But one the island owner
Income per annum:
Two million
The other
his cook/chauffeur
shoeshine boy/butler
gardener/retainer
handyman/labourer

The very same
recently remaindered
by malnutrition

Statistically
It was a rich island
income per capita
one million per annum

From Cecil Rajendra, *Other Voices, Other Places* (New York: Macmillan Co., 1972).

STAGES OF THE DEMOGRAPHIC TRANSITION

Stage 1

The first stage of the model is typified by high birth and death rates. Population growth is stable. It is neither increasing nor decreasing because the number of people born in each generation is roughly equivalent to the number of people dying. While not many countries are at this stage of development today, there was a time just a few hundred years ago when this pattern was widespread. There was no health care or medicine as we know them. Infant and child deaths were common. Nutritional standards were marginal. Disease and death plagued communities. Couples had large families to ensure that at least some of their offspring would survive to adulthood to support them in their later years. Today, the only regions where this stage may exist are in remote areas of the tropical rainforests where there has been little contact with the outside world. Figure 7.6 shows that Sweden, like most of the world, was in Stage 1 of the model in 1740.

Figure 7.7 A hunter from the Wergi tribe in western Australia (stage I)

Figure 7.8 A family in Côte d'Ivoire (stage 2)

Stage 2

This stage is characterized by high birth rates and low death rates. As a result, the rate of natural increase is so large that there is a population explosion. Industrial nations passed through this stage in the nineteenth century. Emerging industrial nations, such as Mexico, entered this stage in the early twentieth century. Many subsistence nations, such as Malawi, are just beginning to enter Stage 2 of the model.

Unprecedented population growth strains the economies of nations at this stage. Western Europe and North America were able to absorb the excess population in the rapidly growing industrial cities. Surplus populations that could not be absorbed migrated to the largely unsettled western territories of the United States and Canada.

The decline in the death rate in Stage 2 is attributed to the expansion of scientific knowledge that occurred in nineteenth-century Europe. Medical advances had a marked influence on public health during this period. For the first time, disease and poor health were linked to unsanitary living conditions. Old superstitions were replaced with proven medical knowledge. Doctors began to gain a better understanding of the nature of many diseases. A vaccine against smallpox was developed in 1801. Over the next 150 years many other diseases were virtually wiped out. Water purification and the development of modern sewage systems eliminated diseases such as cholera and typhus that had plagued the growing urban centres. Pasteurization, refrigeration, and canning led to better nutrition. As a result, people were healthier and lived longer. **Infant mortality rates** dropped as more and more infants lived past their first birthdays.

These scientific advances were extended to the developing world in the latter half of the twentieth century. Improvements in health care caused death rates to drop. But the low standard of living and the high cost of medical technology made it impossible for these subsistence economies to implement health programs at the same level as in developed nations.

As populations in the developing world increased, so did pressure on the environment. Unlike the situation in nineteenth-century North America, there were few new frontiers waiting to be developed. Surplus populations were forced to stay where they were and environments became strained to the point of ecological disaster. Recent famines in Ethiopia, Sudan, and Somalia in sub-Saharan Africa are, in part, the result of too many people living on marginal lands.

In many emerging industrial nations, there has been a massive exodus of people from rural areas to the rapidly growing industrial cities. The low-paid labour these people provide has enabled many Pacific Rim countries, such as China, Taiwan, Thailand, and Indonesia, to enter a new era of economic growth. The living conditions are remarkably similar to those that existed in Victorian London. While these conditions may seem adverse, they are often a vast improvement over areas where environmental degradation, overpopulation, and government disinterest are common. Rural migrants become caught up in the hustle and bustle of the cities. Their children have greater opportunities for education, and family-planning information is more accessible than in remote villages. What remains to be seen is if the giant cities of the developing world can expand their essential infrastructures of water purification plants, sewers, and waste-management facilities to deal with increasing urban populations.

Stage 3

This stage is characterized by low death rates and rapidly declining birth rates. As a consequence, this is the period of greatest natural increase. Nations with diversified economies have already passed through this stage, while many emerging industrial nations are currently in it. While Stage 2 was influenced primarily by technology, Stage 3 is influenced by social and cultural change.

It takes several generations before birth rates start to decline as rapidly as death rates. At the beginning of Stage 3, couples continue to have many children, even though fewer die before reaching maturity. People are now living longer, and for the first time it is possible that they may live longer than they are able to be self-supporting. A large family may ensure that there will be enough offspring to care for and support the parents in their later years. This is still a common characteristic of many developing nations, especially in rural areas.

Another reason for large families is income. In eighteenth-century Europe young children frequently worked as farm labourers. During the Industrial Revolution, children as young as seven laboured in the mills and factories under deplorable conditions. Many were forced to work fourteen hours a day, six days a week. Their meagre earnings helped to support the family, and so the more children a family had, the greater the income. Today many of the emerging industrial economies are undergoing a similar experience. Many children work long hours in sweatshops or spend the day begging, stealing, or scavenging in garbage dumps. Rapid industrialization is outpacing society's ability to cope with change.

In the mid-nineteenth century the diversified economies began to confront the social challenges of the new society. Social welfare programs were introduced to provide for the needs of citizens. Pension plans allowed people to invest a portion of their salary to provide security in their retirement. Unemployment insurance, disability pensions, and welfare payments provided safety nets for those who were unable to work. New laws made it illegal to employ children under the age of sixteen, and children were legally required to attend school. All of these factors eliminated the need for large families. As a result, birth rates began to decline.

Today many developing countries have yet to implement similar programs. They lack the financial resources to initiate such massive reforms. Until they are able to do so, it is unlikely that birth rates will decline, and these countries will remain in Stage 3 of the demographic transition model.

Figure 7.9 A mother with her two children in South Korea (stage 3)

UN SAYS CHILD LABOR RATES HAVE DOUBLED

From the brothels of Asia to the carpet factories of Pakistan, about twice as many children are working full time in developing countries as previously thought, the International Labor Organization (ILO) said Monday.

The latest calculations from the UN labor agency show that 250 million five- to fourteen-year-olds are employed, half of them full time. That's up sharply from earlier estimates of 73 million full-time child workers.

The new figures come after in-depth surveys and interviews in numerous countries. Previous estimates were based almost solely on official statistics.

The ILO report found nearly 153 million children are working in Asia, 80 million in Africa and 17.5 million in Latin America. It called for a new international accord banning the harshest forms of child labor: slavery, prostitution and work in hazardous industries.

Only forty-nine UN members ratified a 1976 child labor convention. Some nations said its limits on paid work were too broad.

ILO director Michael Hansenne said child labor only perpetuates an endless cycle of illiteracy and poverty.

"We all know that . . . many efforts over the years will be required to eliminate it completely," he said. "But there are some forms which are intolerable by any standard. These deserve to be identified, exposed and eradicated without further delay."

Slavery or child bondage still is practised in South Asia, Southeast Asia and West Africa, the report said. Children are either sold outright or rural families are paid in advance by "contractors" who take children away to work in carpet weaving, glass manufacturing or prostitution.

Child trafficking for the sex industry is increasing despite better international awareness of the problem, the ILO said.

In Asia, child prostitutes number about 1 million and rising, the report said. Numbers also are increasing in Burkina Faso, the Ivory Coast, Ghana, Kenya, Zambia and Zimbabwe.

It identified international sex networks that take Latin American children to Europe and the Middle East, and Southern Asian children to northern Europe and the Middle East. Child sex markets were also established in West Africa, Europe and the Arab world, it said.

Among other things, the ILO found that:
• Certain industries are exposing their child workers to pesticide poisoning, lung diseases or even crippling their growing bodies by forcing them to carry heavy weights.
• In Sri Lanka, more children die from pesticide poisoning than from a combination of childhood diseases such as malaria, tetanus and whooping cough.

• Children are exposed to dust and fumes in repair shops, woodwork factories and construction sites in Egypt, the Philippines and Turkey.

In glass factories, children are often forced to drag loads of molten glass from glowing furnaces amid noise levels that could cause deafness. Children as young as three were working in match factories where they were exposed to dust, asbestos and other hazardous fumes.
• Up to 5 million child domestic servants work in Indonesia, including 400 000 in the capital, Jakarta.

From the *Calgary Herald*, November 1996. Reprinted by permission of Associated Press.

Stage 4

By Stage 4, rates of natural increase have stabilized. Birth rates have fallen in line with death rates, so population growth is low. Many nations in western Europe are currently at this stage.

In the future, it is likely that emerging industrial nations such as Brazil and Malaysia will enter this stage. (Appendix A on page 433 provides many examples of countries in which population growth rates are declining.) Brazil's population, for example, is projected to drop from 2.2 per cent per year from the period 1980 to 1985 to 1.2 per cent by the period 2000 to 2005. Countries with subsistence economies will not likely enter this stage in the immediate future. Most of the countries of Equatorial Africa, for example, show little change in population growth projections over the next ten years, so continued growth is indicated for these regions.

The implications of most countries entering Stage 4 of the model in the foreseeable future are significant. If most countries enter Stage 4, world population growth could stabilize by the

end of the next century. But are the projections accurate? There are many variables, including disease and other health-related issues, economic conditions, and social attitudes, that may affect population growth patterns.

The decline in birth rates during Stage 4 is likely the result of a variety of **socio-economic** factors. Consider the trend in Canada since the Second World War. Once peace had been restored in 1945, families were reunited, many couples were married, and the birth rate soared. The country was entering a period of economic prosperity. Families of four or five children were common in the 1950s. Today, however, families with two children are the norm, and a growing number of couples choose not to have children.

It was not the availability of social programs that led to this change, since these had been in place well before the war. Family planning, including contraceptives, offered couples the opportunity to control the size of their families. But there were other important social changes taking place, most notably the changing role of women. During the war, women joined the

Figure 7.10 Some couples choose not to have children. (stage 4)

1960s, and 1970s. But the good times brought with them a high cost of living, so many couples restricted their family size for financial reasons, while some people chose not to have children at all.

Stage 5

In the final stage of the demographic transition model, the birth rate drops below the death rate. There are fewer and fewer births, until the number of births is less than the number of deaths each year. The population at this stage is growing older. As it does, its ability to reproduce declines. The **rate of natural decrease** accelerates. There is a real threat that the population may decline to a level from which it cannot recover. Few countries have reached this stage, but it is beginning to emerge in some European nations, such as Germany.

One of the consequences of an accelerating rate of natural decrease is that the language, culture, and traditions of the people may be threatened and could disappear. On the other hand, reduced population growth rates may have some positive effects. The impact on the environment is reduced. Individual prosperity increases as fewer people share the available resources. And it ensures that Malthus's dire predictions never come true.

labour force to run the industries at home while men were fighting overseas. While many women returned to domestic life after the war, they were not content to remain there. Increasingly, young women sought a good education. They began to enter the workforce in growing numbers. The traditional role of homemaker was now just one of many options available to them. As more women embarked on careers outside the home, the number of children born each year declined and the average age for women to bear children rose. This trend joined forces with the emergence of the **consumer society.** People revelled in the prosperity of the 1950s,

Socio-economic is a term that combines two very broad, interrelated concepts. *Socio* refers to society— the organization of people into a group with specific behavioural characteristics. *Economic* refers to how people obtain the goods and services essential to life. The term *socio-economic* simply means all the specific behavioural and economic characteristics of a group of people.

Figure 7.11 A couple enjoying retirement (stage 5)

Family Allowance benefits were started in Canada after the Second World War as a form of universal welfare to help defray the costs of raising a family. At the time, many economists believed that the increased money in circulation from Family Allowances would help stimulate the economy. In 1992 the program was replaced by the Child Tax Benefit.

CONSOLIDATING AND EXTENDING IDEAS

1 a) In an organizer or a chart like the one in Figure 7.12, summarize the characteristics of the five stages of the demographic transition model.

   b) Identify the stage(s) during which population explosions have occurred.

	STAGE 1	STAGE 2	STAGE 3	STAGE 4	STAGE 5
Birth rate					
Death rate					
Rate of natural increase					
Examples					
Reason for pattern					

Figure 7.12 Five stages of the demographic transition model

2 Read the article "UN Says Labor Rates Have Doubled" on page 85 and summarize it using point-form notes.

3 Consider the effects of the Industrial Revolution on the socio-economic trends of that time. In groups of four, discuss these factors: the Agricultural Revolution; changes in the textile industry; the development of the steam engine, and the iron/coal industries; advances in communication and transportation; new methods of production and their impact in creating a population explosion; urbanization; roles for women in the workforce; the institution of labor laws; and a new social structure.

4 a) Carefully examine the bar graph in Figures 7.13 (a) and (b). With a partner, interpret the graphs to
   i) identify stages in the demographic transition model for each of the regions;
   ii) compare Africa and Latin America with the former USSR, North America, and Europe;
   iii) suggest socio-economic impacts of these youth population trends.
   b) Share your analysis and evaluations with the class.

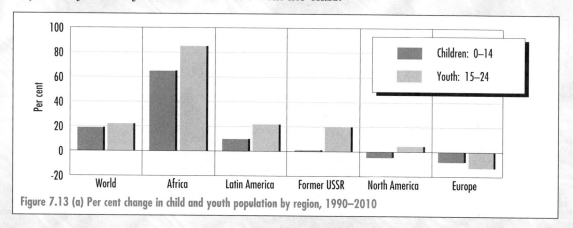

Figure 7.13 (a) Per cent change in child and youth population by region, 1990–2010

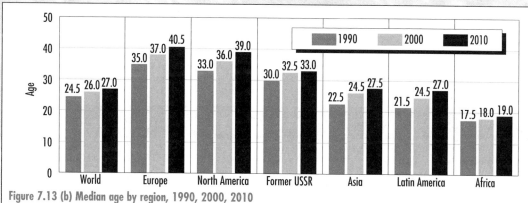

Figure 7.13 (b) Median age by region, 1990, 2000, 2010

5 Explain how Stage 5 in the demographic transition model could be
   i) the beginning of the end for the human race; **or**
   ii) the end of the beginning for the human race.

POPULATION PYRAMIDS

The **population pyramids** in Figure 7.14 are useful analytical tools for demographers. These back-to-back bar graphs set on a vertical axis show the age-and-sex breakdown of a country's population. They reveal a variety of characteristics, including

- a comparison of female life expectancy versus male life expectancy for each age group;
- the number of children (under fifteen years of age) as a percentage of the total population;
- the number of women of child-bearing age (fifteen to forty-nine years) as a percentage of the population;
- the number of people over the age of sixty-five as a percentage of the population;
- the **dependency load**, or percentage of the population under the age of fifteen and over sixty-five;
- the population growth rate;
- the demographic history over the past seventy-five years.

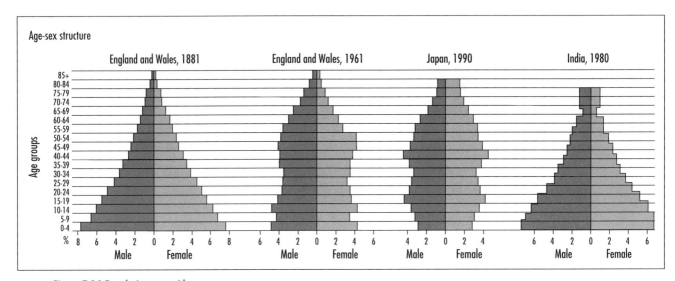

Figure 7.14 Population pyramids

STATLAB

Examine the population pyramids in Figure 7.14, then create your own from the data in Figure 7.15.

1 a) In general, what differences do you notice about the number of males and females for each age group?

b) Which country is the exception? Explain.

c) Which countries show significantly fewer females living into old age? Why do you think this is so?

Figure 7.15 Population by age and sex

CANADA, 1993	AGE GROUP	MALE (%)	FEMALE (%)
	0-4	3.6	3.4
	5-9	3.6	3.4
	10-14	3.5	3.3
	15-19	3.4	3.3
	20-24	3.5	3.4
	25-29	3.8	3.7
	30-34	4.4	4.4
	35-39	4.3	4.3
	40-44	3.9	3.9
	45-49	3.5	3.5
	50-54	2.7	2.7
	55-59	2.2	2.3
	60-64	2.0	2.1
	65-69	1.8	2.0
	70-74	1.4	1.9
	75-79	0.9	1.4
	80+	0.9	1.7

HAITI, 1993	AGE GROUP	MALE (%)	FEMALE (%)
	0-4	7.6	7.5
	5-9	6.7	6.6
	10-14	5.9	5.9
	15-19	5.2	5.1
	20-24	4.6	4.6
	25-29	3.8	4.0
	30-34	3.3	3.5
	35-39	2.6	2.9
	40-44	2.2	2.5
	45-49	1.8	2.0
	50-54	1.5	1.7
	55-59	1.2	1.4
	60-64	1.0	1.1
	65-69	0.7	0.9
	70-74	0.5	0.6
	75-79	0.3	0.4
	80+	0.2	0.3

HUNGARY, 1993	AGE GROUP	MALE (%)	FEMALE (%)
	0-4	3.1	3.0
	5-9	3.0	2.9
	10-14	3.3	3.2
	15-19	4.3	4.1
	20-24	3.7	3.5
	25-29	3.3	3.2
	30-34	3.1	3.0
	35-39	3.9	3.9
	40-44	4.0	4.0
	45-49	3.3	3.4
	50-54	2.9	3.2
	55-59	2.4	3.0
	60-64	2.4	3.0
	65-69	2.1	2.9
	70-74	1.6	2.6
	75-79	0.7	1.2
	80+	0.9	1.9

CHINA, 1993	AGE GROUP	MALE (%)	FEMALE (%)
	0-4	5.0	4.7
	5-9	4.8	4.5
	10-14	4.3	4.0
	15-19	4.0	3.8
	20-24	5.1	4.9
	25-29	5.3	5.0
	30-34	4.5	4.1
	35-39	3.6	3.3
	40-44	3.6	3.4
	45-49	2.7	2.5
	50-54	2.1	1.9
	55-59	1.9	1.7
	60-64	1.7	1.6
	65-69	1.3	1.3
	70-74	0.9	1.0
	75-79	0.5	0.6
	80+	0.3	0.5

CHILE, 1993	AGE GROUP	MALE (%)	FEMALE (%)
	0-4	5.4	5.2
	5-9	5.3	5.1
	10-14	4.8	4.7
	15-19	4.2	4.1
	20-24	4.3	4.2
	25-29	4.3	4.3
	30-34	4.3	4.3
	35-39	3.7	3.7
	40-44	2.9	3.2
	45-49	2.5	2.6
	50-54	2.1	2.2
	55-59	1.6	1.8
	60-64	1.3	1.6
	65-69	1.0	1.3
	70-74	0.7	1.0
	75-79	0.5	0.7
	80+	0.4	0.7

Source: Statistics from *World Resources, 1996–97*.

2 a) Calculate the percentage of the population under fifteen years of age in each population pyramid. Which countries have high birth rates?
  b) What impact would a high population of children have on the economy and social programs?
  c) Which countries have low birth rates?
  d) What impact would a low population of children have on the economy and social programs now and in the future?

3 a) Calculate the percentage of the female population that is of child-bearing age (fifteen to forty-nine) in each population pyramid.
  b) In which countries would you expect the population to decline in twenty years as children reach adulthood?

4 a) Calculate the percentage of the population over sixty-five in each population pyramid.
  b) Why could pension plans be threatened if the population over sixty-five becomes the dominant age group?
  c) Add the percentage of population over sixty-five to the percentage of population under fifteen to determine the dependency load.
  d) Which countries have a large proportion of **dependants**?

5 Identify the stage in the demographic transition model for each population pyramid.

6 Identify those periods in which the number of births grew rapidly (for example, the post-war baby boom) or declined. Determine if there were economic or historic events that affected birth rates during these periods.

7 a) Describe the demographic trends in China. (See Chapter 6, "The Fertility Revolution in China," page 73.)
  b) Why do you think Chinese planners might be alarmed when they analyse the results of the country's population policy?
  c) What do the results imply about Chinese population policies?

8 What generalizations can you make about population pyramids in developing and developed countries? Briefly explain your rationale.

A POLYNESIAN PARABLE

Figure 7.16 The head of a Moai on Easter Island

Easter Island is an enigma. Huge monolithic faces stare out to sea, seemingly looking for the forgotten race that once populated the island's shores. Isolated in the Pacific Ocean 3540 km west of Chile, archaeological evidence suggests that this island once supported a prosperous culture of 7000 Polynesians. Led by Chief Hota Matu'a, they migrated here about 450 CE, probably from the Marquesas Islands. They brought with them domesticated animals and seeds to found a new agricultural society. Once a land of forests, Easter Island provided tree trunks to make canoes and to serve as rollers to move the giant blocks of volcanic stone used to carve the statues.

As the population grew and the island prospered, more and more land was cleared for crops. Eventually there were no trees left. The people were trapped. They could not make canoes to migrate elsewhere. Population pressures mounted. Wars broke out between rival factions. Overpopulation led to environmental destruction and famine, and eventually to cannibalism. This ancient society was destroyed. It was not until 1722 that contact with the outside world was re-established. A Dutch expedition arrived at the remote islands, staying only for one day. By that time, the population had declined to about 4000 inhabitants. Could the lost civilization of Easter Island be a forewarning for the human race?

The first census was conducted in 1822.

Only 150 people remained. External pressures from contact with outsiders had caused the population to drop so dramatically. Many had died from diseases introduced by Europeans. Others had been captured and enslaved. Nevertheless, it is interesting to speculate what would have happened to this forgotten race if they had remained in isolation.

CONSOLIDATING AND EXTENDING IDEAS

1 a) Prepare graded shading maps to show population growth rates for the periods 1980 to 1985 and 2000 to 2005. See Appendix A on pages 433 to 441 for details.

   b) Describe the trends for the major regions of the world. Where do you foresee problems as a result of (i) overpopulation; and (ii) underpopulation?

2 What lessons can we learn from the parable of Easter Island? Explain your thoughts in writing, then discuss them in class.

3 a) Evaluate the theory put forward by Malthus in view of the current world situation.

   b) Do you agree or disagree with his theory? Explain your answer by giving specific examples from this chapter.

4 Many people equate overpopulation with crowding, but, in fact, density is not really the issue. Overpopulation actually relates to the long-term carrying capacity of land. When an area's ability to sustain life is degraded by its human occupants, it is overpopulated. The rate at which developed nations consume resources makes any increase in their populations far more detrimental to the earth than increases in the populations of developing countries. "The whole world cannot sustain the Western level of consumption. If 7 billion humans [projected world population by 2010] were to consume as much energy and other resources as do today's industrialized countries, five planets Earth would be needed to satisfy everyone's needs . . ." (*Caring for the Future* [New York: Oxford University Press, 1996], 51–52).

   Gather statistics that compare the consumption of developed nations with the consumption of developing nations. Your statistics could include annual energy consumption of fossil fuels; resource use (iron/steel, aluminum, paper, timber, etc.); land degradation (soil loss, deforestation, loss of biological diversity). Create pie graphs to compare "the West with the rest."

CHAPTER 8

SOLUTIONS TO OVERPOPULATION

LINKING THEORY TO REALITY

Demographic theories can be useful in assessing population patterns. Demographers use population projections, the mathematics of population growth, and theories of population momentum and spatial distribution to predict future global population growth, urbanization, emigration, and immigration.

Population projections—predictions about how many people live in a certain area—are often inaccurate. There are many reasons for

Figure 8.1 Vancouver has recently become one of Canada's fastest-growing cities.

this. The base data may be wrong. In many developing countries it is difficult to conduct an accurate census. Many people are **illiterate**. They cannot read or respond to census forms, so time-consuming verbal interviews are necessary. This can be complicated by the number of people that must be surveyed. Often, people living in remote villages are suspicious of government workers, so they do not reveal accurate information. In addition, developing countries frequently lack the funds needed for census taking.

Population projections also reflect the assumptions of the demographers. If these assumptions are inaccurate, the predictions will also be erroneous. For example, the reasons that people have children vary from one society to another. There is no simple or clear understanding, and attitudes are constantly changing. Therefore, assumptions made about one cultural group may be completely inappropriate for another. Fertility rates in western Europe dropped as infant mortality rates declined. But the evidence indicates that the opposite has occurred in Africa. Women marry at a young age and bear six or seven children on average. Fertility is viewed as a blessing rather than as something that should be controlled. Will Africa ever enter Stage 3 of the demographic transition model? Or does this stage apply only to developed societies?

Uncertainty about the accuracy of population projections can also result from random events that may dramatically increase death rates. War, environmental degradation, and disease plague many of the world's most heavily populated nations. Somalia, Sudan, Mozambique, Angola, Ethiopia, Afghanistan, Haiti, and Rwanda are countries that have endured civil wars or revolutions. The massive death tolls these conflicts create are obviously not part of any projections. Similarly, as population pressures strain the environment, the carrying capacity is reduced. In the drought-stricken regions of sub-Saharan Africa, the land has been stripped of its vegetation and can no longer support its inhabitants. Thousands of people have died as a result of famine. No demographer can factor the effect of unknown diseases into a projection. Over 15 million people are living with HIV, the virus that causes Aids. The majority live in the developing world. The implications for existing population projections are obvious. Other deadly viruses are being discovered in many jungles of the world. So-called "flesh-eating" diseases, highly contagious respiratory conditions such as **ebola**, and the re-emergence of more virulent strains of cholera, smallpox, and pneumonic plague could cause death rates to increase dramatically in developing countries. Some experts predict that the world's poorest countries could actually regress to Stage 1 of the demographic transition model, with high death and birth rates. While it is hoped that deaths will not escalate, these are factors that need to be considered in any evaluation of population projections.

> The world's population with access to safe drinking water rose from 29 per cent in the 1970s to 43 per cent in the 1980s. This seems like a great improvement. But, in fact, things are not getting better; they are getting worse. The number of people using an unsafe water supply increased by 100 million people over the same period. So while the percentage figures have improved, the actual number of people drinking unsafe water has increased.

THE MATHEMATICS OF POPULATION GROWTH

The way in which data are reported can also be misleading. Consider growth rates. The statistics in Appendix A indicate that population growth rates have been declining steadily and are expected to continue to do so. This should suggest that the number of people is also decreasing, but in fact the number of people is actually increasing. As an example, consider population growth for Malawi. The growth rate was 3.5 per cent a year from 1990 to 1995, but it is expected to drop to 2 per cent for the period 2000 to 2005. The 1990 population of 8.6 million had reached 11.1 million by 1995 and is expected to climb to 14.5 million by 2005, an increase of 59 per cent. As the population increases, the per cent rate of growth will appear to decrease numerically. In reality the population is still increasing dramatically.

It is not only fertility rates that affect high population growth rates. **Population momentum** is a term used to explain the tendency for high population growth to continue long after birth rates have dropped. Suppose a large number of children are reaching adulthood as a result of previously high growth rates. Even though growth rates have decreased, the high number of couples of child-bearing age will result in a large number of births. So the population will continue to grow rapidly even though the birth rate is lower. Latin America is a prime example of this situation. The regional birth rate has dropped rapidly since the 1960s, and it is expected to continue falling until about 2025. But population totals keep rising because the number of fertile couples keeps increasing. In Europe, the fertility rate is below replacement levels, but the population is expected to continue to rise for the next thirty years. Generally, it takes a generation before percentage rates can be translated into actual numbers. So even statistics that measure what is happening today can be deceptive.

CONSOLIDATING AND EXTENDING IDEAS

1 Explain how each of the following factors contributes to the inaccuracy of population projections: (i) the collection of data; (ii) demographic assumptions; (iii) random events.

2 Outline how growth rates can misrepresent what is actually happening.

3 Explain population momentum and how it influences growth rates.

4 a) Copy and complete Figure 8.2 using Appendix A on pages 433 to 441. Africa has been completed as an example.

   b) Rank the continents in order from largest to smallest in each of the four categories.

   c) Which continent is growing fastest in terms of population growth rate? Which continent is growing fastest in terms of actual numbers? How can you account for the difference between actual growth and the percentage rate?

CONTINENT	POPULATION (000 000)			POPULATION GROWTH RATE (%)
	1950	2025	DIFFERENCE	1950–2025
Africa	222	1496	1274	574
Asia				
Europe				
North America				
South America				
Oceania				

Figure 8.2 Population growth rates

d) Prepare a bar graph for one of the continents to show the data from the chart. Compare it to other students' graphs for different continents. What does this comparison reveal?

5 If percentage rates can misrepresent data, what are the advantages of knowing them?

SOLVING THE PROBLEM

Solving the problem of overpopulation is a challenging task. Possible solutions lie in increasing the resource base, improving technology, reducing populations, or perhaps other measures that we have not yet envisaged. In the past, improving technology went a long way to increase the resource base. However, the Green Revolution of the 1950s and 1960s had only limited success because many farmers in the developing world could not afford the expensive requirements of new methods of farming. Today, it would appear that technological changes are not having the same impact. In fact, some regions are suffering the ill effects of poorly planned and badly implemented technological developments. Expanding the resource base also has limited possibilities in today's world. The resource base is shrinking in many regions because of the use of unsustainable practices over the past fifty years. The decline in fish stocks off the coast of Newfoundland is a prime example. The only other solution to overpopulation would appear to be the reduction of the global population. But how can this occur given the complexities of the problem?

Many factors conspire to make population control difficult. Politicians in some countries manipulate the economy of their country for personal gain while many of the citizens suffer. Until people have a safety net against poverty, which includes old age pensions, health care, and subsidized housing, they will continue to have large families to increase family income and to provide security in their later years. The rest of this chapter examines how migration, economic development, health care, and education could help reduce population growth.

CASE STUDY

Large Families
in the Developing World

Overpopulation is causing, without doubt, a continuously detrimental overuse of environmental resources; this fact is accepted by populations in both developed and developing nations. It is important, however, to understand why the peoples of already overpopulated nations continue to have large families. The motivations are many and varied, and some are convincing despite our understandings about the dangers of overpopulation.

Political Pressures

Populations in developing nations are constantly challenged to survive the repercussions of political change. Millions of people flee conflict and war to become refugees. Nationalists struggle to hold on to traditional values in the face of increasing global Westernization. Women encounter resistance as they attempt to gain a voice in the established orders of conservative, male-dominated societies.

Military strategists in nations that wage conventional warfare contend that a large population offers a military advantage. Presumably, the appropriate education and training of the populace will create citizens who will choose to remain in their homeland and fight for their cause rather than flee. Because war takes it toll in terms of human lives, large families are needed to provide the nation with soldiers.

Nations with large, relatively homoge-nous, populations can more easily protect and celebrate nationalistic traditions and goals. Cultural values and behaviours are reinforced because the majority of the population endorses them. Thus, these populations may be able to repulse the influences of other cultures.

Eventually, it is probable that current population trends in many developing countries will result in an even larger proportion of women to men. Perhaps this overwhelming force of numbers will foster and extend the participation of women in decision making, both locally and internationally.

Economic Motivations

Recent years have seen economic uncertainty in developed countries and changes in the economic systems of developing countries. To reduce their dependency on foreign aid, developing nations may allot more agricultural land to subsistence farming. Traditional land-use patterns may be abandoned in response to global market demand for cash crop produce. To compete within the global marketplace, human and natural resources are directed towards industrial development.

This shift towards food self-sufficiency is a motivating factor for having large families. Subsistence farming techniques require a large agricultural labour force; without modern machinery, many hands are needed to cultivate crops and care for livestock. With the expansion of farmland into marginal areas, more labour hours must be spent farther afield in order to gather water and fuel for everyday needs. Children have the ability and the energy to carry out these simple tasks. In countries where infant mortality rates are high and not all children can be expected to reach adulthood, producing as many children as possible is an economic imperative.

Industrialization in developing countries often seems to be synonymous with a child labour force. In India, Jamaica, Morocco, South Africa, Thailand, and the Philippines, children under fifteen are indentured by their parents as carpet weavers, match-stick makers, and waste-pickers. Their contributions to their families' incomes are of immediate and long-term necessity. The more employable children a family has, the better the chances of survival for all members of the family.

Social and Cultural Influences

There are also societal factors that encourage the pattern of large families in some developing nations. The dictates of religion and/or custom, which have established a tradition of high procreation rates, are embedded in many societies. To be fruitful is to be blessed with the wealth of many children. Family planning or birth control methods are considered by the community to be a rejection of one's faith, history, and responsibility. In addition, in many developing countries there is no infrastructure to support elderly people. Consequently, it is both acceptable and desirable to have many children who will care for their parents in later years.

The issue of family size and population growth is intertwined with political, economic, and cultural aspects of life. Since these aspects are entrenched in a society, they are not easily changed nor are they readily responsive to external solutions such as family-planning programs and birth control methods.

CONSOLIDATING AND EXTENDING IDEAS

1 Identify the motives presented in the case study "Large Families in the Developing World." Write a paragraph defending the motive(s) for having a large family that you find most persuasive.

2 Research population data about the children and youth of developing countries and regions. Describe child (0–14) and youth (15–24) populations in China and India, and in one country in *each* of the following regions: Latin America, east Asia, Southeast Asia, south Asia, Africa, and sub-Saharan Africa. Refer to the most recent statistical resources, for example, electronic media.

3 Speculate about the lives of children and youth labourers in developing countries. Create a "Day in the Life" written account that describes how each hour of the day is spent.

OVERPOPULATION AND MIGRATION

Most countries in the world today believe the **spatial distribution** of their population is more critical than either fertility rates or mortality rates. In some countries, there are too many subsistence farmers for the resource base and available technology. In other countries, rural populations are so low that resource development is hampered. As a consequence, many governments encourage people to **migrate** to other parts of the country.

One of the most notable examples of mass migration occurred in Brazil in the 1980s. Farm workers, called *nordestinos*, moved from the economically depressed and overcrowded northeast to the jungles of the western Amazon. The government believed that the vast rainforest would provide an opportunity to establish new farmland. But the soils of the tropical rainforest were unsuitable for agriculture. Unable to eke out even a meagre living, most of the migrants ended up in the slums of the overpopulated cities back east.

People also leave their homelands for other countries. In the nineteenth century, new opportunities in North and South America and Australia lured millions of European immigrants. The availability of free agricultural land that was offered as a migration incentive provided an incredible opportunity for those living on the overcrowded farms of Europe. The cities also had an allure that attracted millions to the New World. These continents were new and prosperous lands where anybody had the chance to be successful regardless of social position or birthright. Employment opportunities abounded in the cities. Today, however, there is little fertile land available for resettlement, and fewer employment opportunities exist in the cities.

Political turmoil can also be a factor in migration. When political freedoms are restricted and people fear political persecution, they move to a safer location. When Saddam Hussein attacked Kurdish villages in northern Iraq in 1995, many Kurds fled to neighbouring Iran and Turkey. Usually, political refugees move to neighbouring countries that offer sanctuary. As explained in Chapter 5, refugees with greater financial means may seek asylum in more prosperous regions of the world.

Still, **emigration** is not a solution to overpopulation. The number of immigrants and refugees who come to countries such as Canada is minuscule compared with the vast numbers who remain in the world's most economically deprived regions. Quotas and regulations established by receiving governments limit the number of immigrants. In addition, most people in developing countries lack the funds needed to emigrate.

> Today, migration to cities in the developing world is significant. In fact, this has become the largest mass migration of all time. Cities in emerging industrial economies are exploding with people. Consider what experts predict could happen by the year 2000:
> - Mexico City could have a population exceeding 25 million—almost as many people as live in all of Canada!
> - São Paulo, Brazil, could be the world's second largest city, with 24 million people.
> - Over 400 cities could have a population greater than 1 million. (There were only 76 cities of this size in 1950.)
> - More than half of these cities will be in developing countries.

Figure 8.3 Settlements in the Amazon near Ariquemes, Brazil

If there were **zero population growth,** in other words, only enough children being born to replace both parents, a farm would not need to be divided. Theoretically, each child would inherit one half of the farm. When a couple married, they would each own half of their farms. In reality, what often happens is that either one daughter or one son inherits all the land and the spouse brings no land to the union. Since half the children are usually boys and half are girls, land inheritance should allow farms to stay intact from generation to generation in a society where there is zero population growth.

If there were many children growing to maturity and all children inherited land equally, the farms would have to be divided into smaller holdings. As this pattern is repeated for several generations, farms would become so small that they would not be functional. It makes more sense if only one child inherits the farm, as was the case during the Industrial Revolution. The other children then had to seek their fortunes elsewhere. Farmers migrated from economically depressed **rural** areas to rapidly growing cities such as London, Paris, and Berlin.

URBAN MIGRATION

Migration into cities, or **urbanization**, is not a new phenomenon. It began over 200 years ago with the Industrial Revolution in western Europe. As the region entered Stage 2 of the demographic transition model (see page 83), the population exploded. There were too many children being born for the subsistence farms to support. With each generation, family farms were passed on to the children.

This pattern of urbanization is being repeated today in parts of the world where populations are exploding. Often, men leave their families, returning after four or five months with supplies bought with their earnings. The women stay in the rural areas, practising subsistence farming and raising the children. Eventually the men usually send for their families, who often end up living in areas that ring the cities. Although limited, economic possibilities are often greater than in rural areas.

The influx of people who migrate to the cities creates a severe strain on urban **infrastructures**. Inadequate housing, contaminated water supplies, lack of sewage facilities, insufficient electrical output, limited garbage collection, and congested transportation grids all contribute to poor health, high crime rates, and an unacceptable

The scene in many cities in the developing world today is not unlike that of the new urban centres of nineteenth-century England. This passage from Charles Dickens's *Oliver Twist* describes London in 1818:

A dirtier or more wretched place he had never seen. The street was very narrow and muddy, and the air was impregnated with filthy odours. There were a great many small shops; but the only stock in trade seemed to be heaps of children, who even at that time of night, were crawling in or out of the doors or screaming from inside.

living environment.

Most cities in the developing world cannot provide the number of jobs needed to support the growing population. Some of the migrants—people who come to the cities usually from rural areas—find work in the sweatshops and factories. Others sell handmade crafts and trinkets. The most desperate of these people, including many children, are forced to beg, scavenge, and steal to survive another day. Despite the incredible economic hardship, more people keep coming to the cities every day.

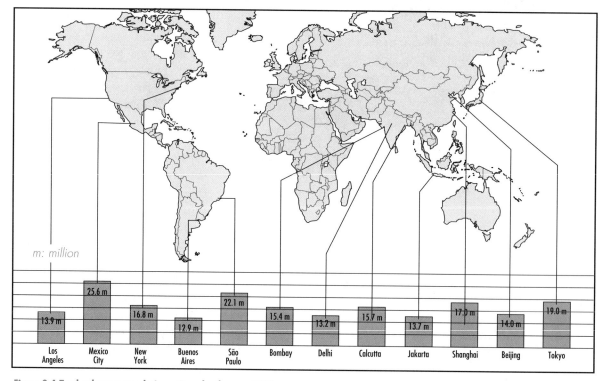

Figure 8.4 Twelve largest population centres by the year 2000

Source: Adapted from *Development*, Special Issue—February 1989. Updated statistics from UN Population Fund, *The 1997 Canadian Global Almanac*.

CASE STUDY: BOMBAY, INDIA

The Perils of Mohammed Hussain

June to October are known to be the monsoon months in India. For Mohammed Hussain, they are the months of peril, because in the part of Bombay where he lives, the streets and sidewalks are flooded with torrential rains.

During the dry season, Hussain and his family live on a sidewalk in a tin shack. When the monsoon comes, they have nowhere to go but under the staircase of a local condominium. Some of the apartment dwellers express compassion for him while others want Hussain, his wife, and four children to get out. But where will Hussain go? Almost half of Bombay's 9 million residents live in slums or on sidewalks. . . .

When he first arrived, Mohammed Hussain belonged to the élite by the standards of slum and sidewalk dwellers. He earned ten rupees (one Canadian dollar) per day as a labourer in a flour mill. His wife earned eight rupees doing bead-work for a garment factory. They could at least eat enough.

But when the first child was born breakfast was sacrificed. With the second child, the mid-day meal of vegetables, pulses, and rice was cut down to dry bread, onions, and a little rice. After the third child, the mid-day meal was given up altogether. Today, with four children and a combined daily income of only twenty-five rupees, Hussain's family eats only once a day. Their hope is that in a year or two the eldest daughter can be sent out as a domestic helper or a beggar to supplement the family's income.

Parents without shelter means children without school. When Hussain's children grow up, they must guard the shack, earn money themselves, or look after the younger ones while the parents work. The chance to earn a few cents today comes at the cost of their earning-power later. By the time they become adults, they walk right into the poverty trap with no skills, no money, and no home. . . .

"The problem of slums in Bombay cannot be viewed in isolation," argues Minar Pimpale, a leader [of the National Campaign for Housing Rights]. "It is intrinsically linked to urbanization in India."

India's urban population is growing at almost 4 per cent per year, compared to the national rate of around 2 per cent. . . .

Most slum dwellers who migrate to the cities, come from economically deprived, underdeveloped, and drought-prone districts. . . . About 52 percent . . . had owned no assets. About 67 per cent . . . had landed in Bombay in search of jobs.

Mohammed Hussain recalls that he first fled to Bombay from his native Aurangabad district to save his new bride's life from the Hindu-Muslim riots. Like everybody else, Hussain decided to stay put once he was in Bombay. "What was the incentive to go back? Eighteen rupees here was more than I was getting back home." . . .

Mohammed Hussain ran away from Aurangabad over a decade ago. Today, existing—not living—on a footpath in Bombay, he is ready to fight for his rights. If Mohammed Hussain is to be secure, his wages must improve, the stripping of forest cover around his village must stop, and rural industries must expand. All this looks like a gigantic task.

S. Waslekar, From *Development* (Winter 1987–88): 18–20 .

CONSOLIDATING AND EXTENDING IDEAS

1 a) Why is internal migration to sparsely settled regions not usually a viable solution to overpopulation?
   b) Why is international migration so limited?

2 Why are cities in the developing world expanding at such a rapid rate?

3 a) Under what conditions would urbanization be a solution to overpopulation?
   b) How does urbanization create more problems than it solves?

4 Read the article "The Perils of Mohammed Hussain" on page 105. Outline in writing the frustrations that Mohammed experiences living in Bombay.

5 As indicated in the graph in Figure 8.5, urbanization is increasing at more accelerated rates in developing countries than in developed nations.
   a) Refer to the *Canadian Oxford School Atlas*, 6th ed., page 132. List the megacities of the developing world in one column and those of the developed world in another column. Write generalization statements that reflect (i) urban populations in the Northern Hemisphere as compared with those in the Southern Hemisphere; (ii) urbanization by continent.
   b) Extrapolate from the graph to predict urbanization trends in African, Asian, and Latin American cities by the year 2020.

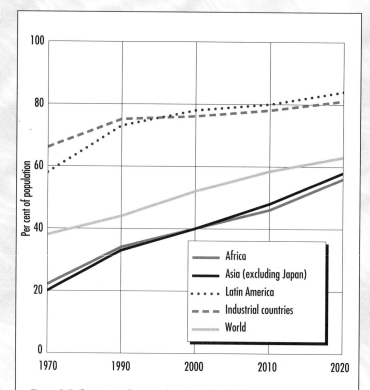

Figure 8.5 Change in urban population, 1970–2020
Source: Copyright © 1994 World Eagle. Reprinted with permission from World Eagle, 111 King St., Littleton, MA 01460 USA. 1-800-854-8273. All rights reserved.

STATLAB

1 Use Figure 8.6 and the population statistics in Appendix A (page 433) to determine the correlation between the fertility rate and urban population as a per cent of total population. Either draw a scattergraph or calculate the linear correlation coefficient for twenty randomly selected countries. (See pages 11 to 18 for details if you have forgotten this skill).
   a) What correlation do you notice?
   b) Based on this study, does urbanization appear to be a solution to overpopulation? Explain.
   c) What other variables could also affect fertility rates?

2 a) Using Figure 8.6, prepare a multiple-line graph showing trends in urbanization.
   b) What percentage seems to be the limit of successful urbanization? Which continents still have largely rural populations?

CONTINENT	1950	1970	1995	2025
Africa	13	30	35	54
Asia	16	28	32	55
Europe	55	72	75	83
Oceania	65	71	71	75
North America	64	74	74	79
South America	48	77	78	88

Figure 8.6 Percentage of the population in urban areas
Source: World Health Organization

3 a) Define the term *infant mortality rate*.
   b) What pattern do you notice in Figure 8.7? Why do you think this pattern occurs?
   c) Calculate the difference between the infant mortality rate in urban and rural areas for each country. Rank the countries from highest difference to lowest difference.
   d) Infant mortality rates are generally considered to be a good indicator of the general health of the population. Taking this into consideration, for which countries does urbanization seem a viable solution to overpopulation? For which countries does urbanization not appear to be a viable solution?

e) What generalizations can you make about the countries best suited for urbanization?
How could these statistics be misleading?

COUNTRY (YEAR)	URBAN	RURAL
Ecuador (1993)	32	41
India (1990)	58	98
Nigeria (1990)	75	96
Pakistan (1991)	69	115
Peru (1993)	65	92
Tajikistan (1993)	54	44
Thailand (1993)	10	7
Zimbabwe (1992)	49	61

Figure 8.7 Infant mortality rates for selected countries (per 1000 population)
Source: *Demographic Yearbook 1994* (New York: United Nations, 1996).

OVERPOPULATION AND ECONOMIC DEVELOPMENT

There has been a strong negative correlation between economic development and fertility rates. (To review correlational analysis, see Chapter 2, page 11.) As **gross domestic product** increases, fertility rates plunge. However, this relationship is not as clear-cut as it seems. Some places with high population densities, such as Hong Kong and Singapore, can become prosperous and increase their industrial development. In other cases, rapid population increase and slow economic development can lead to a deterioration in the standard of living. So the correlation between wealth and population growth is an oversimplification.

In countries where population growth increases faster than economic output, the value of economic gains drops when it is divided by the total population. Although investment capital in new industries may increase dramatically, on a per capita basis it may be even lower than it was before. If economic expansion cannot keep pace with population growth, family incomes decline and unemployment increases. The lower standard of living that results undermines the advances in industrial development, and population growth may increase even more as families struggle to maintain financial security. It is only when they are financially secure that couples do not need the earning capabilities of their children.

Many subsistence and emerging industrial economies are trapped in this vicious circle of poverty. The Philippines is an example. This island nation is located in Southeast Asia, where an economic boom is under way. It has received considerable financial support from the United States. Population growth is 2.6 per cent a year, but the GDP per capita is increasing at only 1.2 per cent a year. The Philippines cannot hope to improve its standard of living and make real economic gains until population growth is controlled.

STATLAB

Study Appendix A (page 433) and Appendix B (page 442).

1 Determine the correlation between the fertility rate and gross domestic product per capita, either by drawing a scattergraph or calculating the linear correlation coefficient for twenty randomly selected countries.
   a) What correlation do you notice?
   b) Why do you think this pattern exists?
   c) Based on this study, does economic growth appear to be a solution to overpopulation? Explain.
   d) What other variables could also affect fertility rates?

2 a) Compare population growth to GDP growth rates.
   b) List ten countries in which the economy is growing substantially faster than the population. Predict what will happen to population growth in the future.
   c) List ten countries in which the economy is growing substantially slower than the population. Predict what will happen to population growth in the future.
   d) Illustrate your findings graphically.

OVERPOPULATION AND HEALTH

There is a strong negative correlation between health and population growth rates. When people receive proper health care, they live longer, healthier lives. An adequate and nutritious food supply also improves the overall health of a population. People who eat well and have access to good health care have longer life expectancy rates. Infant mortality rates and life expectancy are good indicators of quality health care and proper nutrition. Not surprisingly, countries with healthy

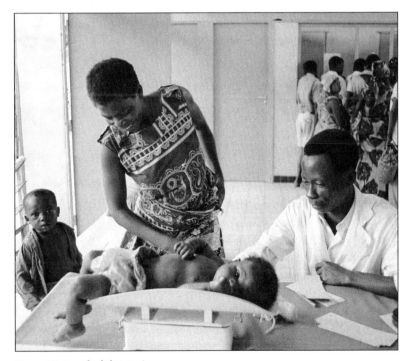

Figure 8.8 A medical clinic in Congo

diets and adequate health care have lower fertility rates. When a society finds that most of its children live to maturity the birth rate usually drops. Fewer children need to be born because more survive to care for their parents in old age. But it usually takes a generation after these standards are achieved before family size begins to decline.

STATLAB

Refer to Appendix A.

1 Determine the correlation between the fertility rate and infant mortality or life expectancy, either by drawing a scattergraph or calculating the linear correlation coefficient for twenty randomly selected countries.
   a) What correlation do you notice? Why do you think this pattern results?
   b) Based on this study, does reducing infant mortality appear to be a solution to overpopulation? Explain.

2 Determine the correlation between fertility rate and average food energy required as a per cent of need per capita, either by drawing a scattergraph or calculating the linear correlation coefficient for twenty randomly selected countries.
   a) What correlation do you notice? Why do you think this pattern results?
   b) Based on this study, does increasing the food supply appear to be a solution to overpopulation? Explain.

3 What difficulties do developing countries with high fertility rates have in implementing improved health care and food supplies?

OVERPOPULATION AND EDUCATION

As the level of education increases, fertility rates decline. The strong negative correlation between literacy and fertility suggests that this may be a key factor in solving the problem of overpopulation. Education results in better nutrition, increased productivity, and higher incomes. All of these factors lead to smaller family size. The problem is that despite massive funding for education, the number of children that have to be educated diminishes the overall results. While literacy rates are improving, less than 10 per cent of children in developing countries receive a secondary-school education.

On the positive side, attitudes towards education are changing in many developing countries. Many children are encouraged to go to school. Yet the necessity of sending children to work to provide income for the family still prevents large numbers of young people from getting an education. In the cities, many of the children living in the shantytowns are disqualified from attending school because their families do not own property or pay school taxes.

The greatest challenge, however, is the education of girls. In many traditional societies

Figure 8.9 A librarian in Zambia

education is reserved for boys. Girls are simply not given the opportunity to go to school. Without education, fertility rates remain high. Evidence indicates that women who complete primary school have fewer children than those who never attend school. A study of fourteen developing countries indicated that in ten of these countries, women with more than six years of formal education married on average 3.5 years later than those who had no education. Delayed marriage usually leads to delayed childbirth. In addition, women with more education are able to enter the workforce, which in turn may contribute to women bearing fewer children later in life.

STATLAB

1 Refer to Appendix A. Determine the correlation between fertility rate and female literacy, either by drawing a scattergraph or calculating the linear correlation coefficient for twenty randomly selected countries.

   a) What correlation do you notice? Why do you think this pattern exists?

   b) Based on this study, does female literacy appear to be a solution to overpopulation? Explain.

   c) What other variables could affect fertility rates?

2 What factors hinder the implementation of education programs for females in many developing countries?

CONSOLIDATING AND EXTENDING IDEAS

1 a) Evaluate how each of the following factors contributes to lower fertility rates: urbanization; economic development; health care and nutrition; female literacy and education.
   b) How do these factors work together to reduce fertility rates?

2 a) Choose a developing country and prepare an organizer or a chart like the one in Figure 8.10. Compare the statistics in that country with those in Canada and with the continent (of the developing country) average. Refer to Appendices A and B.
   b) Study your organizer or chart and circle those indicators that are below the continental average. These are areas where improvement is needed.
   c) Outline a plan whereby socio-economic factors would be improved.
   d) Explain how the proposed changes would help to reduce population growth rates.
   e) How would you implement this policy while respecting the rights of all citizens?

CRITERIA	COUNTRY	CANADA	CONTINENT AVERAGE
Birth rate			
Fertility rate			
Urban population			
Gross domestic product			
Infant mortality			
Life expectancy			
Average food energy available as a per cent of need			
Female literacy			

Figure 8.10 Statistics organizer

3 a) Over a three-week period, gather newspaper stories that highlight life in different continental areas: Europe, Asia, Oceania, Latin America, North America, and Africa.
   b) Select two of the continental areas (one must be developed, for example, Europe; the other must be developing, for example, Africa) for further investigation.
   c) In a concept web that differentiates "developed" from "developing," write the headlines of all the articles you have gathered on the six continental areas.

d) Based on the current news stories, write two paragraphs that assess quality of life in each of your two continental areas. Consider social, economic, and political aspects of life in these areas and compare them to life elsewhere on our planet. Attach the paragraphs to your web.

CHAPTER 9

OVERPOPULATION AND FOREIGN AID

Overpopulation is often the most serious problem facing developing countries. Many developing countries simply have too many people for the resources available. We know that birth rates drop if social services improve. In order for a country to progress to Stage 3 or 4 of the demographic transition model (lower birth rates), living standards need to improve. But it is often difficult or impossible for these nations to improve health care, education, and other social services without aid from foreign sources.

The issue of foreign aid to developing countries is a source of controversy. There are

at least four views about the nature of aid. The **humanitarian view** is that wealthy nations have a responsibility to help countries that are less fortunate. When we see scenes of human suffering, we feel compelled to help in any way possible. The **pragmatic view** argues that the problem of overpopulation must sort itself out. Providing aid promotes dependency. When food is provided free, food production declines and the subsistence economy collapses. The **nationalist view** is that poverty at home must be eliminated before we spend money in aid to other countries. With a growing number

Figure 9.1 An employee of CIDA working with a farmer in Mali

of people finding themselves living in poverty, it is our responsibility to provide for our own citizens first. The **internationalist view** is that overpopulation can only be resolved through external intervention. Developed countries must help developing nations create the necessary social programs that will sustain them over the long term.

FOREIGN AID

After the Second World War, the Western democracies decided it was in their best interest to provide aid to the emerging nations of Africa, South America, and Southeast Asia. This was a decision based on politics, not humanitarian ideals. Soviet-style Communism had spread throughout Eastern Europe. Western democracies wanted to contain Communism from the rest of the world. The Soviet Union and China provided military and financial support to many developing countries, including North Korea, North Vietnam, Cuba, Chile, Angola, and Egypt. In response, the United States and its allies attempted to spread their influence to other developing nations. Much of this aid was in the form of military training and support. Thus, in the suspicious atmosphere of the early days of the **Cold War**, international loyalties were being bought with military aid.

Expansionism was not only of interest to the superpowers. Large corporations had already established themselves in developing countries in the colonial period of the nineteenth century.

> There are three terms that describe the type of aid a country may receive: bilateral aid, multilateral aid, and tied aid. **Bilateral aid** is given by one country to another. **Multilateral aid** comes from a number of countries to a developing country. The third type—**tied aid**—is often the most common, and comes with conditions attached. These conditions are often economic or military. In a sense, the receiving country loses some of its autonomy in exchange for aid.

Many continued to exploit raw materials and low-paid labour well into the twentieth century. Huge development projects were sponsored and financed on the condition that the large corporations in the donor country would receive some benefit in return. For example, a donor country would provide the capital to build a factory. The receiving country then had to purchase the equipment needed to run the factory from corporations in the donor country. Thus, many developing countries were dominated by interests in one of the superpowers, much as they had been dominated by the colonial powers a century earlier.

Far from helping these countries, aid in the 1950s and 1960s resulted in the economic and military domination of many parts of Africa, South America, and Southeast Asia.

THE UNITED NATIONS

The United Nations was formed at the end of the Second World War by the Allied countries. Its role was to prevent the outbreak of any more world wars. Since then the organization has expanded its mandate to include the improvement of living conditions in developing regions of the world. Many aid organizations have been initiated by the United Nations. The World Health Organization (WHO), the Food and Agriculture Organization (FAO), the World Food Program (WFP), and the World Bank are just a few of the UN agencies that offer aid and relief to developing nations.

WHO

The World Health Organization came into effect in 1948. Its objective was "the attainment by all peoples of the highest possible level of health." WHO is the world authority on directing and co-ordinating international health. In collaboration with the United Nations, specialized health agencies, and state governments, it strengthens health services, offers health education programs, improves nutrition and sanitation standards, promotes mother and child care, and provides immunization programs. In addition, it stimulates advanced medical research to eradicate disease and monitors working conditions and other aspects of environmental health.

FAO

The Food and Agriculture Organization was formed in 1945. Its overall objective is to eliminate hunger by raising nutrition levels and living standards, improving food production and distribution, and creating better living conditions for rural populations. To achieve these goals, the FAO promotes agricultural investment, improved soil and water management, and higher crop and livestock yields. In addition, it sponsors agricultural research as well as the transfer of technology to developing countries. Special FAO programs provide relief assistance to countries in times of food emergencies, such as famines. The FAO also sponsors the World Food Program (WFP) with the United Nations. The WFP uses food supplies, cash, and services from member states to support social and economic development and to provide emergency relief.

THE WORLD BANK

The World Bank (formally known as the International Bank for Reconstruction and Development) was established in 1946. Its purpose is to provide funds and technical support to advance economic development in developing countries. The International Development Association (IDA) and the International Finance Corporation (IFC) are affiliated organizations administered by the World Bank. The IDA helps countries to develop their resources effectively, while the IFC encourages the growth of private enterprise in developing countries. In recent years, the World Bank and its affiliates have been criticized for the way in which their loans and debt repayments have been structured. Many developing countries have found themselves simply unable to repay the huge sums of interest that have accumulated on their loans, much less repay the loans themselves. Critics feel the World Bank should focus on improving the quality of life for people in developing countries by meeting the needs of the local people through grass-roots projects rather than through huge megaprojects such as power dams and river diversion. These megaprojects require developing countries to borrow foreign technology and expertise at great financial cost.

NON-GOVERNMENTAL ORGANIZATIONS

Non-governmental organizations (NGOs) represent religious groups, service agencies, and other non-profit organizations. Many of the aid programs provided by NGOs operate at the grass-roots level. Instead of funding government agencies, many NGOs provide direct assistance to the people who will benefit from the project. Government, business, and interested citizens groups combine their unique skills to help people in need around the world. NGOs range from small grass-roots groups that work with individuals to giant transnational organizations that work with governments. Most of the funding for these organizations comes from private groups, unlike government agencies, which are funded by taxes.

Aid or temporary relief in the face of a natural disaster is no longer the primary focus of most NGOs. The goal today is to enable people to give themselves a better life. These organizations would rather give a "hand-up" than a "hand-out." The belief is that empowered people will feel that they are in control of their destinies. They will become self-sufficient and will no longer need aid. To this end, NGOs use local resources, employ local people, and encourage the establishment of self-help agencies run by the people who are being helped. They are often much more successful than government agencies.

The Red Cross is perhaps the best known of the NGOs. An international agency headquartered in Switzerland, it provides emergency medical aid to victims of war and natural disaster. Other groups such as CARE, World Vision, Foster Parents Plan, and Oxfam have done much to improve living conditions in developing regions. One example is World Vision's program to provide drinking water in sub-Saharan Africa. A drilling rig is sent to a local village. The well is drilled and a locally produced pump is used to draw the water to the surface. The organization uses local people, and technology that is appropriate for the community's level of development. When the team leaves, the village has fresh, clean water.

CANADIAN AID AGENCIES

The Canadian International Development Agency (CIDA) was developed by the federal government in 1968. CIDA's main objectives are to establish long-term development strategies and provide short-term humanitarian relief in times of emergency. CIDA divides its aid funding among three major areas: multilateral and international agencies; bilateral aid to individual nations; and financial support for non-governmental organizations. Canadian farmers, technicians, teachers, foresters, nurses, and doctors provide their services to improve the quality of life throughout the world. When disasters strike, CIDA supplies financial aid, emergency supplies, and human

YEAR	MULTILATERAL AID	INTERNATIONAL NGOs	BILATERAL AID	OTHER*	TOTAL
1970–71	1 000 000	510 000	–	–	1 510 000
1973–74	2 000 000	1 500 000	103 600	1 020 000	4 623 600
1976–77	5 000 000	2 770 000	2 900	1 500 000	9 272 900
1979–80	8 000 000	3 715 300	2 760 200	850 000	15 325 500
1982–83	9 500 000	5 563 550	5 942 682	–	21 037 580
1985–86	10 250 000	8 004 100	8 408 345	–	26 662 445
1988–89	13 100 000	10 356 630	23 804 720	–	47 266 530
1991–92	13 400 000	8 305 000	16 446 000	–	38 151 000
1994–95	14 200 000	8 875 000	7 555 000	–	30 630 000

*Contributions to WHO Human Reproduction Program and to OECD's Population Program.

Figure 9.2 CIDA assistance in the population sector, 1970–1995
Source: Population Sector CIDA.

resources to help countries cope with crises.

The International Development Research Centre (IDRC) is a federally funded Crown corporation that supports scientific and technological research into sustainable development in the developing world. The IDRC works with CIDA to provide the most effective aid programs for emerging industrial nations and subsistence economies. Similar organizations are found in most developed countries.

STATLAB

1 Study foreign aid as a percentage of GNP in Appendix B on page 442, then prepare an organizer or a chart like the one in Figure 9.3.

COUNTRY	POPULATION (000 000)	FOREIGN AID (% GNP)	GNP PER CAPITA	TOTAL GNP ($000 000)	TOTAL AID ($000 000)
Canada	29.5	4%	$19 970	$19 970 × 29.5 = $589 115	4% × $589 115 = $2356.5

Figure 9.3 Foreign aid as a percentage of GNP

a) List the countries from highest to lowest for foreign aid as a percentage of GNP. (These are countries with a negative number.)

b) List the GNP per capita for each donor country.

c) Multiply the GNP per capita by the total population to get the total GNP for each country listed in (a).

d) Multiply the total GNP by foreign aid as a percentage of GNP to get the total amount of aid for each country. Show the results in your organizer or chart. Canada has been completed as an example.

2 a) How would the order change if the countries were listed in order by total amount?

b) What patterns do you find surprising?

c) The United States provides only 0.2 per cent of its GNP in foreign aid, yet it gives more money than any other nation. Explain how this is possible.

d) What countries seem to be underrepresented in terms of foreign aid contributions?

3 List the top twenty recipient countries in order from highest to lowest. (These are the countries with positive numbers.) Prepare an organizer or a chart to show these statistics.

a) What patterns do you notice?

b) What do you find surprising?

4 a) Study defence expenditures in Appendix B. Prepare an organizer or a chart comparing aid as a percentage of GNP with defence expenditures as a percentage of GNP and total defence expenditures for the donor countries.

b) Which seems to be more important—defence expenditures or aid? Is this surprising to you? Explain.

c) Which recipient countries spend a significant proportion of their GNP on defence? What advice would you give these countries?

d) What donor countries spend significantly more money on the military than on foreign aid? What advice would you give these countries?

5 Do you think political instability would be reduced if non-military aid increased in regions of conflict? Discuss.

6 a) Prepare a graded shading map to show one of the following:

i) aid as a percentage of GNP

ii) total aid expenditures

iii) defence as a percentage of GNP

iv) total defence expenditures

b) Write an analysis of the patterns your map reveals.

7 Describe the correlation between the following pairs of data:

a) aid received and defence expenditures

b) aid given and defence expenditures

c) aid given and GNP

d) aid received and GNP

CONSOLIDATING AND EXTENDING IDEAS

1 a) Prepare an organizer or a chart showing the pros and cons of each of the four views on foreign aid.

b) Critically access each argument. Which view do you support? Give reasons for your answer.

c) Debate your opinion with other members of your class.

2 Explain how government attitudes towards aid have changed since the Second World War.

3 Some individuals believe that foreign aid should be halted. It is "money down the drain" as conditions in developing countries just continue to get worse; developed nations have needs that must be met first. Others believe that foreign aid to developing countries should be increased; morally, economically, politically, it is the right thing to do. **Should foreign aid be given to developing countries?**

In this position paper you must defend a point of view. Your final paper should include:

a) an introductory paragraph that identifies the issue and alternative positions (points of view).

b) a paragraph that presents one position on the issue as indicated above.

c) a paragraph that presents the other position on the issue as indicated above. Each paragraph (b and c) should argue the advantages of the position, and support the point of view with specific facts and examples.

d) a concluding paragraph that defends your position on the issue.

CURRENT ATTITUDES TO AID

In recent years the nature of foreign aid has changed. Governments have come to realize that overpopulation is a serious global problem. If world population continues to grow, *all* people will be affected. The environment will suffer as greenhouse gases and pollutants contaminate the atmosphere, the hydrosphere, and the lithosphere. Plant and animal diversity will be reduced as habitats are destroyed. Political instability will increase as disadvantaged groups fight for the meagre resources available to them. But more than anything else, people will suffer because they had the misfortune to be born in a country where there are too few resources for the population.

SURVEYING PUBLIC OPINION

People often have misconceptions about an issue as complex as overpopulation. As a class, prepare a survey designed to discover how people in your community feel about the world's population problems and their solutions. Base your survey on a series of multiple-choice questions. (Be sure you have at least ten.) Keep the following tips in mind as you create your survey questions:

• Multiple-choice questions are preferable because the results are quantifiable—that is, they can be analysed using statistical methods.

• Questions should focus on the public's knowledge of population issues and their solutions as well as attitudes towards foreign aid.

• Terminology must be appropriate for a general audience. Avoid references to such things as correlations because people may be unfamiliar with these concepts.

SAMPLE QUESTIONS

The following questions were prepared by one class. They will give you some idea of what is expected. Come up with your own survey using data from this section and from other references such as E-Stat, the Internet, almanacs, yearbooks, and atlases

1 Which of the following solutions would you choose to overcome world population issues?
   a) better education about family planning
   b) enforced sterilization programs
   c) migration to less settled areas
   d) do nothing and allow natural controls to reduce population
   e) provide foreign aid to reduce poverty

2 How much foreign aid does Canada provide per capita?
   a) $40 (0.2% of GNP)
   b) $100 (0.5% of GNP)
   c) $200 (1.0% of GNP)
   d) $300 (1.5% of GNP)
   e) $600 (3% of GNP)

3 Why are families in developing countries often larger than those in developed nations?
   a) The people are ignorant about sex.
   b) The people do not have access to contraceptives.
   c) The people need large families for financial security.
   d) Government regulations encourage large families.
   e) Large numbers of children are necessary to ensure the survival of at least some offspring.

THE SAMPLE

How should you conduct your survey? It is not necessary to have everyone in your community answer your questionnaire. The key is to select a cross-section of people that are representative of your community. This is called the **sample population**. The people in this sample should reveal the attitudes and values of a genuine cross-section of the community.

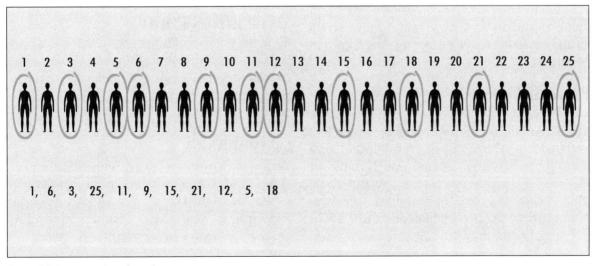

Figure 9.4 A cross-section of people

#	SEX		AGE				CULTURAL BACKGROUND		ANSWERS									
	MALE	FEMALE	<20	20–40	41–60	>60		1	2	3	4	5	6	7	8	9	10

Figure 9.5 A sample population

SELECTING YOUR SAMPLE POPULATION

Select your sample population at random. This will ensure that the information is not biased towards any one group. Try using a list of names and a pair of dice. Role the dice, then count down that number of names from the top of the list. Interview the person you selected.

Try to interview a variety of people—an equal number of women and men, people of different age groups, and people of different cultural backgrounds. Analyse your sample group in an organizer or a chart like the one shown in Figure 9.5.

PROCESSING THE DATA

If each person in a class of thirty students interviews ten people, the total sample will be 300 respondents. The larger the sample, the more accurate the results. The trick now is to put all the results together in a format that the whole class can use as a basis for quantitative analysis. If you have access to a computer network, a simple spreadsheet program can be used for each person/group. Then the data may be merged into one file for use by all students. You should follow this procedure whether you are processing the data manually or electronically.

Each student should summarize the data for all respondents, using the organizer or chart. Then each group of five or six students should summarize the data for the fifty or sixty respondents in their group. Next, the class should then develop an organizer or a chart to show the survey responses.

SYNTHESIZING THE DATA

Total the number of times each answer was chosen by your sample group. Convert these figures to percentages for easy comparison. A computer spreadsheet program will help organize the data.

CONCLUSIONS

Choose two or three survey questions for individual study. Develop a hypothesis of what you expect respondents to say and explain the assumptions you make. From your study, analyse the degree of support for the issue.

- What percentage of the population chose a, b, c, d, or e?

- Do most people have a good understanding of population issues?
- Were there differences based on the sex/age/cultural background of the respondent?
- What generalizations can be made about attitudes to the issue?
- Did the results vary substantially from your hypothesis? How?

COMMUNICATING YOUR FINDINGS

1 Present your findings in a pie or bar graph.

2 Write an analysis of your findings.

3 Develop a strategy to promote public awareness of the issue. Consider writing a letter to the editor of the local paper, preparing posters, or speaking to service clubs.

CONSOLIDATING AND EXTENDING IDEAS

Programs of foreign aid occur through bilateral or multilateral arrangements (see page 115), and may take different forms:

Food aid: Food is sent to countries to feed starving people or to increase the country's available stocks.

Emergency aid: Medicine, food, etc., are provided to deal with natural disasters, such as drought, earthquakes, flooding.

Military assistance: Troops, weapons, and military experts are sent to keep the peace or to make peace.

Project aid: Funds are provided for specific purposes—irrigation-system development, road construction, etc.

Technical assistance: Equipment, and technical trainers and teachers are sent to educate people about new agricultural methods or to help them better utilize their own technologies.

a) In a small group, argue the advantages/benefits and disadvantages/costs of providing countries in need with each or all of these types of foreign aid.

b) Next, select a developing nation in Africa, Asia, or Latin America and identify its foreign aid needs.

c) Each student in the group writes recommendations about which types of foreign aid would be most appropriate to address these needs. Ideas should then be shared with the group.

PART 3

RESOURCE ISSUES: FOOD

CHAPTER 10

CONCEPTS IN RESOURCE MANAGEMENT

As Figure 10.1 suggests, resource consumption has been an issue throughout time for humans as well as other species on the planet. But our needs are more extensive than those of other species. The six basic necessities of all people are food, potable water, shelter, clothing, fuel, and space. Today, as in the past, the earth must continue to provide these life-support systems for the world's population.

NATURAL RESOURCES THROUGH THE AGES

Natural resources include everything in nature that people use. Trees, for example, are a natural resource. First they are processed into **raw materials** such as lumber, timber, and pulp. Then these raw materials are used to manufacture **consumer products** such as furniture, books, and newspapers.

While almost everything found in nature has a use in today's society, this was not always the case. Before our current level of technological development, there was less need for natural resources. Trees did not have as many uses as they do today. During the Stone Age, people burned the wood from trees as fuel and also used it to make weapons and tools. As technology advanced, people worked with tools fashioned from metal to process wood into planks and timber for construction. Wood pulp was first used in the manufacture of paper during the Industrial Revolution (*circa* 1800). The variety of applications for natural resources has increased with our technological development. For example, in the twentieth century a substance found in wood called cellulose was first employed in the manufacture of rayon, a synthetic fibre.

Figure 10.1 Contrasting scenes of traditional and modern resource consumption

The availability of natural resources and the level of technology are key factors in determining the standard of living. Nations with a high level of technological development and abundant natural resources tend to have a higher standard of living than nations without these advantages. One notable exception to this generalization is Japan. For centuries, this island nation has continued to increase its standard of living while constantly meeting the challenges of its very limited natural resource base.

Nations with a low level of technological development are often further disadvantaged by a limited resource base. Some subsistence economies that have historically sustained small populations have been faltering during the 1980s and 1990s. Overpopulation has caused these economies to move to the brink of disaster. Sub-Saharan Africa is a case in point. If either its resource base or its technology were to improve, the region would be better able to support its people. If the people had the technology to draw water from aquifers deep beneath the desert, they would be able to grow more crops. Having a greater resource base—in this case, the food supply—would enable the people to have a higher standard of living. Unfortunately, the solution is not as simple as it sounds. Other factors interfere in the neat theoretical assumption that living standards can increase as technology improves. In the Sahel region of the Sahara, many governments do not have the political will to improve living conditions for subsistence farmers and pastoralists. Furthermore, many groups do not want change; they prefer to continue to live as they have for generations.

CLASSIFYING NATURAL RESOURCES

Resources can be classified into four groups: **renewable resources**, **sustainable resources**, **nonrenewable resources**, and **ubiquitous**, or **permanent**, **resources**. Renewable resources are those that can regenerate once they have been harvested. Agricultural crops, trees, and fish are all renewable resources. Some practices exploit a resource to such an extent that its ability to grow back is severely hampered. The key to renewable resources is to manage them so that they can regenerate. When this happens we say the resources are sustainable. Properly man-

> The cod fishery off Canada's east coast is a renewable resource that has been mismanaged. Cod were caught in these once-rich fishing waters for hundreds of years. Until recently, they were able to regenerate themselves. But advanced technology allowed fishing fleets to catch greater numbers of cod. By 1992, stocks had fallen to such a low point that they could no longer replenish themselves. A four-year moratorium on cod fishing has allowed fish stocks to make a partial recovery.

Figure 10.2 Hauling a cod trap off Fogo Island, Newfoundland, in 1990

aged, renewable resources should never be exhausted; they should last forever.

Nonrenewable resources are those substances that will not regenerate within our lifetime. Once they are gone, they are gone forever! These include metals such as gold and silver, nonmetallic minerals such as salt and asbestos, and fossil fuels such as oil and natural gas.

Ubiquitous resources are those elements we take for granted because they are always available. They include air, soil, and water. The term ubiquitous can be deceiving. When a region experiences some form of environmental degradation, these resources may be reduced to the point that human survival is threatened. We assume that we will always have air to breathe. But when a forest fire breaks out, it consumes so much oxygen that there may not be enough for people to breathe, which can cause them to suffocate. Can we always count on our water supply? Extreme weather conditions can bring drought to agricultural regions. The lack of rainfall may cause harvests to fail and may even lead to famine. And what if there is too much rainfall? Raging flood waters can destroy crops and, even worse, take human lives. It is not only nature that is capable of destroying our ubiquitous resources, but also industrial pollution, which can contaminate lakes and rivers, making the water unsafe to drink. Air pollution can cause respiratory disease. Overcultivation makes valuable topsoil vulnerable to the forces of **erosion**. All of these factors threaten the stability of our ubiquitous resources.

> Probably the best-known nonrenewable resource is oil, an essential element in the operation of industrial society's current technology. Canada is expected to start running out of conventional oil reserves within the next ten years. Of course, there may be new discoveries and there are the enormous reserves locked away in the oil sands of northern Alberta and Saskatchewan. Up until 1996, only 580 000 barrels of upgraded crude oil were being extracted daily from the oil sands north of Fort McMurray. With the federal government's new initiatives, this extraction could easily be tripled (see page 141).

PRODUCERS AND CONSUMERS

In the first half of the twentieth century, Canada was primarily a resource-based country. Our primary industries provided the world with many natural resources, including wood, furs, grain, fish, and minerals. Canada eventually expanded its economic base to include secondary industries, such as the manufacturing of automobiles, communications equipment, and paper products. Today, fewer and fewer Canadians are producers of either primary or secondary products. Instead, we have become a nation of consumers of a wide variety of goods and services. We are more likely to work as bankers, civil servants, teachers, and salespeople. This is the tertiary sector, and the product is a service or a specific expertise offered to the population in exchange for an income. This evolution has been good for our economy and has given us an enviable standard of living. But it has not been without a price. Many Canadians have lost touch with the natural systems that sustain us. We tend to take these systems for granted. In doing so, we fail to recognize the

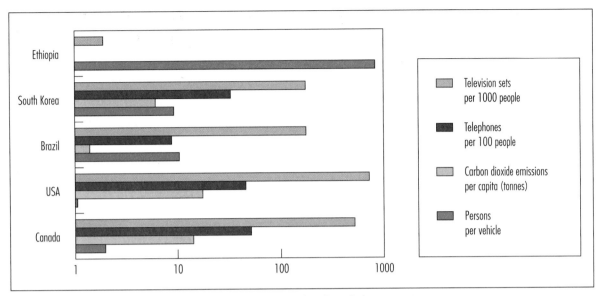

Figure 10.3 Relative consumption of consumer products, and carbon dioxide production per capita
Study the graph and divide the countries into three groups on the basis of consumption characteristics. What characteristics does Canada have? How do they differ from Ethiopia's? How would you account for the differences? Explain why the graph uses a logarithmic scale.
Source: Statistics from *World Resources, 1996–97*.

impact of our own activities on our natural resources. Conservation and moderate consumption are all but forgotten. While it is a pattern that is repeated in most diversified economies, statistics show that Canadians, along with our neighbours in the United States, are among the most wasteful consumers in the world!

Many people contend that this preoccupation with consumption is destroying the planet. The perceived essential needs people have today are much greater than they were in the past. Some people say, "I need a car to get to

The 1996 World Commission on Global Governance recommended that "the time has come when we must . . . consider levying charges on the use of global resources to finance common global purposes." Green taxes are suggested as part of a global scheme called "eco-tax reform," which will encourage sustainable development and discourage destruction of natural resources.

work." This is not entirely correct. If there is public transit or if the place of business is fairly close, a car is not a necessity. It is something we *want*, not something we *need*. Whenever we buy a product, we are using raw materials that were derived from nature. Our homes are often overheated and our lights are left on even when we aren't using them. We waste gas as we drive around our neighbourhood, when we could have walked. As one observer has suggested, we are stealing the natural resources from our children and our children's children.

CONSOLIDATING AND EXTENDING IDEAS

1 a) List the raw materials and consumer products that each of the following natural
       resources provides: trees, copper, wheat, oil, and air.
    b) Classify each of the items in 1 (a) as renewable, nonrenewable, or ubiquitous.

2 A future wheel is a mind map, or concept web, that is used to predict possible outcomes.
   Create a future wheel that predicts the outcomes of current consumption of the earth's natural resources. Develop your predictions using the format in Figure 10.4.

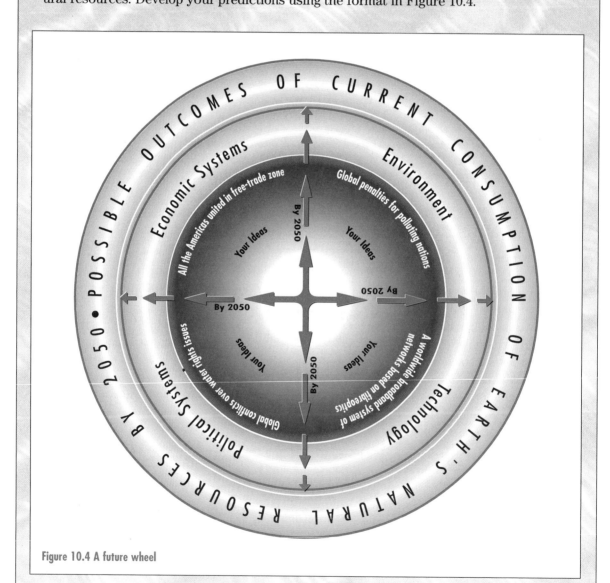

Figure 10.4 A future wheel

NEED	PAST	PRESENT	FUTURE
Food			
types	meat, fish, berries, roots	four food groups, balanced diet	synthetic foods, food substitutes
production methods	hunting and gathering, subsistence farming	large-scale agribusiness, supermarkets	agrifactories, restaurants
quality and quantity	varied greatly	nutritious, varied, readily available	homogenous, shortages of natural foods
cost	inexpensive	expensive	very expensive
Air			
Water			
Shelter			
Clothing			
Space			

Figure 10.5 Human needs—past, present, and future

3 a) Prepare an organizer or a chart like the one in Figure 10.5 to compare human needs in the past, present, and future. The food category has been completed as an example.

b) Evaluate your Future column: circle any factors that seem negative (for example, shortages of natural foods) and develop strategies to overcome these problems.

c) Write a short paper on your "ideal" vision of the future. Use the following format to help you structure your paper:

i) Introduction: Write a statement of your vision in general terms.

ii) Resource usage: Write separate paragraphs for each of the essential human needs, in which you explain how the needs are met in such a way that the environment is not harmed.

iii) Conclusion: Describe what action people need to take today if your vision is to be achieved.

IDENTIFYING RESOURCE ISSUES

There are four elements involved in resource issues: culture, environment, economics, and politics. The role of culture in resource issues is seen in social attitudes towards the earth's natural resources. The consumer societies of North America and, to a lesser extent, western Europe and the eastern Pacific Rim promote the high consumption of manufactured goods. People are encouraged to spend money. In return, the economy is strengthened, jobs are created, and society prospers. This, in turn, leads to even greater consumerism. This paradigm may be advantageous for the economies of the developed world today, but what will happen to future generations when the earth's natural resources have been depleted by the mass consumption of a few privileged nations?

There is a strong link between natural resources and environmental issues. The exploitation of any natural resource affects the immediate environment. Consider soils and agriculture, for example. When we cultivate fields for agriculture, the topsoil is loosened. If proper soil-management techniques are not practised, erosion can carry away valuable topsoil. Without enough topsoil agricultural production declines. In severe cases famine may follow, jeopardizing the lives of thousands, and perhaps millions, of people. This plight occurred in the 1930s in South Dakota. Poor agricultural practices, combined with severe drought, resulted in millions of tonnes of topsoil being lost to wind erosion. Today, this region is a barren landscape of pinnacles, gullies, and sharp-edged ridges known as the **badlands**.

Other resource issues that involve the environment include pollution, deforestation, and waste management. Industrial air pollution in the form of acid deposition, carbon dioxide emissions, and poisonous fumes are clearly environmental issues that result when natural

Figure 10.6 A dust storm over Canada's prairies in the 1930s

resources are processed. In December 1984, the tragic Bhopal disaster in central India resulted when thousands of people were exposed to poisonous fumes from an American-owned pesticide factory. Deforestation of tropical rainforests has been considered an environmental issue, but it is often related to the harvesting (or extraction) of trees in an environmentally sensitive area. The problem of waste management is also considered to be an environmental issue. However, if garbage is considered a resource, then waste management becomes an important resource issue. Far from being just a local issue, waste management has become an international problem as several nations in Africa compete as dumping grounds for garbage from the more affluent North.

Resource issues also involve economics. Natural resources are essential to the economic development of any country. In the past, the developed nations exploited their natural resources with little regard for conservation or preservation. Today, many developing nations are following the same path in the quest to improve their standard of living. In Indonesia forests are being cut down at an alarming rate. The government, eager for foreign exchange funds from the lucrative exports of wood products, and desperate for cleared land on which to resettle urban squatters, appears to have given little thought to sustainability. In West Africa, new plantation crops are being planted in marginal lands that were formerly reserved for pastoral grazing. New irrigation projects have supposedly made these developments viable. But as the soil gives up more and more of its fertility, yields drop, and the land eventually becomes useless. In their quest for short-term gains in employment, standard of living, and fiscal solvency, many countries are using up renewable resources that should be available to future generations.

Politics is yet another element associated with resource issues. Domestically, governments may impose restrictions, create legislation, and

> Deforestation is a complex issue. The harvesting of trees upsets nature's systems. Food webs break down when natural habitats are lost. The land is often clear-cut and a virtual wasteland is left. Yet trees are essential to our lifestyle. We must harvest them because we depend on wood products. Forestry companies in Canada are required to work with governments to ensure that sustainable forestry is practised. But what happens to the animals that depend on the forest? When forests are replanted, many animals return within five to ten years. In other cases, new species take over the habitat.

reach compromises with industries to permit certain levels of consumption. Internationally, governments are confronting issues of equitable resource distribution. (The article "Middle East: Water Rights" on page 134 highlights the issue of resource disputes between nations.)

Natural resources can be a powerful weapon for those countries that possess them. Political **cartels** such as OPEC (the Organization of Petroleum Exporting Countries) regulate resources through pricing arrangements and supply quotas. They wield considerable political influence because they control valuable resources that the world needs. Wars are even fought over natural resources. A prime example is the 1991 Persian Gulf War, which broke out after Iraq invaded its oil-rich neighbour, Kuwait, in late 1990. At issue, at least in part, was Iraq's claim to ownership of the oil deposits that straddled the border between the two countries. This was particularly significant because the region provides 26 per cent of the world's oil and 64 per cent of the world's oil reserves (1995 figures). While Kuwait supplies only 13 per cent of this total (3.3 per cent of the world's total production), stability in the region is considered to be essential to the stability of the global economy. So valued are some natural resources that governments are willing to use military force to gain control of these assets.

MIDDLE EAST: WATER RIGHTS

Event: Israel and the Palestinians have begun secret preliminary discussions on resolving their dispute over water rights.

Significance: Water is one of the paramount issues in the peace process. It ranks in complexity alongside the final status of Jerusalem and the future of Jewish settlements.

Analysis: As their opening position on water rights, the Palestinians are demanding: a water allocation from the Jordan River basin, rehabilitation of the Gaza aquifer, and virtual control over the mountain aquifer and the rights to 80 per cent of the water that emanates from its catchment basin.

Of the three main demands, the issue of the mountain aquifer is by far the most difficult to resolve. It provides 35 per cent of Israel's current water supplies and 100 per cent of the water used on the West Bank [Israeli territory on the border with Jordan]. Moreover, its quality is the best in the region.

Most of the catchment basin lies within the West Bank, but because of the geological structure of the area, most of the underground storage areas lie within Israel's pre-1967 borders in the coastal plain and [Israeli river] valleys.

Although the aquifer can supply 600 million cubic metres of water a year without damage to its basic structure, Israel currently draws 495 million cubic metres annually and the Palestinians have been able to use 105 million cubic metres. . . . The allocation of water between the two sides has remained unchanged since the early 1960s when Jordan controlled the West Bank.

The Palestinians argue that they have not been able to use their fair share. They say that once a Palestinian state has been created, the return of refugees and the increased industrialization will create enormous new demands on water supplies. . . .

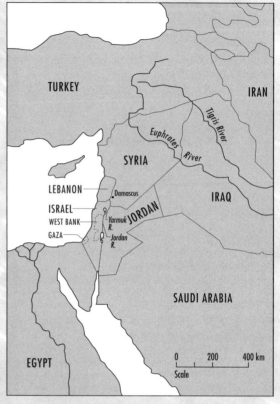

Figure 10.7 The Middle East

The dispute between the Israelis and the Palestinians is not unlike that between the states of the Nile, or Turkey's argument with Syria and Iraq over (water) rights to the Tigris and Euphrates [rivers]. The highland areas claim rights to the rain waters that fall in their area. The lowland downstream users of the water demand free passage of the water to their areas. The only difference in this case is that most of the cloudbursts that land on the West Bank highlands travel underground rather than in open rivers. . . .

The only comprehensive and potentially economically viable solution is a regional approach that has hitherto been considered unthinkable. A proposal is currently being developed by several joint groups of Israeli, Palestinian, Jordanian, and US engineers, hydrologists, and economists. Under this proposal:

- Turkey would release additional water from its recently inaugurated system of dams to supply more [water] to northern Syria and Damascus.
- Syria would release more water from the dams it has built on the Yarmuk River in southern Syria.
- Jordan . . . would use Syrian-controlled water.
- Israel would allow the transfer to the West Bank of water given to Jordan under the peace treaty. Alternatively Israel would use this water and agree to greater pumping by Palestinians from the mountain aquifer.
- Israel would then agree to supply Gaza with more water from its own resources in order to match supply and demand and to rehabilitate the aquifer there.
- Turkey would be compensated by other large investments in its infrastructure, such as a new oil pipeline from the Persian Gulf. . . .

Conclusion: Without an integrated solution to the Israeli-Palestinian water issue, there can be no permanent political solution.

From *The Globe and Mail,* 17 January 1995. Copyright1994 by Oxford Analytica Ltd. All rights reserved.

FRESHWATER RESOURCES

List all of the uses for water and prioritize the list. Estimate how much water your family uses in one day. If you live in an urban centre, the water that comes out of your tap probably travelled a long way and was treated to a great extent before it arrived at the tap. Most municipal water comes from lakes, rivers, or underground water supplies. It is filtered, the sediment is allowed to settle out, and chemicals are added to ensure that the water contains no harmful bacteria. Many Canadians take water for granted. After all, we have more freshwater per capita than any other country in the world. Yet freshwater is not abundant everywhere. Many developing regions of the world lack adequate water supplies. The growing demand results from expanding populations, increasing industrialization, and rising expectations. As more water is needed, surface water reserves and groundwater deposits are used up. This water is replenished by natural processes. Of

Figure 10.8 An aerial view of Scottsdale, Arizona, where backyard swimming pools are a feature of many homes

course, if more water is being removed than natural systems are able to replace, water resources will decrease. This unsustainable use of water will lead to a crisis when the resource is no longer available. Coupled with the problem of decreased freshwater supplies is the issue of water quality. Industrial, agricultural, and urban pollutants in the water often make it unsafe to drink. If water is not purified, it frequently contains microscopic organisms that spread disease. The lack of adequate water resources is a major problem affecting people all over the world. For instance, in Bangkok, Thailand, municipal water is supplied by wells. So much water has been removed from aquifers that salt water from the ocean is flowing into them, thus tainting the freshwater supply.

WATER SHORTAGES IN SUB-SAHARAN AFRICA

Farmers in sub-Saharan Africa are often at the mercy of unreliable annual rains. If these rains fail, disaster is almost sure to follow as crops wither and die, and people go hungry. When rains are steady the countries of North Africa prosper. With abundant grass, the animals that sustain the pastoral peoples of the area grow fat and bear many young. The people prosper from the abundant resources available to them, and many children are born. The hope is that some of the children and some of the cattle will survive when times get bad. Periods of as many as twenty years may go by when the rainfall is below normal. The grasses wither and the land turns into desert. Many of the animals starve and a full-scale famine occurs, causing extreme human suffering. These cycles of drought and plenty have gone on for millennia. But today's modern medical technology allows more children to survive natural disasters, so population pressures are much more severe now than they were in the past.

Even when there is an adequate supply of water, there are problems with water quality. Parasites living in streams and lakes often infect the people who drink the water. Schistosomiasis and Guinea worm are two diseases that afflict people living in North Africa. Efforts to eradicate these diseases through modern water-

treatment facilities have proven successful in many villages, but there are still so many people without clean water that water-borne diseases are endemic to the region.

WATER SHORTAGES IN THE CITIES OF DEVELOPING COUNTRIES

Many cities in developing nations lack adequate supplies of clean water for human consumption. Water is a major necessity for industry, so much of the water supply is allocated for manufacturing. While increasing industrial output is often a priority for many nations, they may lack the funds or the political will to develop clean water supplies for the people living in these cities. A much greater return on investment is obtainable through increasing industrial output. As a result, municipal water supplies remain inadequate for the people who need them.

What are the solutions? Clearly, improving urban sewage systems needs to be a priority. Massive projects to trap water in regions where seasonal precipitation varies have been very successful. The Indus River has an extensive system of weirs, dams, and reservoirs to trap water flowing out of the Himalayas in springtime, when the snow melts, for use in the dry autumn and winter months. But the costs for megaprojects in many developing countries are often prohibitive. Peru has recently abandoned a US $500 million project to drill a tunnel through the Andes Mountains from the wet Amazon basin in the eastern part of the country to the dry desert region on the west coast. Quite simply, the country cannot afford such an expenditure.

WATER SHORTAGES IN THE AMERICAN SOUTHWEST

The southwestern United States is a semi-arid region with a growing population. Originally range country suited to grazing cattle, many parts of the Southwest have become bustling

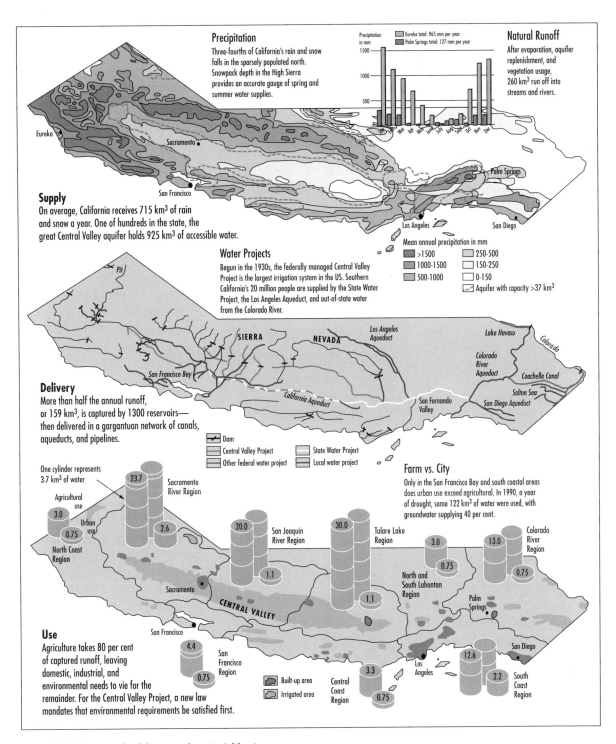

Precipitation

Three-fourths of California's rain and snow falls in the sparsely populated north. Snowpack depth in the High Sierra provides an accurate gauge of spring and summer water supplies.

Precipitation in mm

Eureka total: 965 mm per year
Palm Springs total: 127 mm per year

Natural Runoff

After evaporation, aquifer replenishment, and vegetation usage, 260 km³ run off into streams and rivers.

Supply

On average, California receives 715 km³ of rain and snow a year. One of hundreds in the state, the great Central Valley aquifer holds 925 km³ of accessible water.

Mean annual precipitation in mm

>1500
1000-1500
500-1000
250-500
150-250
0-150
Aquifer with capacity >37 km³

Water Projects

Begun in the 1930s, the federally managed Central Valley Project is the largest irrigation system in the US. Southern California's 20 million people are supplied by the State Water Project, the Los Angeles Aqueduct, and out-of-state water from the Colorado River.

Delivery

More than half the annual runoff, or 159 km³, is captured by 1300 reservoirs—then delivered in a gargantuan network of canals, aqueducts, and pipelines.

Dam
Central Valley Project
Other federal water project
State Water Project
Local water project

Farm vs. City

Only in the San Francisco Bay and south coastal areas does urban use exceed agricultural. In 1990, a year of drought, some 122 km³ of water were used, with groundwater supplying 40 per cent.

One cylinder represents 3.7 km³ of water

Agricultural use
Urban use

Use

Agriculture takes 80 per cent of captured runoff, leaving domestic, industrial, and environmental needs to vie for the remainder. For the Central Valley Project, a new law mandates that environmental requirements be satisfied first.

Built-up area
Irrigated area

Figure 10.9 Water supply, delivery, and use in California
Source: *National Geographic* Special Edition: Water, November 1993: 43

urban centres housing thousands of retired or vacationing northerners. For over four decades, municipalities have relied on the Ogallala aquifer, which stretches from Wyoming to Texas, for irrigation water. But today, advanced pumping systems are removing this water faster than it can be replenished. There is a real danger that the aquifer will no longer be able to supply water in the future.

Precipitation is unreliable in the Southwest. If rainfall is below seasonal averages, the entire water supply is in jeopardy because water from aquifers is drawn out faster than it can be replenished. Yet despite the scarcity in this region, huge amounts of water are channelled into non-essential services. Irrigation systems keep hundreds of hectares of golf courses lush and green in the Arizona desert. In California, thousands of homeowners are able to cool off in their own backyard swimming pools. The solutions to these unsustainable practices lie in educating the population about alternative approaches. People are starting to replace their lawns with attractive desert landscaping, including cacti, sand, and stones. Dryland farming, a practice that utilizes natural processes to reduce water lost through evapotranspiration, is gaining popularity in some sectors as farmers return to farming methods developed years before by their grandparents.

Although Canadians have an abundant water supply, we often do not fully understand its limits. For instance, many municipalities in southern Ontario draw water from underground aquifers. As urban growth spreads in this densely populated region, more of the water supply is consumed. Much of the water in these aquifers was deposited when the last ice age ended about 10 000 years ago. So, in effect, the municipalities are mining ancient waters, much like oil deposits. As consumption continues to increase, this water supply eventually will run out. Canadians may then have to construct expensive **aqueducts** to carry water from northern lakes to southern markets.

However, the long-term prognosis is not good. With the water level in the Ogallala dropping each year, it is just a matter of time until this valuable resource dries up.

FOREST RESOURCES

The world's forests provide essential raw materials for many industries, including construction, furniture making, publishing, and pharmaceuticals. Tropical and temperate forests have been harvested and utilized by humans since prehistoric times. Yet the earth's forests offer much more than bark, lumber, pulp, and vegetable chemicals. They play an important role in the earth's ecosystem. Called "the lungs of the planet," forests absorb greenhouse gases such as carbon dioxide, methane, nitrous oxide, tropospheric ozone, and maintain the world's oxygen supplies. Without forests, our planet's climate would warm, land would degrade and become desertified, and lakes and rivers would dry up.

Deforestation has received an enormous amount of attention in recent years. International recognition of the importance of forests—for their atmospheric and **terrene** benefits, and as sources of raw materials—has led to "debt-for-nature swaps" between developing nations and NGOs. The first "swap" took place in 1992, when Conservation International bought US $650 000

Figure 10.10 Rainforests cover less than 6 per cent of the earth's land surface, but they are the home of up to three-quarters of all known species of plants and animals.

Tropical rainforests provide a habitat for indigenous peoples and thousands of plant and animal species. Many life-saving drugs are derived from tropical rainforest plants (see page 288). These remote forests hide a treasure trove of undiscovered plant species. Could these resources hold the key to important medical discoveries? We may never know! Tropical rainforests are being cleared so rapidly that many unique species may be destroyed before they are ever discovered.

worth of Bolivia's debt to Citicorp, a big US bank, and relieved the Bolivian government of having to repay this money. In turn, the Bolivian government set aside approximately 13 350 ha of forest as a protected reserve. All participants in a debt-for-nature swap benefit. The environmental group is able to preserve an ecologically important part of the world for future generations. The international banks see their debt exposure reduced, at least in part. And the debtor nation gains a nature park and a reduction in debt.

Deforestation also affects ubiquitous resources such as water and soil. In a forest landscape, the natural litter of leaves, roots, and fallen trees soaks up precipitation. When forests are cleared, rainwater flows over the surface of the land instead of percolating into the soil. The impact of the rain causes tiny craters. Multiplied billions of times, these craters loosen the topsoil and free it to the erosive forces of running water. In time, the topsoil is washed away and the groundwater disappears. People who depend on the aquifer that lies under these natural areas may lose their water source if it is not constantly recharged.

Many people believe our forests should be preserved not only for the valuable resources they will provide for generations, but also because of their beauty and recreational value. Stretching from the Atlantic to the Pacific oceans, Canada's forests cover 45 per cent of our nation's total land area. They offer us opportunities for a variety of recreational activities—walking in a picturesque setting, observing plants and wildlife, camping in a natural environment. In recent years, Canadians have come to appreciate that we must protect this fragile resource that enhances our lives.

CONFLICT OVER FOREST RESOURCES

Much of the debate over deforestation is between those who harvest forests and those who want to preserve them. In recent years, the forestry industry in Canada has been required to practise **sustainable development**. Clearly, it is in the industry's best interests to regenerate forests rather than exploit them with little regard for the future. Without forests to harvest, these companies would eventually be out of business! Today, forestry companies use advanced Geography Information Systems (GIS) to determine the age, size, health, and species of forests they intend to harvest. From these studies they can determine the expected yield and the value of the crop to be removed. In addition, maps showing landform characteristics such as slope, topography, drainage, soil type, and rock type are analysed. A forestry company develops a plan that details

Canada has not always practised sustainable development. In the last century most of the great southern forests were cleared. Almost no forest in southern Ontario, southern Quebec, and the Maritimes has been left untouched. Clear-cutting was the rule and reforestation seldom occurred. Only inaccessible regions were spared. Many of the provincial and national parks that we consider wilderness today were completely cleared of forest a hundred years ago. Two of the last remaining stands of old-growth forest are found in Temagami, Ontario, and along Clayoquot Sound in British Columbia.

harvesting methods, reforestation techniques, transportation methods, and remedial actions to protect wildlife and local communities. Provincial ministries of Natural Resources either approve the plan or suggest revisions. Once the plan has been approved, the forestry company carries out its plan under government supervision.

Nevertheless, there have been conflicts between loggers, forestry companies, and governments on the one hand, and preservationists, the tourist industry, and aboriginal peoples on the other. The harvesting of **old-growth forests** in Temagami, Ontario, and Clayoquot Sound, British Columbia, sparked international protests in the early 1990s. Preservationists in Europe boycotted Canadian lumber products. The foresters contended they were harvesting trees that were over 500 years old. Yet it is unlikely that anybody would plant trees and wait 500 years before they were harvested! They were removing these ancient trees as if they were a nonrenewable resource that would take generations to regenerate. Environmentalists argued that these ancient trees must be preserved as monuments to nature and not be disturbed by people. Habitats should remain intact so that plants and animals could continue to live and thrive, and the beauty of nature could be preserved. On the other hand, all living things die. Is it not better to remove the ancient forests so that new trees can be allowed to thrive? The debate goes on.

OIL RESOURCES

One of the most serious resource issues in the world today revolves around oil—a nonrenewable resource. The economies of the developed world are founded on the availability of an affordable and abundant energy supply. Industries rely on energy to manufacture their products. The goods we receive from and send to other countries around the world reach their destinations using the energy provided by gasoline, a by-product of oil. Countless consumer goods, from pens to clothing to carpets, are made from **petrochemicals**.

One way to prolong the life span of current oil reserves is through **conservation**. If we curbed our appetites for so many products, we could conserve some of the oil that is used to generate the energy for manufacturing. We could also reduce energy consumption by car pooling, using public transit, or riding bicycles to school or work. But conservation alone is not enough. Eventually, oil supplies will run out, no matter what efforts we make to use this resource more wisely. But there may be other solutions to this impending crisis.

New technology may enable oil deposits buried in **unconventional oilfields** to be exploited profitably. There is as much oil in the **tar sands** of northern Saskatchewan and Alberta as there is in the Middle East. But the oil is mixed with sand, and extracting and refining this gooey substance is extremely costly. In June 1996, Prime Minister Jean Chrétien announced new tax laws and changes in royalty payments to encourage a regeneration of the Syncrude Project at Fort McMurray in northern Alberta. These changes will allow the oilfield to be competitive with offshore oil producers. An immediate $5 billion expansion is taking place

Figure 10.11 The Syncrude Project in Alberta

and the federal government will increase its investment to $1 billion per year. An estimated 2.5 trillion barrels of oil is thought to be in the sands. About 300 billion barrels are expected to be recovered over the next twenty-five years. The company developing the sands will increase production of high-quality oil to meet Canada's future needs. Offshore oil deposits are also being developed on the continental shelf off Newfoundland and Nova Scotia. These oil deposits are not as extensive as the tar sands, but they are close to eastern markets and hold great potential for economic growth in an economically depressed region of Canada.

There has been considerable research into alternative energy fuels. In Brazil, researchers have been able to refine ethanol from sugar cane to create a petroleum substitute. Experiments with hydrogen extracted from water hold some promise, but the fuel is currently more expensive than oil and does not perform as well. There are many other energy alternatives to oil. Wind and solar energy are sustainable and do not require fuel. While the amount of energy produced by each generator is very low and the start-up costs for production are very high, these two sources are growing faster than all other energy sources.

The generating capacity of wind energy has increased from just over 1000 MW in 1985 to 4880 MW in 1995. The 1995 figures represent a 33 per cent increase over 1994. While wind energy provides less than 1 per cent of the world's energy needs, it is the fastest growing energy source (*Vital Signs* [New York: W.W. Norton & Co., 1996], 56). Solar voltaic cells are thin silicon wafers that directly convert sunlight into electrical energy. While global output is under 600 MW, the sale of these cells is increasing at a rate of about 15 per cent a year. In Japan, a country with few energy resources, the government has initiated a project to encourage people in 70 000 homes to install photovoltaic cells. Although they are used in developed countries, these cells are particularly suited to remote regions that are not on the **power grid**. The World Bank has provided $400 million in funds to developing countries to help them introduce photovoltaic cells to rural villages. Residents will now be able to have refrigeration, and light for reading in the evening. Recent breakthroughs in Australia have improved the efficiency of cells by 21.5 per cent and increased the amount of power they generate by 80 per cent. These innovations make the new technology competitive with coal-fired electricity. Researchers expect that the technology will become widely available to the public early in the next century.

Geothermal power utilizes heat from the earth as a power source. It has been successfully harnessed in Iceland, Russia, and New Zealand, but it has limited potential in most parts of the world because they lack the essential geological formations. Hydrogeologists in New Brunswick and Ontario are providing people with geothermal heat to heat their houses in winter. The process is expensive to install but requires no fuel and minimal power to run water pumps. Water is pumped from aquifers into a heat-exchange system. The heat is removed and the water is returned to the aquifer.

Nuclear energy is cheap and clean but many people fear nuclear accidents such as the one that occurred at Chernobyl. Tidal power is feasible in certain areas where there are sufficient tides, such as the Bay of Fundy between Nova Scotia and New Brunswick. As the water rises, turbines are turned and electricity is generated. The burning of biomass is also a potential source of power. Many large cities have huge amounts of municipal garbage that could be burned to generate energy. As engineers discover more ways to harness energy, further possibilities will present themselves in the future.

CONSOLIDATING AND EXTENDING IDEAS

1 Study Figure 10.9.
   a) Describe the pattern of precipitation in California with regard to latitude.
   b) During which season is there drought in the Central Valley?
   c) What problems exist with these precipitation patterns with regard to human water consumption.
   d) What is the Central Valley aquifer and how does it serve to alleviate the problems of water supply?
   e) How have Californians modified the environment to overcome the problems of water supply?
   f) What difficulties can you foresee if population growth continues in southern California?

2 Study Figure 10.12 and then use this mind map/concept web format to outline the cultural, environmental, economic, and political issues associated with one of the following resources: water, forests, oil.

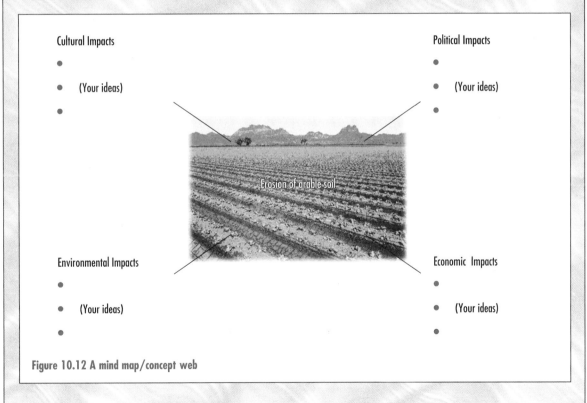

Cultural Impacts
 •
 • (Your ideas)
 •

Political Impacts
 •
 • (Your ideas)
 •

Erosion of arable soil

Environmental Impacts
 •
 • (Your ideas)
 •

Economic Impacts
 •
 • (Your ideas)
 •

Figure 10.12 A mind map/concept web

3 Even though population levels in many developing countries far exceed those of Western societies, their total resource consumption is much less. Why do you think this is so?

4 Deforestation upsets the delicate balance of the ecosystem. Do some research in the school or public library, and present the following information in a pamphlet:
- a world map detailing deforestation
- a flow chart illustrating the relationship between deforestation and flooding
- illustrations of the consequences of flooding
- suggestions for possible solutions to deforestation

5 In groups of three, select one issue associated with natural resource use: (i) habitat and wildlife destruction, (ii) desertification, and acid and toxic precipitation, (iii) destruction of the ozone layer, (iv) water pollution, (v) the greenhouse effect.
   a) Conduct an inquiry using the approach described in Chapter 4 on page 28. Research and gather articles from magazines and newspapers. Electronic retrieval systems, periodical indexes, and vertical file collections will be the most helpful sources in your research.
   b) Compile a bibliography and hand it in to your teacher for evaluation.
   c) Write a synopsis for each of the articles you have gathered. (A synopsis is a brief summary—in this case, a paragraph that outlines the essentials facts presented by the writer of the article.)
   d) Share what you have learned with the rest of the class. Make an oral presentation of six or nine of your synopses.

FOOD, STARVATION, AND HUNGER

DEFINING THE ISSUES

Along with air and water, food is our most essential natural resource. Without it, people cannot survive. Unlike ubiquitous resources such as air and water, food is a renewable resource. It must be either purchased or grown. Many people breathe polluted air or have an inadequate or unsafe water supply. But more people die either directly or indirectly from hunger than from anything else. Hunger and its resulting nutritional diseases kill an estimated 35 000 people every day.

In order to solve a problem, we must first fully understand it. The terms **starvation** and hunger are often used interchangeably when describing the human need for food. Yet, in fact, these two terms have very different meanings. They are caused by different aspects of physical and human geography, and consequently have different solutions.

Most of the food we eat is converted to

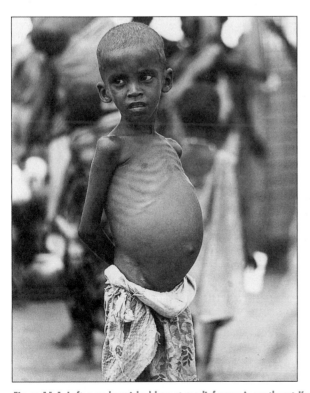

Figure 11.1 *Left,* a malnourished boy at a relief camp in northeast Kenya; *right,* a hungry child waiting for food at a hospital in Muqdisho, Somalia

energy, which is measured in **kilojoules**. The standard measure of hunger relates to national average food availability, stated in kilojoules per person per day (kJ/cap/day). New studies indicate that the world's population has a standard need for adequate nutrition. Called the Standard National Unit (SNU), the suggested figure for national average food availability is approximately 10 350 kJ/cap/day.

WHAT IS STARVATION?

When human beings do not have enough food to sustain themselves, this is known as starvation. People suffering from starvation are emaciated. They have skeletal frames, with thin arms and legs. Their hair may be tinged red and their eyes dull and lifeless. **Marasmus**, often called the disease of starvation, results from a lack of sufficient kilojoules to sustain a person. People with this disease are usually at least 40 per cent underweight. They have hardly any body fat and their muscles are wasted. Their chances of survival are minimal.

Starvation usually occurs in times of **famine**, a natural phenomenon that is common where subsistence farming is practised. The 1984 famine in Africa was blamed on drought. During the previous three years the rain needed for crops did not fall. But drought alone is not the cause of famine. While it may be true that lack of rain contributes to drought conditions, Africa was victimized by this ecological disaster because of inappropriate farming techniques, deforestation and the resulting desertification, and cultivation of its most fertile regions for **cash crops**. (A cash crop is harvested for sale on the world open market; it is not grown to feed farmers and their families.)

These practices placed an intolerable strain on the African continent's agricultural capacity, resulting in famine for millions.

Famines can also be caused by too much rainfall. In China, before the country's most powerful rivers were dammed in the early part of this century, flooding frequently resulted in famine. Floods often occurred when the snows of the Himalayas melted later than usual. If the rivers were swollen with melt water when the monsoon rains of summer arrived, disaster was sure to follow. Those who did not drown in the floods faced starvation because their crops had been destroyed.

Human activities contribute greatly to famine. Traditionally, in the Sahel region of Africa, which includes Ethiopia, the Sudan, Niger, Mali, Burkina, and the Central African Republic, the populations were able to cope with widespread fluctuations of the seasonal rains. People applied traditional strategies to survive cyclical weather changes and periodic drought. In years of sufficient rainfall, cattle grew fat and food crops were plentiful. This abundance sustained the people in years when the rains failed to come.

> In the mid-1980s, millions of well-fed people in Western societies sat in front of their television sets and watched news footage of thousands of people dying of starvation in Ethiopia. The horror they saw unfolding ignited a humanitarian movement unlike anything the world had ever seen. Many NGOs and international aid organizations rushed to the rescue of starving people. Despite massive emergency aid in the hundreds of millions of dollars, 2 million people eventually died in one of the worst disasters of modern time.

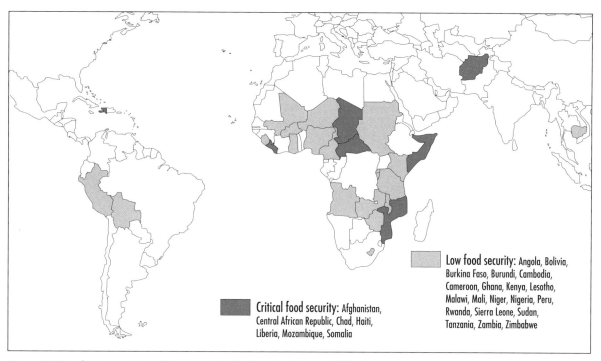

Figure 11.2 Developing countries with low or critical food security indexes, 1990–92

Source: *World Resources 1996–97* by the World Resources Institute. Copyright © 1996 by the World Resources Institute. Used by permission of Oxford University Press, Inc.

Critical food security: Afghanistan, Central African Republic, Chad, Haiti, Liberia, Mozambique, Somalia

Low food security: Angola, Bolivia, Burkina Faso, Burundi, Cambodia, Cameroon, Ghana, Kenya, Lesotho, Malawi, Mali, Niger, Nigeria, Peru, Rwanda, Sierra Leone, Sudan, Tanzania, Zambia, Zimbabwe

Africa is a continent with the potential to feed itself, yet it is known as the "famine continent." Africa is the only continent in the world that exited the 1980s in worse economic shape than it entered the decade.

In 1996, the World Resources Institute noted that the global redistribution of food by public-sector agencies amounted to a record 17 million tonnes of cereals, about 25 per cent of which went to Somalia, Rwanda, and countries in sub-Saharan Africa. Fifteen countries in Africa are facing exceptional food emergencies, and of the twenty-seven countries in the world with household food security problems, twenty-two are located in sub-Saharan Africa (see Figure 11.2).

In recent decades, the politics of land ownership, land use, and policies towards nomads have all contributed as causes of famine. In most

Figure 11.3 Rwandan refugees in the Democratic Republic of Congo

of Africa, Asia, and Latin America, certain cultural groups control most of the land and promote its use for purposes other than feeding the peoples of their continents. Commercial and political élites own the best agricultural land and use their political clout to seize agricultural resources provided by local government and foreign aid organizations. They devote their lands and the labour of itinerant peasants to the production of cash crops, which are then purchased by multinational food corporations. Huge numbers of landless people starve because the regions in which they live do not permit them to become self-sufficient in producing food.

Political conflict also leads to food shortages when food aid is used by governments as a weapon. During the years of the Cold War, wealthy donor nations threatened recipient nations with the withdrawal of food aid if their political ideologies did not match those of the donors. In Ethiopia in the late 1980s, and in Zaïre*, Rwanda, Nigeria, Somalia, and Uganda, rebel groups have been denied food aid by the governments that they oppose.

The politics of food aid have also played a part in creating the human and ecological problems that threaten to overtake the African continent. Nations of the developed world have poured billions of dollars of tied aid into industrial and agricultural projects in Africa. These funds were designated for purposes that suited industries of the donor nations far better than they suited the needs of the recipient peasant farmers. Consequently, over 70 per cent of these projects have failed.

WHAT IS HUNGER?

Hunger is much more widespread than starvation. Hungry people are found everywhere, from the urban slums of both the developing and developed world, to rural backwaters, to suburban neighbourhoods. People suffering from what we call **chronic hunger** are undernourished; that is, they do not get enough to eat on a consistent daily basis. Many hungry people exist on diets that are high in carbohydrates, such as potatoes, rice, corn, and bread. But they lack the essential nutrients obtained from fruits, green vegetables, dairy products, and meat and fish.

The World Resources Institute suggests that the prospects for sub-Saharan Africa are indeed bleak. Current estimates of chronic undernutrition and projections for the year 2010 provide a guide to the nature of the food-security challenge. Sub-Saharan Africa is projected to have a per capita food supply of 9085 kJ per day by 2010—the lowest among all regions of the world (*World Resources, 1996–97*).

Hungry people often suffer from **malnutrition**, a condition in which there is a deficiency of one or more proteins, minerals, or vitamins. Malnutrition is most commonly found among children. Its symptoms include skin rashes, reduced growth, and hair discoloration. The most serious consequence of malnutrition in children is its effect on brain development. Most brain development occurs before birth and during the first two years of life. If either the mother or child is deprived of essential proteins during these critical times, the child's brain development is restricted. The result is that malnourished children may have permanent mental disabilities.

Kwashiorkor is caused by a diet high in carbohydrates and low in proteins. In large families, infants are weaned from their mothers' milk when a new child is born into the family. Malnutrition sets in when this high-protein milk diet is replaced with one based on starchy foods. Children with kwashiorkor may look healthy because they are bloated by the swelling of their tissues. This condition, called **oedema**, leads to round faces and pot-bellies. Other symptoms include pale skin and thinning hair.

*Renamed the Democratic Republic of Congo in May 1997.

Other diseases that develop from malnutrition include **beriberi**, **pellagra**, **scurvy**, and **rickets**. Beriberi is the result of a vitamin B deficiency. It is common among people whose diet staples are rice and other grains. People with this disease have weak leg muscles, which makes walking difficult. As the disease progresses, the arms also become weak. In the final stages, a person is in so much pain that even the weight of sheets may be unbearable. Death usually results from cardiac arrest. Pellagra is a similar nutritional deficiency that stems from a diet high in cornstarch. The symptoms include inflammation of the skin, diarrhoea, vomiting, weight loss, and mental impairment. Scurvy results from a lack of vitamin C. It leads to bleeding gums, bruising,

Figure 11.4 A young boy with kwashiorkor in Mozambique

weakness, and fever. Rickets is caused by deficiencies in vitamin D, calcium, and phosphorous. Common in children, it causes bow-legs, pot-bellies, and incorrectly formed bones. If caught in the early stages, all of these diseases can be cured with a nutritious diet.

Malnutrition also makes people susceptible to other diseases. Diarrhoea, malaria, and anaemia are habitual, and often deadly, diseases among people who are suffering from hunger. Even the common cold can be fatal if the body is too weak to combat it. Children are especially vulnerable to diseases when they are malnourished. Typical childhood ailments such as measles can be combated by healthy children. But measles is the second highest cause of death among children who are malnourished.

CONSOLIDATING AND EXTENDING IDEAS

1 Compare starvation and hunger in an organizer or a chart. Include the symptoms; when, where, and why they occur; and their severity.

2 Study the categories in Appendix A.
  a) List the category that relates to hunger and describe what it measures.
  b) Present the statistics in a graded shading map. What does your map reveal about the pattern of hunger in the world?

3 Research an area that is currently experiencing famine.
  a) Determine the key factors (physical, cultural, political) responsible for the famine.
  b) How are the people of the area dealing with this famine?
  c) What has been the reaction from other groups and countries so far?

PATTERNS OF PHYSICAL GEOGRAPHY: THE DISTRIBUTION OF FOOD

Whenever we examine a global issue, it is important first to understand the physical geography involved. Only then can we develop appropriate solutions. To solve the problem of food shortages, we must begin by looking at our plant and animal resources.

PLANTS

All plant life needs air, space, nutrients, water, sunlight, and heat to survive. Air and space are readily available. The other elements are available only in varying degrees, depending on physical geography.

Nutrients are not uniformly distributed around the earth. The most fertile soils are found in the temperate grasslands of the prairies of North America and the steppes of Ukraine and Russia. This deep, rich, black topsoil is easily cultivated once the overlying grasses are removed. Less fertile soils are found in regions of excessive rainfall. Heavy rains dissolve and remove valuable minerals from topsoil in a process called **leaching.** The tropical rainforests of South America, the humid savannah grasslands of Africa, Australia, and India, and the boreal forests of northern Canada, northern Europe, and Russia all have leached soils. Although artificial fertilizers can enrich leached soils, the soils are often waterlogged and difficult to cultivate. The soils of the mixed forests of the middle latitudes, although slightly leached, are suitable for most forms of agriculture, especially if they are treated with fertilizers and lime to reduce acidity.

Without considerable human intervention, desert soils are unsuitable for most types of farming. The greatest obstacles are deposits of mineral salts, which most plants cannot tolerate. High evaporation draws moisture up through the soil. When the moisture evaporates, the mineral salts remain on the surface. With careful **irrigation** and applications of fertilizer to make up for the lack of **humus,** desert soils can be successfully farmed.

Water is necessary for plants to acquire the nutrients they need, and for **photosynthesis**. Like soil nutrients, the distribution of water is highly varied around the world. Regions that receive abundant rainfall often have lush vegetation. Yet as we have discovered, the soils are not always fertile. Desert and semi-desert regions are unsuitable for agriculture unless other sources of water are available. Some of the most productive agricultural regions in the world are arid. The Imperial Valley in California, the Nile Valley in Egypt, the Indus Valley in Pakistan, and the Tigris-Euphrates flood plain in Iraq all rely on irrigation to compensate for the lack of rainfall. The high temperatures and plentiful sunshine of these desert regions make them well suited to farming once a reliable water supply is obtained.

The elements of sunlight and heat are interdependent. The main hindrance to agriculture in Canada and other countries in high latitudes is the limited amount of sunlight and heat on a year-round basis. The period of time in which there is enough sunlight and heat for farming is called the **growing season**. The growing season determines the length of time crops require to grow from germination to maturity. Many crops grown in Canada have adapted to a short growing season, which can begin as late as June and end as early as September. Other crops such as winter wheat are hardy enough to withstand cold temperatures. Of course, the long summer days in the higher latitudes provide more hours of sunlight than in the tropics. This compensates in part for the short growing season. In lower latitudes the growing season is much longer. This allows plants that require

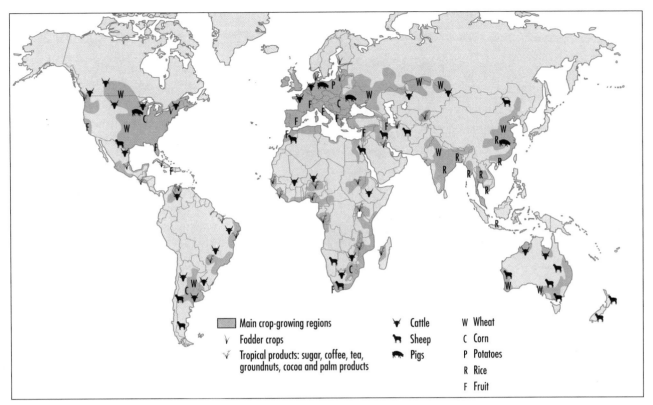

Figure 11.5 Croplands and livestock

Source: *Canadian Oxford Intermediate Atlas,* 2d. ed. Copyright © Oxford University Press 1993. Reprinted by permission of Oxford University Press Canada.

Legend:
- Main crop-growing regions
- V Fodder crops
- V Tropical products: sugar, coffee, tea, groundnuts, cocoa and palm products
- Cattle
- Sheep
- Pigs
- W Wheat
- C Corn
- P Potatoes
- R Rice
- F Fruit

more heat energy time to grow. Tropical regions also have the benefit of being able to grow crops all year round. Two or even three crops can be produced on one field in a year because sunlight and heat are so uniformly distributed. A disadvantage of tropical climates is that pests and diseases are able to thrive from one year to the next. In temperate zones, frosts and cold temperatures ensure that pests and diseases are killed off each year.

ANIMALS

Animals have many of the same needs as plants—air, space, light, heat, and water. But more than anything else, animals need nutrients. **Carnivores** derive their nutrients from other animals. **Herbivores** eat only plant life. Most animals on which people rely for food are herbivores. Many regions that are unsuitable for cultivation make good pastures for grazing animals. The western ranges of North America, the Australian outback, the llanos of Venezuela, and the veldt of South Africa are a few examples. These pastures range from lush to semi-desert. Different animals flourish on different grasses. Dairy cattle need rich grasslands, while beef cattle can survive on less nutritious vegetation. Sheep and goats thrive in semi-arid regions where grass and scrub brush predominate. Camels can survive in the most inhospitable grasslands. Each domesticated animal has its own niche in the pastoral system.

Figure 11.6 A traditional irrigation technique in the Nile Basin

THE EVOLUTION OF AGRICULTURE

Long ago people obtained their food from hunting and gathering. About 10 000 years ago, the incidental planting of food crops occurred more or less simultaneously in Mexico, Peru, Iran, China, and India. Seeds were accidentally dropped on moist ground, where they germinated and grew to maturity. Over time, people came to realize that *they* were, in fact, responsible for the production of certain plants in certain locations, and began deliberately to harvest, store, and plant seeds of their favourite plants. Thus, agriculture was born and sedentary civilizations began to develop. Archaeologists working in each of these birthplaces of agriculture have been able to determine when agriculture started by dating of the artefacts that have been found at sites in these regions.

Eventually, early farmers gained a better understanding of the needs of the crops they were planting. They developed farming techniques that allowed their crops to flourish.

Breaking up or cultivating the soil either with a digging stick or, later, with a plough allowed water and air to get to the roots of growing crops. The removal of weeds reduced the competition for water, sunlight, and nutrients. The fact that many early agricultural societies developed in arid regions meant that irrigation was necessary for seeds to germinate. The harvesting process was refined as people learned to winnow the grain from the useless straw of grain crops. Replanting the seeds of plants with favourable characteristics improved the natural selection. Plant species that became domesticated in this way acquired the characteristics people wanted—good taste, adaptability to climatic variations, and resistance to diseases and pests. Over the years, a variety of different crop strains, each with its own genetic characteristics, was developed.

Around this same period of time, animals were domesticated. Cross-breeding enhanced those traits that were beneficial for human use. For example, beasts of burden such as dogs, yaks, horses, llamas, and camels became more passive in their association with people. The more wild individuals escaped and the more docile ones were bred. In time, these animals had totally different dispositions from their ancestors. Goats, alpaca, and sheep that had thick coats were preferred over animals with thinner coats. So these animals gradually grew thicker wool coats. Similarly, creatures that were bred for food, such as chickens, turkeys, pigs, and cattle, developed more meat as fatter animals were bred and the gene pool of leaner specimens was allowed to disappear.

With a reliable food supply, life became easier. Freed from their nomadic existence following the herds and fresh pastures, people could adopt a sedentary lifestyle. Since people now stayed in one place, they could build permanent dwellings and acquire other material possessions. The development of agriculture required land ownership and cooperation. Political systems in these rapidly growing societies evolved to govern land ownership, provide protection against enemies, and regulate trade. Of course, taxes had to be levied to support government services. With taxes and trade came the development of mathematics, currency, and writing. While this pattern occurred universally among agricultural peoples, it is especially evident in the Nile Valley. Hieroglyphs on tombs and archaeological remains trace the evolution of political systems as a result of the domestication of grains and animals. As civilizations grew in power and affluence, large temples and other monuments such as the Sphinx and, much later, the pyramids were erected. And so it was that the discovery of agriculture led to civilization. Writing, religion, currency, urbanization, and technology all became possible because of the discovery of agriculture.

Agricultural scientists in Israel have modified ancient water-conservation techniques in the Negev Desert. The sparse rainfall is collected from across a huge highland area where **orographic precipitation** exceeds the rainfall of the area in the rain-shadow lower in the valley. Channels direct the water to one small centrally located greenhouse.

A system similar to hydroponics, utilizing drip irrigation technology, enables each plant to get its moisture requirements with a minimum of salt contamination. Unlike greenhouses in Canada, the building is not used to increase heat. The roof is made of opaque polyethylene, a semi-transparent plastic shell that protects the crops from the harsh desert sun. Moisture from **evapotranspiration** forms as condensation on the plastic ceiling. This raises the humidity level and recycles the moisture.

EXPANDING AGRICULTURE TO MARGINAL LANDS

When agriculture was developed around 8000 BCE, the world's population was 4 million. Today, the earth's population increases by 4 million people every ten days! If population growth continues at this rate, we will have to produce as much food by the year 2020 as we have produced since agriculture was invented 10 000 years ago. Is this possible? Scientists believe it is, in spite of the fact that most of the best farmland is already being cultivated.

The key is to expand agriculture into **marginal lands**—that is, lands that are not naturally suited to farming. There have been many attempts to convert marginal lands to productive use over the years, with varying degrees of success. In arid regions there has to be sufficient water to irrigate the land without the accumulation of harmful salts in the soil. Cold regions often need greenhouses or special hardy hybrid crops that can survive periods of frost. Many developing countries have used and still are using rainforests as potential agricultural lands, with very limited success. Even though the crops' require-

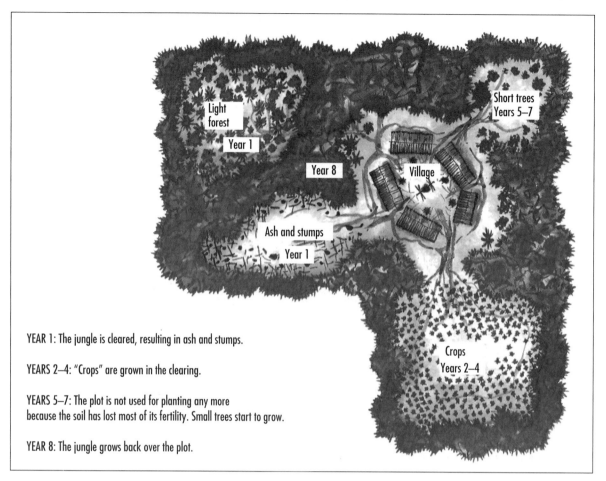

Light forest
Year 1

Short trees Years 5–7

Year 8

Village

Ash and stumps
Year 1

Crops Years 2–4

YEAR 1: The jungle is cleared, resulting in ash and stumps.

YEARS 2–4: "Crops" are grown in the clearing.

YEARS 5–7: The plot is not used for planting any more
because the soil has lost most of its fertility. Small trees start to grow.

YEAR 8: The jungle grows back over the plot.

Figure 11.7 Slash-and-burn agriculture
Source: *Earth Dynamics: Studies in Physical Geography* By Ron Chasmer. Copyright © Oxford University Press Canada 1995. Reprinted by permission of Oxford University Press Canada.

ments are adequately met, rainforest soils are characteristically very poor. Heavy rainfall leaches nutrients away, and erosion is a major problem once the protective forest has been cleared away.

TRADITIONAL FOOD PRODUCTION

The most successful farmers in marginal lands are often those who rely on traditional methods of intensive agriculture. It is possible to grow enough food on relatively small parcels of land to support a large family and provide an income from the sale of cash crops such as peanuts, sugar, or coffee. In the mountainous regions of Peru, Southeast Asia, and East Africa, **terraces** are carved into the hillsides. Rainfall flowing down the mountain slopes is channelled for distribution from one terrace to another. Steep walls hold the soil in place. This system enables crops to be grown in areas that are not naturally suited to agriculture. Another farming technique is **intercropping,** in which several crops are grown together in a single field. Low-lying vegetation that requires shade can be protected by larger plants. Insects are easier to control because they cannot move between plants of the same species. One crop may be harvested at the same time that another is being planted, which creates a steady harvest and income. While intensive cultivation is ill suited to large-scale mechanized farming, it is appropriate for subsistence agriculture because crop yields are high and it is labour intensive.

There are other traditional farming methods that work well in sparsely populated areas but are not as successful today because of increasing populations. **Slash-and-burn agriculture** has been practised in tropical rainforests for thousands of years. A farmer clears a plot of land by girdling trees (removing a strip of bark) and letting them die, then burning off much of the undergrowth. Crops are grown among the roots and burnt-out tree trunks. After several years, crop yields are so low that the plot is abandoned. It is then reclaimed by the rainforest, while the farmer moves on to a new plot of land. However, this process cannot sustain the growing populations in the outer limits of many wilderness areas such as Rondônia in central Amazonia, Kalimantan and Irian Jaya in Indonesia, and New Guinea in Papua New Guinea.

In the savannah of East Africa, farmers practise **shifting cultivation**. Scrub brush and grasses are burned off in the dry season. The resulting ash infuses nutrients into the soil. Grains such as sorghum and millet and other crops such as cassava are harvested over the next few years. When yields decline, the area is abandoned. The natural vegetation returns and soil fertility is gradually restored. In the meantime, farmers graze their cattle on the grasslands, waiting to cultivate the land once it has fully recovered. As with slash-and-burn agriculture, this traditional farming method is not possible with an expanding population.

Alley cropping is another form of sustainable farming in the savannah. Acacia trees have nitrogen-fixing bacteria that grow on their roots. The bacteria allow the trees to thrive even in poor soils. The nitrogen can also be utilized by field crops growing near the trees. So farmers plant acacia trees in between rows of crops. The trees also provide a valuable cash crop—gum arabic, a substance used in the manufacture of waxes and other products. Thus, traditional techniques have enabled intensive sedentary farming to be practised in regions that are naturally unsuited to permanent agriculture.

MODERN FOOD PRODUCTION

While many places in the world experience food shortages, there is a surplus of food in some regions. Most of this surplus is produced in developed countries such as Canada, the United States, and France. Specialized, highly mechanized farming operations extend over hundreds of hectares. In places such as the Canadian prairies, fields of wheat stretch as far as the eye can see. These bread baskets of the world produce grains and other crops for consumption at home and for export.

Many farmers in North America earn less for their grain than they did thirty years ago because there is more grain produced than local economies can absorb. Of course, grain prices fluctuate greatly as market conditions change. In 1997, a global shortage of grain, resulting from the economic restructuring of Eastern Europe and Russia, led to higher grain prices. The removal of railway subsidies in Canada further complicates the picture—wheat boards now have to pay more to ship grain from prairie farms to ocean ports. The problem lies not so much in what farmers can produce, but in political and economic structures that limit the flow of foods.

There are two possible long-term solutions to the problem of food supply. Hungry people either must be given the opportunity to earn enough money to buy the food they need or they must grow more food themselves. By providing farmers with the tools, seeds, and knowledge to grow more food, hunger can be alleviated. Unfortunately, there are often local factors that interfere with grass-roots development plans. War, famine, disease, and limited access to good farmland all contribute to patterns of hunger and poverty in rural regions of the developing world. People can earn the money to buy the food they need in a number of

Bradford, Ontario, is the largest carrot-producing region in the country. Here, livestock farmers can buy carrots to use as animal feed for ten dollars a truckload. However, a Bradford resident returning from vacation on the Caribbean island of Martinique reported that carrots grown in Bradford were selling in this tropical resort for five dollars per half kilogram!

ways. For example, they can move to urban centres with millions of other people and work to support themselves. However, the competition for jobs in the cities of the developing world is fierce. There are too few jobs for the millions of rural workers who migrate to the cities everyday. If education programs and local industries were established, there is a chance that people could have a better life. Which solution do you think is more sensible? Or are both solutions appropriate depending on specific circumstances?

STATLAB

1 Study the statistics in Appendix B on pages 442 to 449. Find the correlation between GDP per capita and the percentage of GDP earned from agriculture. (See pages 16 to 18 to review how to calculate linear correlation coefficients.)
   a) What patterns do you notice?
   b) What strategies might increase GDP per capita in countries with subsistence economies?

2 Study the statistics in Appendix B.
   a) List the twenty countries with the most crop land per capita, from highest to lowest.
   b) Multiply the crop land per capita figure by the total population of each country to determine total crop land.
   c) Compare your lists from (a) and (b). This comparison should show you where most of the world's food is produced.
   d) Why could these regions be considered the bread baskets of the world?

3 Study the statistics in Appendix B.
   a) List the twenty countries with the least crop land per capita, from lowest to highest.
   b) Multiply the crop land per capita figure by the total population of each country (Appendix A) to determine total crop land.

c) Compare your lists from (a) and (b).

d) Why could these regions be considered the hungriest nations in the world?

4 a) List the continents/regions in Figure 11.8 in order by land-use category. For example, crop land from greatest to least would be Asia, North America, (former) USSR, etc.

b) What patterns do you see?

c) Which continents/regions appear to have maximized their land use?

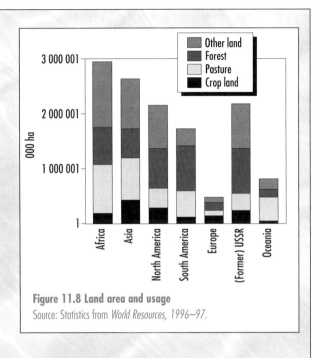

Figure 11.8 Land area and usage
Source: Statistics from *World Resources, 1996–97.*

CONSOLIDATING AND EXTENDING IDEAS

1 Summarize the environmental limits to agriculture in an organizer or a chart similar to Figure 11.9. The tropical rainforest has been completed as an example.

2 a) Identify the advantages and disadvantages of traditional food production.

b) How do population pressures make the continuation of some traditional practices tenuous?

3 a) What problems exist with modern farming?

b) Discuss the feasibility of giving food away to developing countries.

4 What potential solutions can you foresee regarding the alleviation of hunger?

ENVIRONMENT	LIMITATION
Tropical rainforest	Limited soil fertility due to extensive leaching
Temperate grassland	
Mixed forest	
Savannah	
High latitude	
Desert	

Figure 11.9 The environmental limits to agriculture

PATTERNS OF HUMAN GEOGRAPHY: WHY ARE PEOPLE HUNGRY?

Human geography also plays a key role in the problem of hunger. The main cause of hunger is poverty or lack of opportunity. If the resource base expands or technology improves, opportunities open up that reduce the effect of too many people living on too little land. It is only when political and economic systems allow people to participate in economic growth that hunger can be alleviated.

ENVIRONMENTAL DETERIORATION

The strains people place on the environment cannot be underestimated. Certain forms of environmental destruction can seriously hamper people's ability to produce food. This is particularly common in subsistence economies where, in some desert regions, the environment is deteriorating at an unprecedented rate. **Desertification** is transforming once arable land into desert. But is this an element of human geography? Many people might think that this phenomenon is the result of natural processes. In the past, this may have been true as climatic shifts created major changes in rainfall patterns in the Sahara. But a major contributing factor to desertification today, and indeed for the past 3000 years, involves human land-use patterns.

The Sahara has not always been a desert. Cave paintings discovered in the middle of the desert depict grassland animals that today are found only thousands of kilometres to the south. Rhinoceros bones have been discovered in parts of the northern Sahara, and yet these animals too are found only much farther south in more humid regions. During the time of the Egyptian Empire, historians referred to the Sahara as a rich agricultural land. All of this evidence suggests that the Sahara had a much wetter climate

Figure 11.10 Cave paintings in central Sahara, 6000 BCE

Figure 11.11 Overgrazing in North Africa has led to erosion of the land.

as recently as 3000 years ago. Today the region is the driest it has been in 3000 years. What human activities took place to make the Sahara the world's most formidable desert?

Pastoralism is the dominant land-use pattern in the Sahara today. Peoples throughout North Africa raise cattle, sheep, goats, and camels, mainly for their milk but also for their meat and fur. The animals graze on the grasses and small bushes. A large herd is a status symbol, so people keep as many animals as possible. These herds are also insurance against periods of drought when pastures dry up. Some of the animals will survive even if most of the herd is wiped out. These hardier animals can regenerate the herd when the period of famine is over. Traditionally, the people simply moved to wetter regions when the rains failed. As a result, they were able to thrive even in difficult times. Today such nomadic movement is restricted. The pressures of a growing population and the increase in the number of small farms and large plantations on the peripheries of desert lands have limited the people's access to traditional grazing lands.

The survival of domesticated livestock hinges on the

Human geography was not solely responsible for transforming the Sahara into a desert. Physical geography also came into play. As the glacial ice of Europe and Asia started to melt 15 000 to 12 000 years ago, the region began to warm up. As it did, evaporation rates gradually increased. Higher temperatures and rates of evapotranspiration slowly changed the ecosystem to one where **xerophytes** predominated over trees and grasses. Over time the region became semi-arid and then desert. Farmers migrated south to river valleys where they could use irrigation as a substitute for rainfall. The only people left in the dry lands were nomadic pastoralists. They raised animals and moved from place to place to get the fodder and water their animals needed. This less intensive land use was well suited to the dry lands.

availability of fresh pastures and an adequate water supply. Without these, ecological disaster is inevitable. The animals overgraze the land, leaving the soil exposed to the erosive forces of wind and rain. Once the land is eroded, the soil's fertility is diminished. It can no longer support the grasses that sustain the herds. Humidity is reduced, so there is less water vapour in the air and therefore less rainfall. When it does rain, the ground is so hardened by the hoof prints of too many animals competing for too little land that the water runs off the surface without soaking into the soil. What valuable surface soil may have remained is washed away, making the land even more desolate. Thus begins a vicious cycle of drought, reduced vegetation, and higher evaporation rates, followed by more drought.

Ponds, springs, rivers, and wells provide the water needed by both animals and people. Oases occur where **artesian wells** allow water to seep to the surface. An **oasis** can be thought of as a tiny island in a sea of sand, rock, and stone. Life abounds for one reason only—the presence of water. Some oases are dusty little villages with little plant life, but they have access to a deep well that may be thousands of years old.

Others may have relatively large populations depending on the amount of water that is available. Population pressures in these areas have strained these valuable water resources. The concentration of animals around these resources also threatens the water quality and contributes to poor sanitation.

It is not only the animals that are responsible for the destruction of the landscape, but also the people who cut down trees and bushes to use as firewood for heating and cooking. The ground is then exposed to the searing heat of the sun. The surface of

The effects of desertification in the Sahara can be seen elsewhere in the world. Cultivated lands bordering the Sahara desert are often buried under layers of wind-blown sand. In Algeria, a giant sand dune has been moving continuously through one small village for over 100 years. Wind-blown Saharan sands have even been found as far away as the Amazon Basin in South America!

the ground is baked hard. Microclimates change as the humidity close to the earth's surface drops and temperatures increase. Ground-cover plants such as grasses and herbs will not grow if exposed to the full force of the sunlight. When it does rain the water tends to flow over the surface of the land without soaking into the soil. Flash flooding causes damage in **wadis** and low-lying areas. Upland regions experience gullying and other forms of erosion. The soil is washed away and the possibility of revegetating the region is diminished.

THE TRAGEDY OF THE COMMONS
COMMON SENSE OR OVERSIMPLIFICATION?

The tragedy of the commons is a parable that some individuals have used to explain patterns of environmental deterioration that exist today. In feudal Europe peasant farmers were allowed to graze their animals on pasture land owned by the lord. Since they did not own the land, it was to each farmer's advantage to graze as many animals as possible. The farmers realized that if they all grazed their animals in this way, the pasture would eventually be destroyed and no one would benefit. Despite this, no one wanted to give up using the pasture land, so everyone lost.

This parable has been applied to pastoralists in Africa to explain how the Sahel is being destroyed. At first glance it does seem to apply. But further study reveals that African pastoralists often have far greater political sense and understanding for the environment than many political theorists assume. Migratory herders usually have sophisticated systems to manage natural resources. For example, the FulBe elect a manager to regulate the use of pasture lands. This official decides on migratory routes, determines when the herds move, and settles conflicts between herders and farmers. The *Barabaig*, another North African tribe, regulate pasture use and water supply. Families that do not follow the rules are punished by the tribal elders.

Figure 11.12 The tragedy of the commons
Overgrazing destroys pasture land and leaves the soil exposed to erosive forces.

THE KALAHARI DESERT

Desertification is found in other parts of the world, including southern Africa, the Middle East, southern Asia, and South America. The Kalahari Desert in southern Africa is undergoing changes similar to those that took place in the Sahara 2000 years ago. Traditionally, the indigenous San people were hunters and gatherers. They practised a sustainable lifestyle, taking from nature only what they needed to survive. Living primarily on nuts, tubers, roots, berries, and meat from the hunt, the San migrated from place to place. In their wake, they left little or no impact on the fragile ecosystem of the Kalahari.

In the past, the Kalahari had few sand dunes. Surface water was abundant in huge wetland regions. Natural vegetation was plentiful because the plants, which adapted to the dry winters, flourished during the brief summer rainy season. Today, animals are being kept in domesticated herds, and the cycle of overgrazing, erosion, and **desiccation** is beginning.

Recently, the San have come in contact with urban society and with the Bantu-speaking peoples who live on the edge of the desert. Now their traditional lifestyle is changing. The San are no longer nomadic.

They herd cattle and work on farms and in mines and live in sedentary communities. San children are going to schools to learn curriculum dictated by bureaucrats in the cities. They are losing the bush lore that their ancestors discovered during 20 000 years of living on this unforgiving land. With a more reliable source of food from the animals they raise, and with better health care and standardized education, most people would agree the San have a higher standard of living than they had as nomads. But is the change really for the better? It is impossible for the San to return to their traditional lifestyle. Most do not want to and, in any case, could not because they have forgotten the old ways.

With animals overgrazing the land and microclimates changing, there is evidence that the Kalahari Desert is becoming drier. If the human population increases during the good years, there may be too many people for the resources to support when the rains fail and the herds are unable to survive. The San are a remarkable people who have lived in harmony with their environment for thousands of years. Their sustainable use of resources allowed them, in the past, to thrive as an optimum population. But with a growing population, will the Kalahari be "grazed to death" by the increasing number of cattle needed to support the San? How long will the San be able to live in the Kalahari as pastoralists?

CONSOLIDATING AND EXTENDING IDEAS

1 a) Summarize the practices that led to environmental deterioration in the Sahel region of Africa.
   b) Describe the specific negative effects.

2 Read the article "Don't Fence Me In" on page 163.
   a) What evidence of bias is there in the article?
   b) Prepare an organizer or a chart comparing pastoralism to farming.
   c) What is the government's view of the nomadic lifestyle of the FulBe?
   d) What conflict exists between the FulBe and the Mourides?
   e) Who do you think is responsible for the destruction of the Senegalese savannah?
   f) Develop an argument either for or against the nomadic lifestyle and debate the issue in class.

3 Desertification is only one type of environmental deterioration that results from resource mismanagement. Research another environmentally sensitive area—possibilities include the Maritime fishery in Canada; deltas, for example, the Mississippi or the Nile; wetlands, for example, The Everglades in Florida or Gran Chaco in South America; and tropical rainforests. Prepare a report outlining the causes of the problem, the pattern of environmental abuse, and the prospects of solving the problem.

DON'T FENCE ME IN

In the spring of 1991 hundreds of brightly painted buses suddenly approached Senegal's Mbegué Forest from all directions. Wearing their distinctive patchwork garb, members of the country's Mouride Islamic Brotherhood filled the air with singing and chanting. Following the orders of their leader, Serigne Mbacké, within a matter of weeks they cut down more than 5 million trees and brutally expelled more than 6000 FulBe pastoralists and 100 000 cattle from the forest. Two months later, the rains began: the once-lush pastures were ploughed under and the land planted with a vast plantation of peanuts.

The Mbegué reserve lies in the Sahel, the transitional area between the vast ocean of desert in the Saharan north and the more fertile agricultural zones to the south. The Sahara and the northern Sahel have traditionally been the domain of nomadic herders. . . . The southern Sahel and the tropics beyond are occupied largely by farmers whose carefully tended crops create a patchwork of fields on land passed from generation to generation. . . . The pastoralists and farmers . . . often compete for the same land and water. Pastoralists need open spaces for extensive grazing. Farmers enclose their land, intensively cultivating their fields and taking great pains to ensure that animals are kept far from tempting grain or vegetable crops.

A Windswept Landscape

This competition for space is as old as the occupations themselves. But now it is intensifying as farmers expand into areas once exclusively used by pastoralists. Nowhere is this more evident than in the Ferlo region of northern Senegal where FulBe pastoralists have lived with their hump-backed cattle since the fifteenth century. The Ferlo lies in the heart of the Sahel and covers the northern quarter of Senegal. To an outsider many of its wide expanses appear as desolate wastelands. Rainfall is both sparse and highly capricious. One hamlet may receive a drenching downpour while another 5 km away remains bone dry. A village may have good rains one year followed by three or four years of drought.

The windswept Ferlo landscape is painted in shades of brown and grey and many of the trees are spiny, drought-resistant species easily mistaken for stunted bushes. The grasses beneath their spindly boughs appear dark and brittle. But the region's stark appearance belies the richness and complexity of its ecology, best understood by the FulBe.

The FulBe who live in the Ferlo . . . grow a few subsistence crops but live mostly from the milk and meat of their cattle, sheep, and goats. For them, the Ferlo has been a land of opportunity, rich in a multitude of tree and grass species that make excellent fodder for their animals. The key to their success has been their mobility. By practising a degree of nomadism, the FulBe have been able to adapt to whatever conditions nature serves up. Rather than waiting helplessly for rain, they seek out the productive pastures where rain has already fallen. Underlining the advantages of nomadism, a young FulBe herder wryly asked: "Does a field have legs to walk when it gets thirsty?"

At times when the entire Ferlo experiences drought, the FulBe move out of the region altogether, driving their cattle farther south into areas of higher rainfall. These more southern zones are inhabited by the Wolof and Serer people, who make their living primarily from agriculture. These areas are densely populated and the FulBe would have trouble finding grazing land for their cattle were it not for several protected forest regions. These reserves were created by the French colonists and maintained by later governments in order to preserve trees and protect the land from soil erosion. Farming was prohibited in the reserves, though herders were allowed to graze their animals. In years when the rains don't come, these reserves are havens for the FulBe as they move south, serving as a kind of insurance policy for those who make their homes in the riskier climates to the north. The Mbegué forest, until it was decimated in 1991, was one of the most important of these reserves.

The FulBe are far less mobile now than they were fifty years ago. In the past, herders had to move their families out of the Ferlo each dry season when the watering holes dried up. Then, in the 1950s, the colonial government started to drill boreholes 200 to 300 m into the aquifer. There is now a grid of wells, with pumping stations at 30 km intervals. Water sloshes into giant cement tanks and then is siphoned into barrels for household needs. Nearby, animals . . . take their turns drinking from the long cement troughs fed from the tanks.

Elegant Straw Igloos

Since the installation of these boreholes, many FulBe have set up semi-permanent camps or villages around the water sources. Most people remain year-round while one or two men head out to accompany the cattle, guiding them from borehole to borehole depending on the availability of water and the quality of pastures in different areas. Others practise "micro-nomadism"—moving their camps of . . . straw igloos to more remote pasture areas when water is widely available, then returning to the boreholes during the dry season. . . .

It is the mobility [of the FulBe] that is threatened as the grazing lands of the Ferlo are compressed on all sides. . . . To the north the Ferlo is bordered by the Senegal River. Traditionally, the river's annual flood nourished the fields and pastures of the FulBe who lived along its banks. In the 1980s two dams . . . were built to support large-scale irrigation schemes. Because the land could now be irrigated, it was taken over by wealthy farmers. As a result, the FulBe were pushed deeper into the Ferlo.

To the west former grazing lands have been turned into orchards and gardens whose high walls keep out both people and animals. Water pumped from the shallow aquifer along Senegal's coast helps grow green beans, mangoes, and strawberries. These lush gardens are owned by powerful religious leaders, Lebanese business people, and other members of an élite based in the capital, Dakar. . . . To the south the issue is encroachment of so-called "insurance pastures." . . . In 1991 the government granted the . . . Mouride Islamic Brotherhood permission to put to the plough more than 60 per cent of the Mbegué reserve. A year later (shortly before the election) the Mouride leader urged his followers to support Abdou Diouf in his bid for re-election as president of Senegal.

Where rich pastures once flourished, the Mbegué forest is now carpeted with vast peanut plantations. The original Mbegué reserve had thirty-eight water points suitable for watering animals; thirty-five of those were on fertile lands taken over by the Mourides. The Mbegué was the last important reserve remaining in the agricultural zone immediately to the south of the Ferlo. . . .

Erratic Climate

Bureaucrats in the capital, Dakar, and others argue that concern for the FulBe's nomadic system is a sentimental attachment to an archaic way of life. And they point to the FulBe themselves, many of whom have aban-

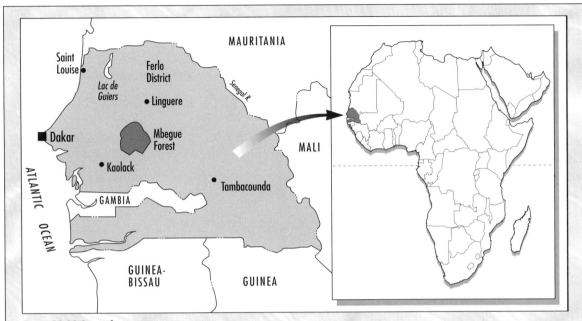

Figure 11.13 Senegal

doned pastoralism in favour of farming and ranching. But this misses the point. Nomadic grazing systems have evolved in response to erratic climates with limited rainfall. By moving from place to place in relation to the available resources, the FulBe make optimal use of the limited water resources and spread the impact of their activities more evenly over the environment.

The pastoral footprint is light on these fragile ecologies. Peanut farming is not. . . . When it is harvested the entire plant is uprooted, leaving neither organic matter to enrich the soil nor ground cover to protect the topsoil during the long dry season. . . .

Peanut Damage

Small farmers with a few hectares make scrupulous efforts to combat the negative side of growing peanuts. . . . The powerful religious leaders and their followers are much less conscientious. Their farming practices mine the soils for short-term profit. . . . When the soils have been exhausted the Mourides move on. The area is a wasteland where even grasses take years to regenerate.

Each new farm contributes to the relentless compression of grazing lands and reduces the FulBe's options for mobility. . . . Both the human and the environmental costs of the transition are high. When the FulBe can no longer move their cattle in search of better pastures, they will be driven to a different kind of migration. A path which will take them to the crowded slums of Dakar.

Pastoralists like the FulBe play a critical role in the sustainable management of fragile ecosystems. Both . . . governments and aid agencies need to recognize this. The key is to work with local people to develop policies to protect grazing lands *and* nomadic rights where pastoralism is both more sustainable and more environmentally appropriate than farming. In Senegal they need to act quickly— or there won't be any pastures to protect.

Karen Schoonmaker Freudenberger. From "Don't Fence Me In," *New Internationalist* April 1995: 14–16. Copyright © New Internationalist. Reprinted by permission of Guardian News Service Limited.

CHAPTER 12

GLOBAL SYSTEMS AT WORK: POLITICAL AND ECONOMIC REALITIES

More than enough food is grown today to feed the world's 6 billion people. Despite Malthus's theory, food production has continued to increase as technology improves, and will continue to do so as more advances are made. Why, then, are so many people without an adequate food supply? There are four global systems that reduce the availability of food: the legacy of colonialism, modern economic imperialism, militarism, and international debt.

THE LEGACY OF COLONIALISM

Some authorities believe that the problem of world hunger is historically rooted in **colonialism**. Starting in the late fifteenth century, Britain, France, Spain, Portugal, and other European nations battled for supremacy by establishing colonial interests throughout the world. European powers colonized Africa, southern Asia, North, Central, and South America,

Figure 12.1 Bean and banana crops in Uganda

THE TRIANGULAR TRADE ROUTE

During colonial times, millions of Africans were forcibly taken from their homelands and placed into slavery on the plantations of European settlers. These enslaved people were taken to the Caribbean Islands, Spanish and Portuguese colonies in the Americas, and the British Thirteen Colonies. Conditions on the ships were deplorable; people were treated like cargo. In the New World colonies, they were forced to work as farm labourers on the numerous plantations that had developed there. Many people died as a result of the severity of their living and working conditions.

Tobacco, cotton, sugar, and other crops were deliberately cultivated for export to the home countries in exchange for manufactured goods needed in the colony. A triangular trade route evolved. Sailing ships used the prevailing winds—the westerlies—to transport plantation crops such as tobacco, indigo, cotton, and sugar to Europe. These winds blow from the southwest all the way across the Atlantic to Europe. Manufactured goods such as guns and machinery, luxury items such as wine and silk, as well as cash were picked up in European ports. From Europe, the ships followed the Canary Current down the coast of West Africa to ports such as San Pedro, Port Harcourt, and Porto Novo. Portuguese, Dutch, Spanish, and English traders travelled up the many rivers along the Guinea coast of West Africa to capture or purchase indigenous peoples in the European colonial towns. They were held in prisons and auctioned off to merchants who sold them to the sea captains.

From the west coast of Africa, the ships again took advantage of the winds to travel west. The easterly trade winds blew them from the African ports across the Atlantic to the Caribbean Islands, where the enslaved people were traded for plantation goods that were sold in Europe. And so the cycle of the triangular trade route continued.

Even though most countries had banned the slave trade by the early 1800s, slavery continued for many years as the children of enslaved people were born into captivity.

Figure 12.2 Enslaved people being taken to America, 1830

and the South Pacific. Europeans took control of the most productive lands. Precious metals and gems, agricultural commodities such as tobacco, sugar, and cotton, and other valuable raw materials were exploited by the colonial powers and exported back to the home coun-tries. Indonesia, for example, was a treasure house of unique spices for Europe's chefs. Egypt provided cotton for the clothing mills of north-ern England. Mexican silver financed Spanish wars back in Europe. Over the years, millions of dollars were siphoned out of the colonial economies. In turn, the colonies became the markets for the manufactured prod-ucts of the colonial powers.

European companies, pri-marily interested in the extrac-tion of raw materials such as rubber, diamonds, gold, cop-per, iron ore, and other miner-als, exploited the rich natural resources of their colonies with impunity. The best agri-cultural lands of Africa, Latin America, India, and Southeast Asia were used by the colonial powers to grow cash crops for export. Commodities that were in particular demand during the age of imperialism included tea, coffee, cotton, sugar, cocoa, tobacco, palm oil, peanuts, jute, and silk. Encouraged by their colonial rulers to produce export crops, farmers turned to grow-ing commodities that could be sold on world markets. As a result, the amount of food grown to sustain the indige-nous populations declined. Forced from prime agricultural lands to marginal regions ill suited to the growing of food crops, millions of people in India died during the famines of the 1800s. One legacy of imperialism was the depen-

Figure 12.3 A cacao plantation in West Africa

dence of entire populations on the production of a single export commodity. This commodity was produced to the detriment of food self-sufficiency and was subject to extreme price fluctuation in the global marketplace

The military and political power of the European nations ensured that the colonial system benefited the home countries at the expense of the colonies. The colonies were prevented from developing diversified economies because this was not in the best interests of the ruling powers. It became difficult for European colonists and impossible for indigenous peoples to improve their standards of living. Eventually, people living in the colonies refused to be subjugated any longer. As their wealth, numbers, and military strength grew, they became dissatisfied with the one-sided arrangement. They wanted to be free of the overbearing domination of the ruling countries and to develop their own political and economic systems. One of the first nations to demand independence from Great Britain was the United States. After a bloody revolution the Thirteen Colonies united to form a new nation. Later, Haiti, Mexico, Brazil, and many other Latin American colonies declared their independence. By the mid-twentieth century, other colonies in Africa and Asia became independent, sometimes through revolution, sometimes through peaceful negotiation. Canada is an example of a colony that separated from a ruling country with very little hostility. Over a period of nearly 100 years, it

evolved into a fully sovereign state.

The end of colonial rule did little to change the economic systems that were so deeply entrenched in these newly liberated nations. The same economic infrastructure based on the export of one or two commodities remained in place. The most productive land and other natural resources were owned by foreign companies, expatriates, or the wealthy élite within the country. In many colonies, the home country was supplanted by public companies whose shareholders lived in the former home country. Local officials often took over existing economic structures for their own financial gains. Farmers and former enslaved people were left out of this reorganization. They continued to work on plantations for paltry wages, or they became **sharecroppers**—subsistence farmers who paid a portion of their crops to the landowner in exchange for rent. Unable to get ahead financially and to feed their families adequately, these people were doomed for generations to a cycle of poverty, hunger, and injustice.

ECONOMIC IMPERIALISM

Today, political colonialism has been replaced by **economic imperialism**. Multinational corporations headquartered in Europe, the United States, Japan, and other developed nations dominate the economies of many developing countries. Some of these corporations specialize in all aspects of agricultural production and marketing. These **transnational agri-**

> In Ghana, 35 per cent of all arable land is controlled by four multinational corporations. Each year these corporations purchase between 60 and 80 per cent of the world's cocoa from this small nation in Africa. Since these corporations control the price of cocoa, they can pay any amount they want for the crop. Ghanaian farmers who produce it get only a small fraction of the profit made on the finished product, chocolate. Most of the profit goes to traders, processors, and merchandisers.

businesses control much of the world's agricultural trade. Alleviating hunger is not their priority. Like all companies in free-enterprise systems, they are in business to make money. Unfortunately, making a profit in agriculture can jeopardize the opportunities of many people in developing countries to obtain the food they need.

For example, if a company has a monopoly on a product, it can control the price. If prices are low, it can limit the amount of the product in circulation to drive up the price. If prices are low because of too much competition, the company can flood the market and lower the price further. While a giant company can afford to sustain losses over a period of time, small local suppliers are unable to compete. Farmers must sell their crops at the lower market price, but they must still pay high market prices for seeds and fertilizer. As a result, they go out of business or are ripe for take-over by a corporation.

Multinational corporations often obtain government subsidies and financial support from government agencies. Rather than funding local ventures that could benefit their own peoples, many small developing countries support the multinationals because so much of their export industry is dependent on one or two companies doing business in their countries. In order to increase exports to earn foreign currency, developing countries finance the multinationals even though their citizens are being exploited. Multinationals can then continue to buy plantation crops at bargain prices. If a multinational were to withdraw from a developing country, it

> Historically, the term "banana republic" referred to nations that were practically owned by American banana importers. The countries of Ecuador and Honduras relied so much on the export of their banana crops that they had to comply with the companies' wishes. Today, the term pertains to any country that relies on one or two agricultural products.

could result in financial ruin for the country.

In return, many multinationals provide food aid by distributing surplus food supplies in developing countries. But this does not help the people of these nations to provide for themselves. Many of the corporations' practices would not be tolerated if they occurred in developed countries, but they are able to avoid prosecution because international laws do not adequately govern their activities.

Multinational corporations also sell food products to developing nations. Unfortunately, this does little to relieve hunger because of the nature of these products. Most are expensive luxury foods such as soft drinks, candy, white bread, and baby formula. Far from improving the situation, these products may actually intensify the problem. Consider baby formula, for example. Traditionally, infants were breast-fed. Not only is breast milk free of bacteria and other harmful germs, but it also contains antibodies, nutrients, and other substances that nourish infants more than baby formula can. The water that powdered formula is mixed with must be clean—free of bacteria and parasites. Once the formula is made up, it must be kept refrigerated. More often than not, it is impossible to meet these essential demands in rural districts of many developing countries where potable water is rare and refrigeration is practically non-existent. In addition, nipples and bottles need to be sterilized before each feeding. Without adequate sanitation and education about the importance of sterilized equipment, the chance of infant malnutrition and disease is very high.

Cargill Incorporated of Minnesota, a multinational corporation, is active in sixty countries around the world. It has annual sales in excess of US $48 billion in fifty different commodities. It is a private company, without public stock-holders, and as such Cargill does not have to make its business ventures public. Among its varied interests, Cargill is the world's largest producer of barley malt, used in brewing beer. As the world's biggest oil-seed producer, it controls the price of sunflower seeds in India. It is the third largest flour miller in the United States as well as a major cattle feed supplier. Among Cargill's many farming operations is an orange grove in Brazil that has over 1.4 million trees. The operation includes processing plants, refineries, barges, ships, packagers, retail distributors, and even bio-genetic research facilities.

Figure 12.4 Advertisements for Western food products outside a market in Perinet, Madagascar

Figure 12.5 Nairobi, Kenya, is the East African headquarters for many multinational corporations.

MILITARISM

Another contributing factor to world hunger is war. In Africa, some countries have experienced continued political upheavals since they gained independence. At the heart of this instability is the colonial past. The nations that make up Africa today were created by European powers. Their boundaries do not consider cultural, linguistic, or other social characteristics that frequently bind nations together as a common entity. As a result, a country may contain several traditional enemies within its borders, while people with a shared background and language may be divided among several nations. The artificial boundaries arbitrarily created by the colonial powers have led to political unrest in modern times.

It is impossible to grow crops and raise animals in the midst of war. Furthermore, war is extremely destructive to environments and people. In Bosnia, millions of land-mines have been buried without any apparent reason. It has become impossible to farm fertile land because of the fear of triggering these treacherous bombs. Even if the Bosnian people win the war, they will never win back their land unless they can determine how to find and dis-

Damage assessments of the war in Bosnia indicate that almost two-thirds of the houses, one-third of the hospitals, and half the schools have been destroyed or damaged. Fields and vineyards are in ruins; rivers are contaminated with toxic wastes from destroyed manufacturing plants; roads, water supplies, and power plants are unusable. The World Bank estimates that the economic recovery of the region will cost US $5 billion over three years.

The civil war in the former Ethiopian province of Eritrea has caused similar economic disruption. This conflict has decreased the labour force and forced farmers to avoid some areas of agricultural production. During the last years of the 1980s, 40 per cent less land was cultivated, resulting in crop reductions of one-third to one-half. This has led to Eritrea's present-day food shortages.

arm the mines that plague their once bountiful land.

When war strikes, many things created by humans or nature are destroyed. Soils are eroded and ecosystems ruined. Even climates change. When the retreating Iraqis set fire to oil wells in Kuwait during their 1991 occupation, the local climate was disrupted for months by black clouds. The destruction of irrigation works, dams, fields, storage and transportation facilities, farm equipment, fertilizer stores, and fuel depots needed to run tractors damages agricultural infrastructures. All this devastation exacerbates the hunger of people in war-torn nations. Today, many former colonies spend more on military hardware than on education, health care, and economic development. It is estimated that the total amount spent on weapons each year by developing countries is over $25 billion. Just a fraction of this amount would eliminate hunger in many of these nations.

INTERNATIONAL DEBT

In spite of the vast amounts of financial aid distributed to developing countries, more money flows out of these countries than enters them. The reason is international debt. Developed nations such as Canada and the United

States receive interest payments on loans from developing countries such as Bangladesh and Peru. This situation has its roots in the 1960s. As the global economy began to expand, many developing countries took out huge loans for economic development projects. These included such megaprojects as the construction of the Aswan High Dam in Egypt and the creation of the new capital city of Brasilia in Brazil. The loans were also used to finance smaller projects, such as irrigation systems, road networks, and educational facilities. All of these projects were financed with foreign loans from international organizations such as the World Bank and other financial institutions. As with all loans, interest was charged on the principal amount. In the 1980s interest rates soared, and in the 1990s the global economy faltered. As a result, developing countries found it increasingly difficult to pay the annual interest payments on their debt, let alone repay the principal.

The impact of international debt on hunger means that developing countries have no more credit available, so they cannot borrow money to improve social services, education, or health care. Any revenues they receive must be used to make interest payments on loans taken out many years ago. When countries are unable to make their payment schedule, they may be aided by the **International Monetary Fund (IMF)**. But this help comes at a price. Debtor nations must agree to implement an austerity program. Spending on social services is reduced and food subsidies are eliminated. These measures invariably have the greatest impact on people who depend on the "safety nets" of government-provided social services.

CONSOLIDATING AND EXTENDING IDEAS

1 Summarize the colonial practices that contributed to the problem of hunger prior to the twentieth century.

2 a) How do multinational corporations control world agriculture today?
   b) How does this domination contribute to the problem of hunger?
   c) What can be done to solve this problem?

3 Explode some myths! For each of the following, write a paragraph to explain the facts that refute each myth:
   a) The main cause of hunger is that not enough food is being produced.
   b) The main cause of hunger is drought and other natural disasters.
   c) The main cause of hunger is overpopulation.
   d) The main cause of hunger is ignorance of modern agricultural practices.

4 Make a list of commodities you use that are cash crops of developing nations. Then determine which countries' land is growing your coffee, pineapples, cocoa, tomatoes, bananas, peanuts, tea, and cotton, etc.

5 Use the statistics in Appendix B on pages 442 to 449 to find the defence expenditures for the countries of one of the following continents: Africa, South America, Asia, Europe.

a) Divide the figures for defence expenditures by the total population to obtain per capita figures for each country. These calculations will allow you to compare countries.

b) Prepare a graded shading map showing defence expenditures per capita. Compare your map to the one in Figure 12.6. Which countries spend a large amount of their income on the military?

c) What conclusions can you make about how certain countries spend their money?

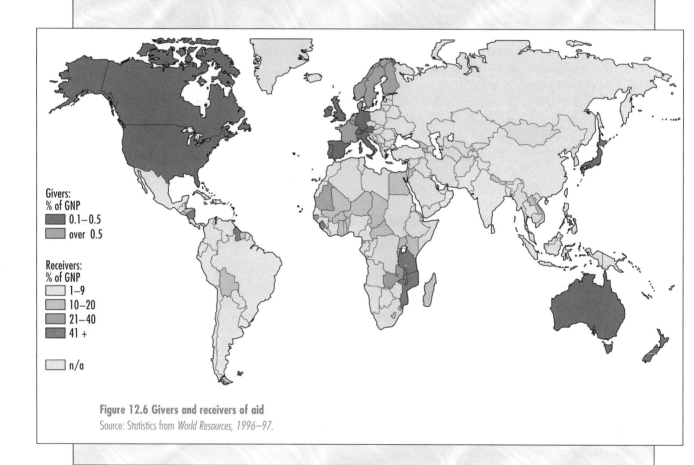

Figure 12.6 Givers and receivers of aid
Source: Statistics from *World Resources, 1996–97*.

COUNTRY PROFILE:
DEMOCRATIC REPUBLIC OF CONGO

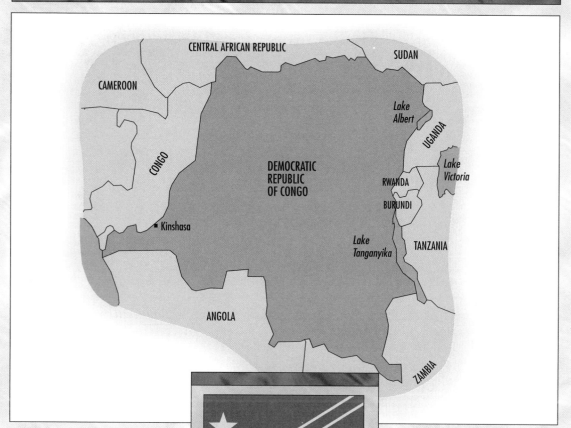

CONGO—FACTS AND FIGURES

GNP per capita	n/a
Population	37 400 000
Birth rate*	48
Death rate*	15
Infant mortality**	93
Female literacy	61%
Defence expenditures	n/a

*per 1000 population
**per 1000 births

Figure 12.7
Source: Statistics (1990–95) from *World Resources, 1996–97.*

Formerly known as Zaïre—and until 1971 as the Belgian Congo—this country brings together about 200 different cultural groups, which have experienced long-standing tribal animosities and isolationism. The original forest dwellers were forced into the interior centuries ago when Bantu-speaking tribes from the south and west and Nilotic groups from the north invaded the region.

The Belgian Congo was formed as a European colony with no consideration for the indigenous peoples. This is the principal reason that the country has been in turmoil since its creation in 1885. In the colonial days, Belgium exploited the land for its rich mineral resources,

Figure 12.8 Protestors marching through Kinshasa in defiance of Laurent Kabila's ban on public demonstrations

especially copper. After independence in 1960, the country was ripe for rebellion as smaller cultural groups tried to break away and form their own countries. The United Nations worked to support the fledgling government in the belief that stability in the country was vital if the entire continent was to develop into self-governing nations.

In 1965, Mobutu Sese Seko staged a military coup and remained in power until 1997. He siphoned an estimated $5 to 7 billion dollars out of the country while the people of Zaïre endured one of the lowest standards of living in the world. Border clashes escalated in 1995 and 1996 as tribal groups on both sides of the Zaïrean border struggled for survival. The worst situation was along the border with Rwanda. Thousands of Hutu refugees were starving to death or dying from volcanic fumes on lava beds as Zaïrean forces prevented the United

Nations from sending in aid. When UN peacekeeping forces moved into the region in late 1995, a measure of stability was restored and refugees started to return to their homes.

In May of 1997, rebel forces walked into the capital city of Kinshasa and took over the government in a bloodless coup. Their leader, Laurent Kabila, declared himself ruler of the country, which he renamed the Democratic Republic of Congo. The rebels' easy take-over of Kinshasa marked the end of their seven-month campaign that vowed to sweep away the political system known as "Mobutism." However, there is uncertainty about Kabila's commitment to democracy and human rights stemming from his Marxist background and the alleged killings of Rwandan Hutu refugees by his Tutsi fighters. As for elections, Kabila has promised a multi-party vote no later than June 1998.

CHAPTER 13

SOLVING THE PROBLEMS OF FAMINE AND HUNGER

Figure 13.1

The developed world produces more food than it can consume, while other regions of the world continue to endure hunger.

Famine and hunger are two distinct problems with different solutions. While it is much more dramatic, famine is less widespread than chronic, persistent hunger. Finding solutions to both problems poses a strong challenge. However, the statistics in Figure 13.2 suggest that the situation is better today than it was over twenty-five years ago. There is reason to be optimistic that solutions to famine are attainable and an end to hunger is possible.

ENDING FAMINE

There are a variety of ways in which we can eliminate famine. These include short-range and mid-range strategies that can work together to produce long-range solutions. In the short term,

Region	1971		1981		1991	
	MILLIONS OF UNDER-NOURISHED	% OF TOTAL POPULATION	MILLIONS OF UNDER-NOURISHED	% OF TOTAL POPULATION	MILLIONS OF UNDER-NOURISHED	% OF TOTAL POPULATION
Africa	101	35	128	33	168	33
Asia	751	40	645	28	528	19
Latin America	54	19	47	13	59	13
Middle East	35	22	24	12	31	12
Total developing regions	941	36	844	26	786	20

Figure 13.2 Prevalence of chronic undernutrition in developing regions

Which regions have shown a marked improvement? In which regions are the numbers of malnourished people growing?

Source: *World Resources, 1994–95* by the World Resources Institute. Copyright © 1994 by the World Resources Institute. Used by permission of Oxford University Press, Inc.

the crisis can be managed by supplying food to those who are victims of famine. While food aid is readily available, the main problem is usually the distribution of food. Sometimes governments withhold food supplies from groups they consider to be their political enemies. A solution, although a costly one, is to fly food directly into those regions where it is needed. But supplying food on a continuous basis is not a long-term solution. It would be misguided to provide food aid as an ongoing policy. People would become reliant on this aid and local farmers would not be able to compete with the free or inexpensive food. Food aid must only be used as a measure to prevent people from dying in an immediate crisis. After the emergency has passed, other strategies need to be implemented.

As a mid-range solution, adequate infrastructures must be established to help people survive the next drought. In India, estimates suggest that a third of all grain stored in the country is either contaminated or eaten by rats and mice. New storage facilities must be built to

house food surpluses in times of plenty. In addition, governments should not sell surplus food to increase their export revenues. If this food were stored instead, it would go a long way to warding off the next crisis.

Another mid-range solution is to modify human land-use activities to ensure that land remains a sustainable resource. While nomadic herding was economically sustainable in the past, it is impossible to return to the old ways now that population pressures have reduced the amount of land available to herders. The nomads used intricate systems of migration to reach forested, more humid regions during periods of drought. These regions have now been largely cleared and developed as plantations that grow cash crops, so herders are forced to remain in semi-arid regions even when the rains do fail. In Chapter 12 you read about the nomads of Senegal. Trapped between a semi-desert, which cannot always provide adequate water for their herds, and huge agricultural projects, where their animals are not welcome, they have nowhere to go. Should they be resettled on

Figure 13.3 CIDA workers distributing food to victims of famine in Bangladesh

farms and made to adopt a lifestyle contrary to their age-old traditions? Should they be allowed to starve because they cannot pasture their cattle on traditional grazing lands? Or should the new plantations that were built on traditional pasture lands be returned to them? None of these "solutions" is viable. The first two are not humane. The latter solution is improbable since the new owners of the plantation land have political power and are not likely to give up their land even though their practices are unsustainable. Nomads throughout the world are faced with a similar scenario. They are losing traditional pastures to new agricultural developments. Governments need to take action to ensure that all people, including nomads, farmers, and urban residents, are allowed access to food or are provided with opportunities to earn enough money to buy the food they require.

> Attempts to resettle nomadic herders in central China have met with some success. Nomadic Mongol herders were given huge ranches of 20 to 30 km². Because winter pasture is not available, they have to grow hay and store it for winter feed just as Canadian ranchers do on the prairies. Rather than living off a subsistence economy, the herders now sell cattle as a cash crop.

Over the long-term, the environment must be modified to reduce the number and intensity of famines. When there are too many people living in marginal lands, the populations must cope by undertaking new approaches—relocating some of the population, implementing better land-management strategies, and increasing the resource base through new agricultural products or practices. Returning the land to its former healthy state is a monumental task, but is necessary in many semi-arid regions where the natural vegetation has been removed. In Ethiopia, many mountain forests have been destroyed by browsing goats and by people seeking wood for fuel. The loss of vegetative cover has caused the climate of the region to change. Less rainwater is stored in the ground because the mat of humus and litter is gone. The soil is exposed to the desiccating rays of the sun, which increases erosion. While the mountain forests could be regenerated, it takes many years for a forest to reach maturity. And it may be much more difficult to restore climatic patterns. Some countries, such as Israel, are establishing reserves where development is limited and ecosystems are able to recover from unsustainable practices. Another example of a country restoring its ecosystems is Costa Rica.

Costa Rica abolished its army forty-five years ago. The money that would have been spent on the military has been directed instead to human services, such as health and education, and to the preservation and cautionary development of the largest national park system in the world. Ten per cent of Costa Rica's land area is devoted to natural conservation. The country is home to 5 per cent of the world's known plant and animal species, and its soils can grow almost any kind of fruit or vegetable. The restoration of Costa Rica's ecosystems has involved agricultural experimentation, adoption of policies of sustainable development, and a national commitment to land reform. TATIA, an association funded by all Central American countries, has led and supervised experimentation aimed at increasing agricultural production while protecting the life systems on which agriculture depends. Agroforesty (the harvesting of gums, resins, waxes, medicines, and food from trees) is replacing cash cropping, so forests are protected and maintained rather than cut down. Reforestation occurs hand-in-hand with harvesting. Costa Rica is also trying to break its pattern of growing export market crops such as bananas, coffee, cotton, and sugar. Instead, it is producing new, unusual crops such as ornamental flowers. It has become national policy that all farmers participate in education and training programs that foster sustainable development. Land reform is returning some of the country's agricultural land to family farmers. And Costa Rica is also using its forests and national parks to generate foreign income through tourism, which in turn encourages conservationism and further restoration of the ecosystems.

One method of modifying the environment involves water management. **Catchment basins,** or cisterns, are large storage tanks often located underground. These have been used successfully in India. In some parts of the country the monsoon rains of summer cause extensive flooding, but during the winter there is no rainfall at all. Collecting the summer rainfall for irrigation during the winter has enabled Indian farmers to extend their growing season and produce more food. Such a scheme could be possible in other areas that experience famine, such as the Sahel in Africa where summer rainfall is plentiful but winter droughts can be severe.

New technology enables wells to be drilled into aquifers deep beneath the desert sands of

Figure 13.4 Colorado River water maintains this golf course in Lake Havasu City in the Arizona desert.

the Sahara. Many villages can now obtain abundant supplies of freshwater. While aquifers offer a solution over both the short- and medium-term, they are not a long-term solution. If more water is withdrawn from these wells than can be naturally replenished, the resource will dry up.

STATLAB

1 Prepare a graph that illustrates the data in Figure 13.5 (a). Note that the values are integers, so zero should appear in the middle of the graph (see Figure 13.5 (b)).

2 Identify periods of drought and potential famine and periods of abundant rainfall.

YEAR	INDEX*	YEAR	INDEX*	YEAR	INDEX*
1960	0.8	1973	-0.9	1986	-0.3
1961	1.2	1974	0	1987	-0.5
1962	1.1	1975	0.5	1988	0.5
1963	0.7	1976	0	1989	0.1
1964	1.5	1977	-1.0	1990	-1.0
1965	1.0	1978	0.5	1991	0.1
1966	0.7	1979	-0.1	1992	0.4
1967	1.1	1980	-0.3	1993	0.1
1968	-0.2	1981	-0.3		
1969	1.0	1982	-0.9		
1970	-0.1	1983	-1.5		
1971	-0.2	1984	-1.4		
1972	-1.0	1985	-0.7		

Note: Values of the sub-Saharan precipitation index are derived from April to October rainfall data from stations located in Burkina Faso, Cape Verdi, Chad, Gambia, Guinea-Bissau, Mali, Mauritania, Niger, Senegal, and Sudan.
*The Precipitation Index is based on a calculation of deviations from rainfall in 1977.

Figure 13.5 (a) Rainfall patterns in the Sahel and Sudan
Source: *Rainfall Variability and Drought in Sub-Saharan Africa since 1960* (FAO Agrometeorogy Series No. 9). Reprinted by permission of the Food and Agricultural Organization of the United Nations.

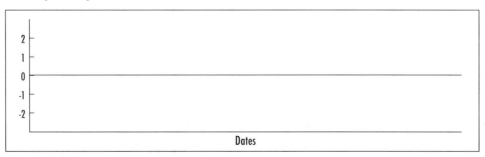

Figure 13.5 (b) Graph outline

3 a) What trends can you identify in the magnitude (size) of the deviation and in reliability of rainfall?
  b) What could be the consequences of these trends?
  c) What predictions could you make about rainfall in the region in the year 2000?

ENDING HUNGER

The problem of hunger is more difficult to solve than that of famine. It is more widespread and is the result of many different factors, as we have seen in Chapter 12. Determining a solution is indeed a challenge.

One possible answer is to increase local food supplies. But the options are limited. There is little value in increasing the food supply in the developed world. This region already produces more food than it can consume, yet still other regions of the world remain hungry. Another popular option is to use technology to expand the resource base in regions where hunger is a problem.

In the 1950s and 1960s, the technology of the **Green Revolution** allowed dramatic increases in crop yields. New hybrid grains developed through the cross-breeding of wild and domestic varieties of rice and wheat were especially successful. The new technology produced higher yields and stronger, more resilient, grain varieties. Shorter and sturdier stalks on plants reduced breakage during heavy rains and strong winds. Resistance to many diseases and pests was enhanced.

Ironically, while these "miracle grains" of the Green Revolution have greatly boosted the production of wheat and rice in countries such as Mexico, India, Indonesia, and the Philippines, the problem of hunger among their populations has worsened over the past three decades. The new crop varieties revolutionized large-scale agriculture, yet they did little to help small farmers. The new seeds required ideal growing conditions. They thrived in good soils with abundant rainfall or adequate irrigation, but they did not do well in marginal lands. In addition, they required a substantial financial investment, not only for the seeds but also for fertilizers and pesticides. Few farmers could afford the cost, and so they continued to rely on the old techniques. Family income dropped when the local markets became glutted with excess grain produced in the large agricultural operations. In the past, the farmers could sell grain that they did not need for their own consumption for a modest profit. Now, higher yields have resulted in an abundance of grain, causing prices to drop. With a shrinking income, many small farmers have been forced to sell out to large landowners and move to the city.

Unfortunately, the Green Revolution had nothing to offer these subsistence farmers. It did not work to improve local seeds indigenous to dry lands, nor did it concentrate on improving age-old traditional methods of agriculture. Those who gained through this agricultural development were wealthy landowners in developing countries who could afford mechanized farming. The increased yields resulting from the "miracle seeds" went to well-off people in urban centres, to cattle, and to affluent countries for development of luxury food products.

STATLAB

1 Study the statistics in Figures 13.6 (a) and (b).
  a) Determine which of the three columns best shows how the Green Revolution affected agriculture.
  b) Prepare bar graphs to show the results from part (a).

c) How can you explain the fact that while wheat production increased dramatically, yields showed only modest gains?

d) Identify the countries that benefited most from the Green Revolution. Describe the advantages these countries enjoyed.

2 What are some of the disadvantages of growing high-yield varieties of wheat and rice?

COUNTRY	INCREASE IN RICE AREA ha (000 000)	%	INCREASE IN RICE PRODUCTION t (000 000)	%	INCREASE IN RICE YIELD t/ha	%
China	3.1	10	67.9	73	1.76	57
India	4.0	11	31.5	69	9.66	51
Indonesia	1.6	21	18.8	122	1.72	83
Bangladesh	1.4	15	7.0	46	0.44	27
Thailand	2.8	43	5.2	42	-0.02	-1
Burma	0.1	2	7.0	97	1.45	94
Vietnam	0.9	19	5.3	60	0.66	75
Philippines	0.3	10	3.8	93	0.99	75
South Korea	0.1	0	2.3	45	1.96	47
Pakistan	0.6	43	3.2	160	1.20	83
North Korea	0.2	33	2.7	117	2.47	66
Total (11 nations)	**15.0**	**14**	**154.7**	**74**	**1.05**	**52**

Figure 13.6 (a) Rice production during the Green Revolution

COUNTRY	INCREASE IN WHEAT AREA ha (000 000)	%	INCREASE IN WHEAT PRODUCTION t (000 000)	%	INCREASE IN WHEAT YIELD t/ha	%
Egypt	47	9	728	57	1.08	45
Morocco	318	19	874	80	0.34	52
Afghanistan	294	13	800	36	0.20	21
Bangladesh	405	405	960	813	0.96	81
China	4 106	17	51 671	173	1.63	135
India	9 685	72	25 900	156	0.61	50
Iran	2 000	50	2 582	66	0.10	10
Pakistan	2 052	39	6 284	103	0.54	47
Turkey	1 574	22	3 264	33	0.12	9
Total (9 nations)	**20 481**	**39**	**93 063**	**131**	**0.82**	**66**

Figure 13.6 (b) Wheat production during the Green Revolution
Source: Palacpac; US Department of Agriculture and FAO.

3 Figure 13.7 indicates how wheat yields increased as a result of the introduction of new hybrid varieties.

a) What implications do these statistics have for alleviating hunger in the regions listed in Figure 13.7?

b) What do the statistics fail to consider?

REGION	INCREASE IN WHEAT YIELDS 1977–1990 (%)	TOTAL VALUE OF PRODUCTION INCREASE (US$)
Sub-Saharan Africa	57	31 000 000
West Africa/North Africa	43	515 000 000
South Asia	17	1 822 000 000
Latin America	53	662 000 000

Figure 13.7 Estimated effects of spring wheat breeding on production in developing regions

THE GREEN REVOLUTION IN THE TWENTY-FIRST CENTURY

The Green Revolution is not over. Researchers in **genetic engineering** are continuing to develop new crops that can be grown in marginal agricultural lands. In the past, much of the technology of the Green Revolution originated in the laboratories of the developed nations. Today, more research is being conducted by geneticists and other scientists in the countries in need of this new technology. Local plant geneticists are also doing research into tropical grains and other crops that are not normally used in the developed world. The first Green Revolution concentrated on grains such as wheat, rice, corn, barley, and oats. Little research was done on crops such as manioc, sorghum, or millet because they were not common in the countries that were carrying out the research. Both sorghum and millet, as well as many other local grains, grow best in subtropical climates with irregular patterns of rainfall. Indigenous grasses are often used to develop new hybrid grain varieties. These hardy domestic strains have developed resistance to unreliable precipitation patterns and pests over thousands of years of evolution. By breeding these plants with grains that have high yields, new varieties are being produced that can withstand some of the environmental constraints of the region and still produce high yields. This technique is especially useful in developing countries where farmers do not have the money for irrigation or chemical pesticides. With the development of new hybrids, yields are improving for local farmers. Another reason local scientists are having a greater impact than scientists from the developed world is that they are more readily accepted than outsiders. They speak the same language, they understand regional traditions, and they are trusted.

While food production in developed countries is declining as a result of depressed global markets, it continues to increase in most developing nations. Countries in Asia and South America in particular are increasing their food production. Moreover, transportation facilities are being modernized, storage facilities are better built, and economic infrastructures such as capital funding, commodities markets, and economic cooperation among countries are improving. As a consequence, the number of malnourished people in the world has continued to decline since the early 1970s (see Figure 13.2). People are eating better, not only because of increased food production but also because of better distribution facilities and transportation networks.

This is not the case in Africa, however. High population growth rates negate any increases in total food production, so food production per capita is actually declining. In an effort to help alleviate the problems many African countries

> From 1961 to 1985, global investment in agricultural research increased at an average annual rate of 4.2 per cent. Only 1.7 per cent of this amount was invested in developed countries, while investment increased at a rate of 6 per cent annually in developing nations. A recent survey by the Food and Agriculture Organization of the UN showed that 68 per cent of agricultural researchers were working on crop research, 19 per cent were involved with livestock research, and the rest were working in the fields of forestry and fisheries. These figures are encouraging because they indicate that research in developing countries is growing faster than in the rest of the world.

are experiencing, the Consultative Group on International Research, a group made up of scientists from around the world, has turned its attentions to increasing food production in the Sahel. As explained in earlier chapters, famine occurs regularly in the Sahel, causing tremendous suffering among the people. This group examines problems related to environmental deterioration, overpopulation, food supply, and famine, and proposes strategies to lessen their effects.

LAND REFORM

Improved agricultural practices are key to increasing food production and eliminating hunger. But without land reform and redistribution new technology cannot solve the problem of hunger. In order for rural farmers to become self-sufficient, they need access to land. In much of the developing world, most of the land is held by a small but wealthy, politically influential élite. Many political leaders are themselves the owners of much of their nations' productive land. It is in their personal interests to maintain the status quo. The vast majority of the people do not own land and are sharecroppers or paid labourers who work for very low wages.

CASE STUDY

Land Reform and Migration in the Philippines

With a population in excess of 200 people per square kilometre and a population growth rate of 2.5 per cent a year, the Philippines must implement major land reform if its population is to be self-sustaining. The country has fertile lowland coastal plains and forested uplands in the interior. Most of the productive lowlands are divided into large farms that are either held by wealthy landowners or leased to multinational exporters of coconuts, bananas, and pineapples. The combination of inequitable land tenure and population growth has contributed to a large increase in people who do not own land. Without land they cannot get the food they need. From 1975 to 1990, the percentage of farm workers in the agricultural labour force who did not own land rose from 40 to 60 per cent. In the 1970s many workers migrated to the capital city of Manila and other coastal cities. When economic downturn and high unemployment made migration to the cities no longer a practical solution, they moved back to the country. They worked on sugar and coconut farms for wages that could not adequately support them. When the sugar industry collapsed in the mid-1980s, many workers migrated to the undeveloped forests of the interior mountains. Today about one-third of the Philippines's population lives in these marginal regions.

The impact of this migration has been devastating for the fragile ecosystem of the rainforests. Already threatened by unsustainable lumbering operations, the tropical

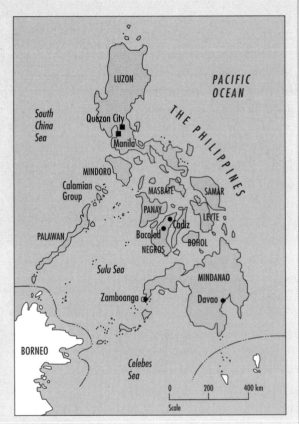

Figure 13.8 The Philippines

forests are now being depleted by subsistence farmers. Crop land increased in the mountains from just under 600 000 ha in 1960 to 4 million ha in 1990. As the forests are cleared, the soil is exposed to the elements. Soil erosion of 122 to 210 t/ha annually has occurred where the forest has been cleared compared to less than 2 t/ha when the forest canopy has been left intact. As

Figure 13.9 A temporary settlement in the rainforest of the Philippines

more and more people migrate to the rainforests, the resource potential is being stretched to the point where the environment will be destroyed and the people will not be able to survive.

The country was ripe for political extremism because of the population pressures it was experiencing. When populist president Corazon Aquino led the non-violent overthrow of Ferdinand Marcos in 1986, many people believed it would be the beginning of a land redistribution program. Aquino promised land reform to this island nation. For years Marcos and his friends had controlled large tracts of land. Despite Aquino's good intentions, she never fulfilled her promises. Revolts in the mountain areas, coup threats from Marcos's supporters, debt repayments, and economic restructuring took all of her efforts. Aquino spent her entire term in office trying to remain in office.

In 1992 Fidel Ramos, an Aquino supporter, was elected to succeed her. He started the process of land reform by giving approximately 3 million ha of government-owned and public lands to farmers. But none of the huge private estates have been dismantled. The Congress, which is controlled by a political élite, wants to exempt commercial farmland from redistribution because of the valuable cash crops it produces. With the Philippines's runaway debt and stagnant economy, there is little wonder these politicians sound so convincing. The country is expanding its production of cash crops and importing inexpensive grain from Thailand. Will this policy help the farm workers of the Philippines?

FINANCIAL AND TECHNICAL SUPPORT

Once land reform has been enacted, rural farmers must have access to low-interest loans to purchase seed, fertilizer, and tools. In much of the developing world, farm labour is performed by women. While women run the farms, they have little say in the home, are excluded from politics, and are almost always refused loans by banks, which are usually run by men. The men often leave rural areas in search of work in the towns and cities or in the mines. They may be gone for months at a time, and often do not come back at all. Women remain at home to support and raise their families.

Technical support must be appropriate for the level of development. Intensive agricultural methods are success-

In Bangladesh, specially trained local women travel through the country advising women about child care, health care, and nutrition. They also provide loans to women to help them become self-sufficient. These respected women are well received by villagers. The loans are important because they allow female farm workers to be successful without feeling they are accepting charity from an aid agency. They are used to buy tools, seeds, fertilizer, and basic health-care aids. In many villages these women are the only government representatives the residents encounter.

ful in most developing countries. The small size of the farm plots means that providing tractors and other large automotive equipment would be impractical. In addition, subsistence farmers could not afford the gas to operate this machinery. More appropriate support would lie in the development of labour-intensive techniques that improve yields and reduce costs. The introduction of new hybrid crops that do not need expensive fertilizers or pesticides is one way yields could be increased. The development of **cooperatives** would provide seed, fertilizers, and other equipment at low cost. Members could lease expensive implements without buying them, and the cooperative could be the vehicle for selling surpluses at a profit. Subsistence farmers must have a say in the policies that affect their lives.

CONSOLIDATING AND EXTENDING IDEAS

1 Explain how each of the following factors contributes to the problem of famine:
   a) irregular rainfall patterns
   b) food aid
   c) land-use patterns
   d) population pressure

2 Create a "Quotable Quotes" visual display that highlights the comments of various people who have written or spoken about hunger as a world issue. Gather your quotes by using

electronic media research tools and/or by checking resources in your ⌐
library's reference sections. Here are some suggestions: *Bartlett's Fami╷*
(1988), *The Oxford Dictionary of Modern Quotations* (1991), *The New Ⓠ*
(1992), *The Penguin Dictionary of Twentieth Century Quotations* (1993).

3 a) Explain how each of the following factors contributes to the problem of ╷
    i)   colonialism
    ii)   food aid
    iii) cash crops
    iv) inequitable land-tenure systems
    v)   environmental deterioration
    vi) overpopulation
    vii) militarism
    viii) international debt
    ix) the Green Revolution
  b) Explain how each of these factors could be improved to help eliminate hunger.

PROBLEM SOLVING IN GEOGRAPHY: SELECTED CASE STUDIES

In order to understand and find solutions to global issues, we need to examine, on a global scale, diversity and disparity, development, and quality of life issues. This involves the following steps:

- gathering and organizing information
- analysing and evaluating this information
- generating alternative solutions to address the issues
- assessing the appropriateness and feasibility of these solutions
- resolving the issues through action

So far in this textbook, you have been mainly involved in the first three steps of decision making—researching various topics and consolidating your findings into position paragraphs and papers; using statistics to explain, support, or refute these points of view; and, through the use of future wheels and mind maps, speculating about possible future outcomes.

As you proceed to the fourth step of decision making, you can employ several strategies to assess appropriateness (suitability to environmental, social, economic, and political conditions) and feasibility (can it be done?) of alternative solutions. One strategy is **cost-benefit analysis**, where the costs (monetary, environmental, human) are "weighed" against the benefits (financial gains, social/political advantages, etc.) for each proposed solution to an issue. A second approach is **needs assessment**, an innovation designed in the business world to improve productivity. Like cost-benefit analyses, needs assessments can be used to make effective decisions about issues.

NEEDS ASSESSMENT
Step 1

Determine the present situation. Suppose we are asked to analyse the lifestyle of a Canadian family. The first column in Figure 13.10 lists a variety of aspects of family life.

Step 2

Consider how you would like things to be. Make an outline copy of Figure 13.10 and list your as in the fourth column entitled Vision. Be realistic! You might like to win the lottery, but this is an unrealistic expectation. Keep your visions to what is truly achievable through your own actions or those of others.

Step 3

Now determine how you can fill the gap between the reality of the present situation and the possibilities of the future. Remember that your goals are interrelated. This is your "action plan." It outlines what needs to be done to achieve your vision of the future. Since your action plan will likely take several years to implement, it would be a good idea to divide it into smaller, manageable steps. Keep in mind, of course, that nothing is static. Everyone experiences both good fortune and bad. This means your action plan may need to be revised to take into account unforeseen circumstances.

CRITERIA	PRESENT SITUATION	ACTION PLAN	VISION
Employment	Wife: part-time registered nurse. Husband: unemployed construction worker.	Wife: i) continue working part time until children reach school age. ii) Then start working full time. Husband: i) retrain as an electrician. ii) work full time for an uncle who owns an electrical company. iii) start a business.	Wife: Continue part-time job. Husband: employed, with a good trade.
Monthly Income	$1400 (net)	Extra income from new jobs. Loss of income-tax benefits.	$3500, increasing to $4000
Expenses per month	Rent: $800 Food: $300 Loan: $100 Car: $100 Day care: $0 Savings: $0 Other: $100	i) Save the extra income for 3 to 5 years for down payment on a house. ii) Buy a small house. iii) Save money for vacations and new consumer goods.	Mortgage: $1200 Food: $500 Loan: $150 Car: $400 Day care: $500 Savings: $500 Other: $250
Family	Daughter: 4 years old. Son: 18 months.	Children in day care until they reach school age.	Children in day care. No plans to expand the family.
Health	Good.	Improve personal fitness.	Good.
Leisure	TV, walking in park, local hockey team.	Join a fitness centre.	Working out, annual vacation, TV, park, hockey.
Car	4-year-old Mazda in fairly good condition.	Buy a late-model used car so that the husband can get to work.	Buy a second car.
Investments	None.	Put aside a portion of income.	House. Savings account.

Figure 13.10 Lifestyle of a Canadian family

CONSOLIDATING AND EXTENDING IDEAS

Using the following inquiry model, plan how to eliminate hunger in a region of your choice.

a) FOCUS:

You have been asked to head a United Nations Task Force charged with increasing the food supply in a country where hunger is a problem. You have been allotted $300 million to spend over the next ten years. Prepare a strategy to improve the food supply in the region you have chosen.

b) ORGANIZE:

Prepare a needs assessment using the model in Figure 13.11.

CRITERIA	PRESENT SITUATION	ACTION PLAN	VISION
Arable Land			
• amount			
• soil quality			
• water resources			
• erosion			
• irrigated land (% total arable)			
Climate			
• temperature			
• precipitation			
• reliability of rainfall			
• growing season			
Natural Vegetation			
• types of ecosystems			
• cleared land			
• forested			
• pasture land			
• unusable land			
Farming Practices			
• land tenure			
• fertilizing			
• pesticide use			
• terracing			
• equipment			

• technology (e.g., tractors, harvesters)			
• crop rotation, intercropping			
• agricultural research			
Crops Produced			
• cash crops			
• subsistence farming			
• livestock			
Environmental			
• natural disasters			
• insect infestations			
• environmental destruction			
Food Infrastructures			
• storage			
• distribution			
• processing			
Food Aid			
Economics			
• foreign debt			
• rural credit plans			
• public agricultural research expenditures			

Figure 13.11 Needs assessment

c) LOCATE/RECORD: .

Use the library resource centre, the Internet, encyclopaedias, year books, electronic periodical retrieval systems such as SIRS, or other resources to obtain information on the region you selected.

d) SYNTHESIZE AND EVALUATE:

Rank each criterion based on its significance in increasing the food supply. Decide which criteria should be selected as part of the development project.

e) CONCLUSION AND APPLICATION:

Create a development plan that will achieve maximum results over the ten-year period. Include both short- and long-term projects in your plan. Determine what other countries could benefit from a similar strategy, and explain why. Remember to consider the background and beliefs of the cultural groups under study.

f) COMMUNICATION:
Communicate the results of your study to your class or group. Describe the decisions you reached, including how and why you came to these decisions. Be prepared to defend your position. Consider using a common graphic of long- and short-term projects for each presentation (see the example in Figure 13.12). This approach will allow you to display the logic in your sequence of choices.

YEAR									
1	2	3	4	5	6	7	8	9	10
FOOD AID									
	LITERACY EDUCATION								
		TECHNICAL EDUCATION							
					RESEARCH AND DEVELOPMENT				

Figure 13.12 Timeline of development priorities

CASE STUDY

Agriculture in India

Most experts agree that India will surpass China as the world's most populous nation sometime in the next century. India currently has a population of 936 million and a population growth rate of almost 2 per cent a year. Not only is the population growing rapidly, it is also incredibly diverse, with over 700 languages and dialects, and several major religions. While India has many large modern cities such as Bombay, Calcutta, Delhi, Hyderabad, and Madras, most of the people live in small villages where subsistence farming is the predominant economic activity. By 2025, India's population is projected to reach 1.4 billion. Since the resource base cannot expand dramatically, how will the country be able to provide food for all of its people?

Farming has always been important to India's economy. The country contains several of the best agricultural regions in the world. Most of these regions grow crops all year round. In fact, the majority of farming activities take place in winter, when temperatures are high enough for most crops and frost is unknown except in the mountainous north. The summers are often extremely wet because of the summer monsoon. Heavy rains cause flooding, which makes farming difficult. The winters are usually dry, but river water and water stored in tanks in the

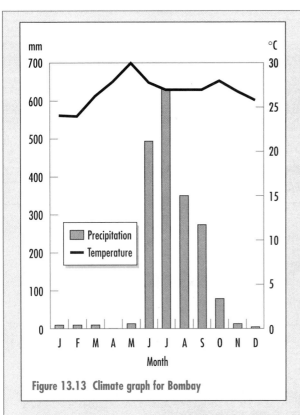

Figure 13.13 Climate graph for Bombay

The Green Revolution has helped to increase productivity, and India is now more able to feed its population. Agriculture in India has been a great success story. In the early 1950s, the country had to import much of its food, famines were common, and millions lived without adequate food. Increased fertilizer use, storage of water, irrigation technology, and high-yield grains have even allowed this populous country to export surplus food to other countries. The government provides grass-roots training programs to teach scientific farming methods to peasant farmers. Low-interest loans are available to farmers so that they can purchase seeds, fertilizer, and other agricultural necessities. Unlike many other developing countries, India is a democracy. The government can gain popular support by helping the largest sector of the population—peasant farmers.

Will India be able to increase food production in light of its high population growth rate? While six out of ten workers are employed in agriculture, the percentage of people working in agriculture is declining in the wake of urbanization and industrialization. As more and more urbanites look to the countryside for food, it is possible that food production may not be able to keep pace with the burgeoning population. At the present time, most farmers in India use the food they grow for their own consumption, but also grow some cash crops so that they can purchase the items they cannot produce. Half of all Indian farms are less than 1 ha in area and only 4 per cent are over 10 ha. The pressure on farmland to produce adequate food for all Indians is rising each year.

For India, there are essentially three solutions to increasing productivity: expand food production in marginal areas by improv-

soil make up for the lack of precipitation.

India's soils are generally excellent for farming. Rich farmland in the Indo-Gangetic Plain and the Punjab is the result of **alluvium** deposits. When the snow melts on the slopes of the Himalayas, mountain streams carry silt down into the heavily populated valleys of northern India. Heavy rains add to this runoff. In the past, annual floods resulted in the deposition of these rich silts on farmland. The annual floods are now controlled by reservoirs and dams. Consequently, loss of life and property has been eliminated and water can be stored for use in the dry season. Unfortunately, since the rich silts no longer fertilize the soils, the only way crop yields can be maintained is to add organic or artificial fertilizers.

ing soil fertility; reduce the production of cash crops for export in favour of food crops; and rely on food imports, which would be paid for by exports of manufactured goods (see Import Substitution, page 254).

The redevelopment of marginal lands is the current strategy favoured by the government. Minister of State for Agriculture, Dr. U. Venkatewarlu, an expert in agricultural finance, announced in June 1996 that the government is preparing for a second Green Revolution by converting "grey arable regions into green areas." This is to be accomplished by restructuring fertilizer prices and urging banks to use guidelines created by the Reserve Bank of India. These policies establish that 18 per cent of all loans should go to the agricultural sector. The current level is less than 14 per cent. In 1996, only 32 per cent of arable land in India was being farmed, 7 per cent was being restored to full use by local schemes, and more than half the potential farmland in India was not being redeveloped. It is hoped that the improved availability of funding for agriculture will help small farmers in marginal farming areas.

In addition to funding, the agriculture minister is initiating a program to promote a more balanced use of artificial fertilizers. Prior to 1991, the price of fertilizers containing potassium was regulated so that farmers could afford to use them. After the price was deregulated, it became so high that farmers reverted to cheaper urea fertilizers, which play havoc with soil fertility. Improved training and access to funding should help expand food production nationally.

Almost 50 per cent of India's farmland is used to grow cereal crops such as wheat and rice. Much of the remainder is allotted to the growing of cash crops such as tea and sun-

flowers. Most of these cash crops are exported so they do not add to the national food supply. Should these lands be used to produce food for the people or to produce exports? In the Punjab, 7.4 per cent of the land is devoted to cotton production, 4.2 per cent is used for growing peanuts, and 3 per cent goes towards sugar cane. This is over 15 per cent of the total arable land in the region. There is a strong argument in support of cash crops. All countries need valuable foreign-exchange income from the export of crops to help balance trade deficits. The Punjab is especially suited to the production of cash crops. Sugar cane, for example, requires very precise growing conditions not found in most places: rich soil with high water-retention properties, and a frost-free period with heavy rainfall and high temperatures. But this crop also needs a great deal of sunshine and dry weather for harvesting. Because the Punjab meets all these conditions, sugar grows well there. Millet and sorghum would also thrive, but the value of a grain crop on the same land is very low in comparison. It stands to reason that a farmer would grow a crop that produces the highest rate of return; in this case, sugar. On the other hand, the argument in favour of increasing food production is also strong. People living in an agricultural region should be able to obtain food that is produced locally. From a humanitarian standpoint food production should be a priority. Another factor relates to world commodity markets. Often, multinational companies manipulate the prices of cash crops to their own advantage. In recent years, prices for many agricultural goods (such as cotton) have dropped to such an extent that they are no longer profitable to produce.

The third strategy—the export of manufactured goods in exchange for food—is popular among many people living in the industrialized cities of India and other countries. Export income from manufactured goods is substantially higher than it is for exported agricultural products. It makes economic sense to produce manufactured goods and export them in exchange for the money needed to import food from elsewhere. Japan has employed this strategy in recent times. With a huge, low-paid labour force, India is able to manufacture goods inexpensively.

The consumers of these goods are mostly Indians. The country is so big that the domestic market absorbs almost everything that the country's factories produce. Without export income, India would not have foreign funds to import needed food supplies. As the country expands its economy into consumer durable goods, its ability to import food will grow. Nevertheless, this strategy could result in an unequal distribution of wealth, with those members of society most in need remaining hungry.

COMBATING THE SALT OF THE EARTH

Mundir Lal knows the price of a vote.

For his support during the state elections, the Uttar Pradesh [a state in India] government gave him free water and a patch of desolate, marginal land. . . . That patch of land has meant a change in his status and prospects since a new state project has helped thousands of small farmers . . . convert their marginal land to a productive farm belt. . . . Over the next seven years the government hopes to reclaim 45 000 ha of land this way.

The new fields came from 1.2 million ha of farmland that had fallen prey to **sodicity**, a condition that occurs when salt rises to the surface and renders soil unproductive. Since 1993, the state-run project has reclaimed more than 6000 ha of parched, white soil that previously sprouted nothing but weeds. Now, the new cropland produces enough rice and wheat—$1.3 million last year [1994]—to pay for the work in three years.

For Mundir Lal, the additional land has given him 500 kg of rice each year, which he

can cash in for about $100. . . .Officials say the project will not only reach the disempowered of northern India; it will bring a new sense of food security to India's biggest farm belt—a state of 130 million that is growing by 4 million people a year. . . .

The main reason [for sodicity]: large irrigation schemes raise water tables and bring sodium to the surface. [As the water evaporates, the salts that are left render the topsoil useless.] The salt problem can be solved rather easily, though. By flooding the salt-laden land with water and gypsum, the chemical balance is restored to the alkaline soil and the salt is pushed well below the surface to make way for productive soil. As long as the soil is kept active, and the soil is nourished by planting trees, the salt tends not to return. The cost: about US $1000/ha.

In wealthier states . . . big commercial farmers took care of the salt problem on their own, buying gypsum and applying it to their land. But in Uttar Pradesh, small landown-

Figure 13.14 A woman cleaning grain near Agra, Uttar Pradesh

interest, long-term loans of $55 million from the World Bank and US $30 million from the European Union, the project managers then recruited groups of small farmers to reclaim their salty land. . . .

As some of India's poorest farmers became productive landowners, the only objections seemed to come from the high-caste farmers. They feared that development would cost them their cheap field labour—and they were right. . . . The daily wage for a field worker has risen in the past year to 22 rupees ($1) from 15, an increase of about 50 per cent. Most of the workers now come from other villages. Such a protest might have stymied the project in an earlier decade, but the election of a lower-caste government in Uttar Pradesh in 1993 . . . has started to change the nature of development in the state. . . .

At first, many farmers saw the land project as another boondoggle, and they refused to dredge their common drains. But as the former salty land produced fresh crops, and new incomes, many farmers took the work more seriously. . . .

In Karrayya village, where 43 ha of land have been reclaimed, the farmers formed savings groups to buy irrigation pumps and pooled their money to dredge drainage canals. With productive land as collateral, the farmers now find themselves eligible for loans for water pumps and fertilizer.

"We can look after the problems," said Ayodhya Prasad . . . surveying the 1 ha of paddy land he has reclaimed. "After all, we are the ones who stand to benefit."

ers—most of whom make less than US $330—did not have the same resources. . . . As a result, previous land reclamation projects in Uttar Pradesh showed little success. Farmers refused to level their land and enclose it with the bunds needed to capture water. Even if they did, most failed to drain the surplus water, which served only to bring back the salt.

Finally, in 1990, the state . . . recruited non-governmental organizations such as the Society for Promotion of Wasteland Development to organize small farmers. With low-

John Stackhouse. From *The Globe and Mail*, 21 August 1995. Reprinted with permission from The Globe and Mail.

CONSOLIDATING AND EXTENDING IDEAS

1 Study Figure 13.13.

a) How are precipitation patterns different from those in Canada?

b) How are temperature patterns different from those in Canada?

c) What benefits and difficulties does the climate of India pose for farmers?

d) Synthesize the information in Figure 13.15 with the table in Figure in 13.16 to create a new map that illustrates where different grains should grow in India based on climate considerations.

e) How have farmers in India overcome the monsoon climate?

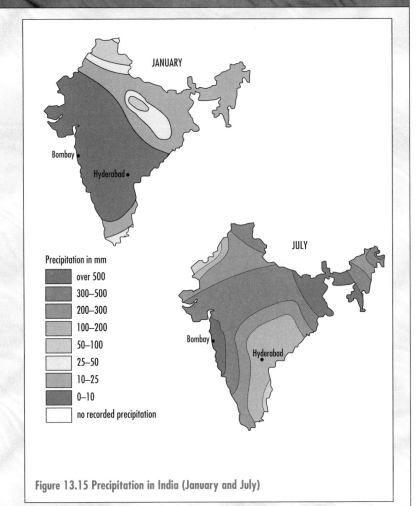

Precipitation in mm

- over 500
- 300–500
- 200–300
- 100–200
- 50–100
- 25–50
- 10–25
- 0–10
- no recorded precipitation

Figure 13.15 Precipitation in India (January and July)

CROP	SUMMER RICE	WINTER RICE	WHEAT	MILLET	SORGHUM
Temperature minimum	24°C	24°C	16°C	24°C	24°C
Total water needed (mm)	965	1057	356	508	635

Figure 13.16 Temperature and water requirements for different grains

Source: Reproduced with permission from *Decision-Making Geography* by N. Law and D. Smith. © Stanley Thornes (Publishers) Ltd.

2 Study Figure 13.17.

COUNTRY	YIELD 1961*	YIELD 1990*
China	600	3200
Albania	800	2800
India	800	2200
Egypt	2500	5300
South Korea	1800	4300
Ireland	3400	8200
Saudi Arabia	1500	4900
Paraguay	800	2200
Chile	1350	3000

* kg/ha per year

Figure 13.17 Rates of yield gains in wheat in selected countries
Source: Dr. D. L. Plucknett, Worldwide Yields and Crop Productivity (research program), 1993.

a) Calculate the percentage increase in yield from 1961 to 1990 for each country (divide the 1990 figure by the 1961 figure, subtract one, and multiply by 100 per cent).
b) How does India compare to other countries?
c) What factors could account for the differences?
d) Describe how the Green Revolution changed food production in India.
e) Why is it essential that crop yields continue to increase?

3 a) Consider the following options for alleviating hunger in India:
   - Increase the amount of land under cultivation by clearing rainforests, draining and reclaiming wetlands, irrigating semi-deserts, and terracing mountain slopes.
   - Increase food production by reducing cash crops intended for export.
   - Reduce population by encouraging emigration and developing new family-planning strategies
   - Import more food and develop more industries.
   - Intensify cultivation through land reform, increased use of fertilizers and pesticides, more irrigation, and further mechanization.
   - Generate new food sources by introducing new crops, genetic engineering of existing crops, developing aquaculture, and manufacturing more and a greater variety of food products.
   b) Prepare an organizer or a chart to investigate each of these alternative strategies. Include such criteria as:
   Preconditions: Is there enough capital and technological ability to implement the strategy?

Environmental repercussions: Would the modification of wildlife habitats be supported by preservationists? Would new developments be environmentally sustainable?

Economic repercussions: Would funds be available for implementation of the strategy? Would displaced farm workers be given new opportunities?

Political repercussions: How would various interest groups support the idea? Would people accept change willingly?

Social repercussions: Is the strategy compatible with religious and moral values?

c) In light of your evaluation, rank order the solutions from most desirable to least desirable.

4 Read the article "Combating the Salt of the Earth" on page 196.
   a) What elements of physical geography result in soil infertility?
   b) How have unsustainable agricultural practices exacerbated the problem? What solutions are being proposed?
   c) What part did politics play in the project?
   d) Evaluate the proposed solution and state why you support or do not support the project.
   e) What are the possible repercussions of this scheme—to higher-caste land owners, to lower-caste farmers, and to the government of the state?

5 a) Research the majority of Indians' attitude to cattle.
   b) How could this attitude affect food production?

CASE STUDY

Agriculture in Botswana

The situation in Botswana is much different than the one in India. Botswana is a considerably smaller country, both in land area and population. Its subsistence agricultural economy is generally much less productive than the agricultural industry in India. Located in southern Africa, the landscape of Botswana is primarily savannah and tropical steppe, with the Kalahari Desert dominating the eastern third of the country. The economy is based on subsistence pastoralism, much like that of the sub-Saharan region of North Africa. However, there are regions where farming and commercial cattle production flourish.

The government has chosen two different agricultural development strategies to improve pastoral practices and increase food supply. Your job is to evaluate these plans and decide which one you believe is the best model for this country.

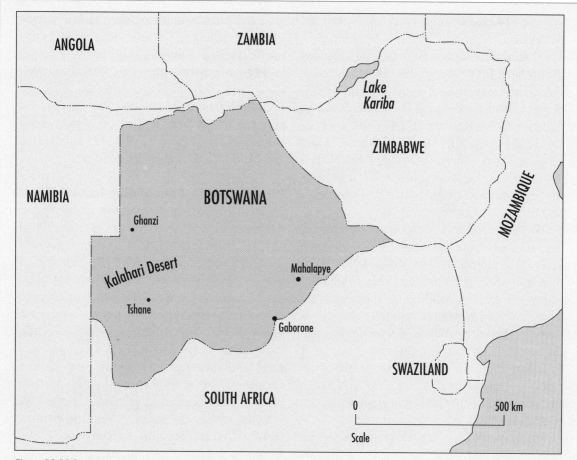

Figure 13.18 Botswana

Figure 13.19 A cattle drive in Botswana

Plan A: The Range and Livestock Management Program

In 1974, six technicians from Texas, backed by American foreign aid and the Botswana Ministry of Agriculture, came to Botswana as part of the Range and Livestock Management Program (RLMP). Their mandate was to modernize livestock production practices in Botswana. Because Texas and Botswana share similar natural environments, researchers believed that Botswana could benefit from agricultural methods practised in Texas. Two experimental ranches modelled on the Texas system were planned for Botswana.

One of the ranches, Sebelalo, was located in southern Botswana where the range country is similar to that of south-central Texas. Sixteen families occupied the area, maintaining a herd of 360 cattle. In 1975, members of the RLMP met with the Botswanans. They proposed that the cattle owners establish a 6400 ha ranch. Wells would be dug with financial support from both the Botswanan and American governments.

With the support of the Botswanan families, the area was developed between 1975 and 1977. Boreholes, or deep wells, were drilled at strategic locations. This eliminated the need for the traditional migration for water in the dry winter months. The ranch was delineated by 37 km of fencing. Smaller corrals for bulls and weanlings were also built to help manage the herd effectively. All of the work was performed by local people, with financial and technical support provided by the RLMP.

In July 1977 the ranch was ready. The Botswanans managed the ranch, with the advice of an advisory group from the RLMP. A committee of local people was formed and met monthly to prepare reports and register the herd. A respected member of the community was elected as herd manager.

From 1977 to 1979, the RLMP offered advice to the Botswanans based on the Texan model of ranch management. The Botswanans learned how to build fire breaks to prevent grass fires from depleting pasture stocks; to rotate pastures so that one pasture could be used while another was recovering from previous grazing; and to wean heifers from their mothers by isolating the babies so that they would graze, leaving the milk available for human consumption. Other advice included selling off substandard bullocks for meat, thereby allowing only the best young bulls to reach breeding age. This improved the quality of the herd through selective breeding. The Botswanans also discovered how to control ticks and other insects using pesticides, and how to dehorn animals to reduce injury during confrontations.

Plan B: The Tribal Grazing Land Policy

At about the same time as the RLMP began its work in southern Botswana, the Tribal Grazing Land Policy (TGLP)—a local government initiative—was formed. Its aim was to assist small livestock owners in acquiring land and establishing their own commercial ranches by helping them obtain leases on grazing land. The program operated on the basis of three underlying principles: democracy, self-reliance, and social justice. The people would be responsible for the project. They would perform the work themselves, with some financial support from the government. No one cultural or religious group would be favoured over another.

The Nkange-Maitengwe Lands Enclosure Project was one of the schemes supported by the TGLP. It involved two villages, both of which owned small herds of cattle and goats and practised subsistence farming. The villagers wanted to fence in their land to contain the herds, which in the dry season concentrated near the rivers close to the villages, where they damaged the crops. When the people asked for assistance from the TGLP in 1979, local officials conducted a land-use survey. The results revealed that the soil was highly fertile and could be used more intensively than simply as pasture. As a consequence, the TGLP provided 90 per cent of the cost of the fencing, but the villagers had to contribute the remainder and perform all of the work. The government built a 1 km long demonstration fence to display the five types of fencing that were available. A fencing committee was formed to select the best fencing for the project. Over the next year a fence of about 100 km was built. The cattle were contained and field crops were introduced close to the village on the more fertile land. Those who participated in the project were charged a fee, and by-laws on herd control and fence maintenance were established to resolve problems.

CONSOLIDATING AND EXTENDING IDEAS

1 Consider the two case studies of agricultural development in Botswana. The first project failed, while the second locally developed initiative was a great success.
   a) List the objectives of each plan. In what ways were they similar? In what ways were they different?
   b) Why do you think the second plan succeeded while the Texan model failed?
   c) What does this suggest that we have to learn about foreign-aid projects?

2 Create an illustrated storyboard that details the significant issues, individuals, and events in either the article "Combating the Salt of the Earth" *or* the case study Agriculture in Botswana. Try to give your completed storyboard the appearance of an extended "cartoon strip," by using visuals (drawings, photos, etc.) and written text.

3 In a paragraph, suggest what we can learn about providing aid and managing change from the Botswana case study.

PART 4

ECONOMIC ISSUES: INDUSTRIAL DEVELOPMENT

CHAPTER 14

BASIC CONCEPTS IN ECONOMIC GEOGRAPHY

The Oxford *Dictionary of Geography* defines economic geography as "the study of the creation of wealth and income as it affects the geography of the landscape; the analysis of the spatial distribution of the transportation and consumption of resources, goods, and services."

The term *economics* is derived from ancient Latin and Greek words relating to the management of money in a household. While all of us are involved in managing our own money, economics has come to mean a great deal more in the twentieth century. In fact, economics is increasingly on everyone's mind as the world continues to grapple with economic problems on a day-to-day basis. Pick up any newspaper and there is bound to be at least one front page article pertaining to economics.

ECONOMIES AND PRODUCTION

Traditional economies rely on one or two ways of

Figure 14.1 Daily newspapers include business or financial sections that focus on domestic and global economic issues.

producing the essentials of life. Hunting and gathering, subsistence agriculture, and pastoral economies concentrate their energies on primary production, extracting resources directly from the natural environment. As societies develop, production becomes more elaborate. Raw materials such as grain and wood are processed by specialized manufacturers. This manufacturing stage is known as secondary production. Because much more labour is involved in making a finished product such as furniture or bread, its value is considerably greater than that of the raw material.

Today's diversified economies include primary and secondary production, but the greatest share of the labour force is engaged in tertiary and quaternary production. In Canada, for example, people are still involved in primary production of such raw materials as wheat, nickel ore, fish, and so on. In addition, many workers are employed in secondary production, in industries such as steel, housing, and automobile components. Tertiary production includes the many services needed to support primary and secondary production. Everything from transportation services to teaching and health care are included in this category of economic production. In Canada, tertiary production has become the mainstay of the economy. More people are employed in offices than in mines or factories. Quaternary production deals primarily with ideas, information processing, and the control of other industrial sectors. Research and information processing are becoming increasingly important sectors of the economy.

With growing numbers of people in diversified economies working in sales, accounting, public health, and other service professions, economic development has shifted away from traditional industrial expansion into communications systems. Many people now sell ideas rather than products. A sales representative for a major textbook publisher does not go to her employer's office premises unless she has to. She does all her work at home, 150 km away. She meets with clients, and then sends orders to head office via the Internet. Another of this new breed of entrepreneur develops communications software in his home in the Cayman Islands. Since the product can be transmitted electronically around the world, this software developer does not need to live near the people who will purchase it. Another individual utilizes Geography Information Systems (GIS) to reduce transportation costs for her company. Working as a dispatcher for a courier service, she can determine where every truck in the fleet is located at any time because of satellite positioning technology. Each truck is identified on a master digital map. When a client requires the company's service, the dispatcher sends the vehicle that is closest to the client. Each package can be tracked electronically and located instantly through electronic technology. Many courier companies are able to provide twenty-four-hour delivery service because of advanced communications technology.

In nations of the developed world, the nature of work has changed from basic production to the provision of services. Typically, well above 50 per cent of the employed labour force works in the services sector. Industries that are projected to grow rapidly within the next decade (into 2005) include health, education, social services, personnel supply, computer and data processing, amusement and recreation, management, public relations, and child care. The value of these highly sophisticated services is often far greater than the value of manufactured consumer goods. With this increased affluence, living standards have risen even over the high levels of the 1970s and 1980s.

The nature of production is changing as well. Industry worldwide is currently dominated by the onslaught of technology, particularly in computers, telecommunications, robotics, and medicine. Of course, workers have to be more sophisticated, better educated, and more able to adapt to new ideas than ever before. The industrial worker is now often expected to be part of the design team and has to have advanced skills in communications, computer analysis, and problem solving. For many, it is a challenge to develop the skills needed to get a good job. Consequently, more adults are going back to school for further education; students are choosing more specialized fields of study; and ministries of education are developing programs that redefine basic education and literacy, and are allocating more funding for technological development and instruction in schools.

Most of these advances are taking place in OECD countries (members of the Organization for Economic Cooperation and Development); very few non-OECD countries are able to adjust to rapid technological change. Thus, the technological gap between developed and developing countries continues to widen. The growth of quaternary industries in developing nations is an issue for the future.

Economies that specialize in exporting high-technology products enjoy better employment opportunities and higher wages. Countries that export primary products or low-technology manufactured products are worse off—they experience slow growth and have difficulties in creating enough new jobs.

> Production results from a process, effort, action, or product. Shelley and Clarke's *Human and Cultural Geography: A Global Perspective* defines production as "the identification and extraction of raw materials, the transformation of raw materials into finished products, the distribution of consumer products to consumers, and the development of new products and techniques through research."

ECONOMICS THROUGH THE AGES

There are three stages of economic development that evolved from major paradigm shifts or technological advances. The first major shift occurred thousands of years ago when people started to use tools. For the first time, they were able to hunt and gather foods more effectively than other animals. Human populations increased as people developed a **competitive advantage** over other animals. Populations soared and the species thrived in isolated pockets of the African savannah and the Chinese plains. Hunters and gatherers developed very simple economic systems. They collected enough food to last them through the winter, or dry season, or any other period of privation. Because they were nomadic—following the animals—they did not have many material possessions and virtually no accounting was necessary. A few weapons, gourds, or bladders for holding food and water, transportable shelters, and some clothing were all these people owned.

Today, most people have moved on past this stage of development. However, hunting and gathering economies are still viable in marginal regions such as northern Quebec and central Asia where agriculture cannot be practised because of environmental limitations.

The second series of technological advances involved the development of agricultural practices—the cultivation of plants and the husbandry of animals to produce food. Archaeological evidence suggests that **domestication** (the intentional planting of crops and the taming and raising of animals) as an agricultural practice occurred simultaneously in the foothills and upland valleys of the Middle East, in present-day areas of Iraq, Iran, Syria, Lebanon, and Israel. From sites in the Shanidar area of Iraq, archaeologists have recovered sickles for harvesting grains and hand milling stones (mortars and pestles) for grinding grains into flour, which date to 8000 BCE. The remains of a domesticated sheep at one site was carbon-dated to 8900 BCE. Wild variants of modern grains, the predecessors of wheat and barley, still grow in mountain regions. Wild relatives of domesticated pigs, sheep, goats and cattle continue to roam in isolated regions. The settlement patterns of the peoples in these times were sedentary, not nomadic; hunting and gathering was being replaced by agriculture.

Archaeological evidence also indicates that agricultural practices developed at about this same time in the river valleys of coastal Peru and in the southwestern United States. Cultivation derived from the replanting of stems and roots in the soft mud of river banks. Some people believe that this practice predated the technique of growing seeds to produce crops.

The cultivation of crops and animal husbandry encouraged populations to settle down in areas that could provide freshwater, rich soils, and a temperate climate. Regular, adequate supplies of food, the accumulation of surplus food, and the division of labour led to the development of civilizations. With agriculture and sedentary settlement patterns came the luxury of leisure time, which allowed certain individuals to specialize in activities not directly tied to daily survival. Mathematics, writing, and artistic endeavours enriched societies and contributed to the origin and growth of economic and political systems. Simple tallying—to keep track of the numbers of animals raised and the

Figure 14.2 The three stages of economic development: stage 1 (*left*), simple tools are used to hunt and gather food; stage 2 (*middle*), agriculture replaces hunting and gathering; stage 3 (*right*), improved agricultural technology increases food production.

amounts of grain produced—led to mathematics. Increasingly sophisticated systems of barter developed until money was invented. Rudimentary habits and rules for living in groups paved the way for political systems, which required financial support. Governments needed funds to build public works and monuments, to provide courts and systems of justice, and for military expansion. This resulted in taxation, for which sophisticated accounting systems were devised. The economic system became integral to society.

The third great stage in economic development began in the mid-eighteenth century in western Europe and slightly later in nineteenth-century North America. Major changes took place in agriculture. Common lands were enclosed so that new farming practices—crop rotation and experimentation with new crops—could be employed. New technologies in the form of the seed drill, iron ploughs, mechanical reapers, and threshers greatly increased the amount and variety of food being produced. More efficient means of agricultural production meant that fewer people were needed to work the land. Farmers and peasants uprooted their families and migrated from the countryside to the emerging and rapidly growing industrial cities, where they would become part of a large new labour force for the factories and workhouses.

Unlike the two previous paradigm shifts in technological advancement, which were related primarily to the attainment of food, the Industrial Revolution centred on secondary industrial activities of manufacture and increased production. Prior to the Industrial Revolution, products were hand-made by skilled craftspeople usually within their homes or in buildings adjoining their living quarters. Within this **domestic system** of manufacture, products tended to be expensive, inconsistent in quality, and limited in quantity. By the mid-1700s, new methods of production were being developed in the textile, coal and iron, transportation, and communications industries. The invention of the steam engine, the use of fossil fuels to generate power, the mechanization of manufacture, and the advent of mass production through assembly lines were all advances of the Industrial Revolution. They increased production speed and efficiency, standardized quality, and reduced the cost of products to the consumer.

Industrialization contributed to population growth and urbanization, and transformed the economic structure of Europe. During the nineteenth century, the middle class rose in prominence. Its status in society was associated with financial wealth. The most powerful members of this new class were the wealthiest factory and mine owners, the bankers and financiers, and the owners of the resources and means of production. They adopted practices of free-market enterprise and advocated the benefits of unfettered capitalism. A second new social class—the industrial workers—also emerged during industrialization, and they too came to be defined by economics. This large sector of the population was largely untrained, landless, and occupied the lowest ranks of society. It was from this class that future demands for reform and social services would come.*

Some people argue that this third stage of economic development is over. This may be somewhat true for many of the nations that we call "developed," since industrial development has given way to a new **information age**. Fewer people are producers of durable products; they have instead become producers of ideas and information. But in developing countries the Industrial Revolution is very much a part of everyday life. Industrialization, growth of urban centres and infrastructures, and the evolution of social, economic, and political systems associated with economic development are evident in Asia, Africa, and Latin America today.

*All references to socio-economic status are presented in a historical context; the terms are not intended to label societal groups.

STATLAB

1 Refer to Appendix B on pages 442 to 449. Compare percentage of GDP from agriculture, percentage of GDP from industry (manufacturing), and percentage of GDP from services.

   a) List ten countries that obtain most of their GDP from service industries and state the percentage of GDP from services for each.

   b) List ten countries that obtain most of their percentage of GDP from manufacturing and state the percentage of GDP from industry for each.

   c) List ten countries that obtain most of their percentage of GDP from agriculture and state the percentage of GDP from agriculture for each.

2 a) Find the GDP per capita for each of the thirty countries.

   b) What patterns do you notice between GDP per capita and each production category?

3 a) Prepare a scattergraph for twenty randomly selected countries, showing the relationship between GDP per capita and percentage of GDP from services.

   b) Prepare a scattergraph for twenty randomly selected countries, showing the relationship between GDP per capita and percentage of GDP from manufacturing.

   c) Prepare a scattergraph for twenty randomly selected countries, showing the relationship between GDP per capita and percentage of GDP from agriculture.

   d) What correlations are apparent for each category of production?

   e) Account for any anomalies that you observe.

4 a) Assume that you are an economic planner in a developing country. In light of the patterns revealed by the previous activities, what development strategy would you suggest to the government to increase the country's GDP?

   b) How could the interpretation of statistics be an oversimplification that might lead a developing country down the wrong path?

ECONOMIC THEORIES

Economists and philosophers such as Adam Smith in the eighteenth century, Karl Marx in the nineteenth century, and John Maynard Keynes in the twentieth century developed strategies to make the most of economic development. Now, more than 200 years after it all started, people are starting to re-evaluate the different economic theories and modify them to make them relevant to economic situations today.

From 1500 through the mid-eighteenth cen-tury the prevailing economic theory was **mercantilism**. In practice, governments attempted to augment their national wealth and prestige by increasing income from exports and keeping import costs low. When the value of imports was subtracted from the value of exports, the resultant **trade balance** was in favour of the exporting country. European nations were very successful in maintaining favourable trade balances because they were able to exploit the natural resources and cash crops of their colony countries. For example, Great Britain would import a raw material such as cotton from India at very low cost, and export back to its colony finished cotton goods, which would be sold at moderate to high prices. European nations consistently encouraged the production of raw materials in their colonies and repressed attempts by colonists to industrialize. They did not want their colonies as competitors in markets for manufactured commodities. Instead, they wanted to create markets for more expensive finished products manufactured by their own labour forces.

In the late 1700s the Industrial Revolution sparked new ideas about the importance of human rights. The economic system endorsed by governments and ruling élites gave way to increased democratization of the marketplace. In 1776 Adam Smith wrote *An Inquiry into the Nature and Causes of the Wealth of Nations.* Smith's basic premise—that economic growth would increase naturally and rapidly if individuals were free of government restrictions to act in their own self-interest—found immediate and long-lasting support in the societies of late eighteenth-century France and America. Political rights of the individual had taken centre stage in both the American and French Revolutions. Ideals of liberty, equality, and freedom for the individual meshed well with Smith's philosophy of individuals participating in a milieu of free competition. **Laissez-faire economics**, the system in which entrepreneurs and farmers were permitted to function freely without government intervention, encouraged the rise of every person. The American **free-enterprise system**, based on the writings of Adam Smith, encouraged individual investment, innovation, and industrialization, and allowed the United States to become a great economic power in the nineteenth and early twentieth centuries.

This paradigm shift to the market economy was not universally endorsed. European powers such as Great Britain did not want to lose the markets and sources of raw materials they had developed though imperialism and colonization. Their economic success was dependent on continued exploitation of foreign markets, either within their own colonies or through advantageous trade arrangements with other nations. Most emerging industrial nations also rejected the American system. In Canada, it was essential that the government control economic development to protect fledgling industries from being overwhelmed economically by their powerful southern neighbours. Railroad, steel, and construction industries were funded by the Canadian government and were protected by trade barriers and preferential taxation. Unprotected, Canadian manufacturing could not have competed with American producers. American industrialists had to build factories and plants in Canada if they wanted to have access to Canadian markets. Thus, Canadian companies such as Alcan, Stelco, and Dofasco were born. This national policy in Canada during the 1870s provides a model that developing countries unable to compete in the international free market could consider today.

With the onset of the Great Depression in the 1930s, governments began to realize that many of their ideas about free enterprise might be flawed. John Maynard Keynes wrote a response to the economic problems of the "Dirty Thirties" in his *General Theory of Employment, Interest and Money.* Keynes argued that a government should be actively engaged in creating jobs to reduce unemployment during times of economic stagnation or downturn. Unemployment, hardship, and destitution, provoked by the economic crash and depression that began in 1929, caused the paradigm to shift once more. The British and American governments intervened to establish a greater role in economic policy. In France, Scandinavia, Britain, and the United States, governments tried to stimulate their economies through currency devaluation, price stabilization, and make-work projects. Despite their best efforts, governments saw little economic improvement until increased industrialization brought on by the Second World War provided employment for millions of workers. After the war, Keynesian economics fit in well with new socialist ideas. Governments in booming developed nations spent millions improving infrastructures and providing social services unheard of before the war. Airports were built, dams provided new hydroelectric power, and highways connected rapidly growing cities. Universal health care, pension plans, unemployment insurance plans, and expanded educational opportunities made people in these developed countries more prosperous and healthy than ever before.

While much of the Western world was experimenting with Keynesian economic views, Communist governments in eastern Europe and Asia were developing their own form of government intervention in the economy. Within the public-enterprise economies of these nations, governments played the primary role in the production and provision of goods and services. All resources—

Hong Kong has become one of the world's most successful examples of free-market enterprise. It is/has the world's

- third major economy;
- eighth largest trading entity;
- largest exporter of clothing;
- busiest container port in terms of throughput;
- third largest gold bullion market;
- eighth largest stock market (second largest in Asia);
- fourth busiest airport for international passengers and second busiest for cargo;
- highest per capita ownership of Rolls-Royce and Mercedes Benz.

Hong Kong reverted to China on 1 July 1997, at the end of a ninety-nine-year lease to Great Britain. It will be interesting to observe the impacts of Communist China's centrally planned economics on this "mighty-mite" of capitalism.

human, capital, and natural—and all means of production were owned and controlled by the state. Central-planning boards—not entrepreneurs or heads of industry—decided which sectors of the economy would be developed in the best interests of the state. In countries such as the former USSR, China, and Cuba, economic development within the framework of the **central-planning system** has encouraged high rates of employment and more equal distribution of goods. It has reduced the wasting of resources, and avoided the boom-and-bust fluctuations of the free market.

The post-war boom years are over! For the past decade or so Western economies have been in the doldrums. Imports outpace exports, government debts are growing astronomically, and the values of Western currencies are dropping. Keynesian economic policies encouraged overspending and overborrowing. The pendulum has swung back again. Today, economists are returning to the ideas of free enterprise originally espoused by Adam Smith 200 years ago. Planners look to the developing nations of Southeast Asia. Hong Kong especially has been an entrepreneur's paradise. Limited government interference and strong competition have made it the most dynamic economy in the world. Japan, Taiwan, and South Korea also show vigorous economic growth. The current trend in Western governments is to slash government spending. Thousands of civil servants are losing their jobs and government services are being cut back. As governments scramble to balance budgets and improve trade balances, a new era of entrepreneurship is emerging.

As nations compete in the international marketplace, there is a sense that only the strongest will survive. This is a return to the doctrine of Adam Smith's free-market economy, in which only the "strongest" (the wealthiest, the most entrepreneurial, the hardest working) will succeed and survive. This has meant that some nations, already successful in terms of annual growth in GNP, per capita annual income, etc., continue to thrive economically at the expense of other nations. **Protectionism**, trade barriers, quotas, and **tariffs** have fostered the success of some economies and have denied other economies the opportunities of free-market global competition.

Fortunately, participation in regional and international trade arrangements has encouraged cooperation between countries. Trade organizations have tempered some of the "survival of the fittest" aspects of competition in the global marketplace. GATT (the General Agreement on Tariffs and Trade), established in 1948, prescribed rules for international trade among its more than 100 member nations. Its ongoing goal is to sponsor negotiations aimed at progressively liberalizing the world's trading system.

Historically, individual nations have sought advantage in the global marketplace though participation in a variety of preferential trade arrangements. Four types of trading blocs have evolved: the free-trade area, the customs union, the common market, and the economic union. Within each type of arrangement, member countries seek the elimination of barriers that hinder trade and economic growth.

Free trade is the new paradigm for economic growth. A free-trade area is a type of preferential trading arrangement in which member countries reduce or eliminate tariffs and other trade barriers among themselves. But they maintain their individual national trade barriers and tariff schedules on goods from outside the area. The development of free-trade areas is seen by some as a vital component in boosting national competitiveness. NAFTA, the North American Free Trade Agreement of 1994, economically linked Canada, the United States, and Mexico by gradually eliminating tariffs and other barriers to the movement of goods, services, and investment among the three countries. Ultimately, the pact will create a massive open market—over 360 million people and over $6 trillion in annual output.

A customs union goes a step beyond a free-trade area by reducing or removing tariffs, quotas, and other trade barriers among themselves and establishing a common set of trade barriers for goods from countries outside the union. SACU (the South African Customs Union), which has existed since 1910, is one example of such an arrangement. A common market is a customs union that has eliminated all barriers to the movement of labour and capital among member countries, and has a common external system of tariffs. Common markets exist in the Middle East (the Arab Common Market since 1964), Latin America (the Central American Common Market since 1960), and Europe (the European Community, from 1957 to 1993). In 1993, the EU (European Union) came into effect to move countries of the European Community from membership in a common market to participation in a true economic union. In this type of trade arrangement there is full economic integration—in trade; in the movement of goods and services, and labour; and in fiscal policies (including a common currency).

CONSOLIDATING AND EXTENDING IDEAS

1 Economic theories basically come down to two divergent ideologies: government control versus free enterprise. Develop an argument to support one of these viewpoints. Debate the issue with your group or class.

2 Develop a flow chart to show the sequence of steps in the process of prehistoric/historic economic development. Use illustrations to enhance your presentation of anecdotal data.

3 Identify the twenty-four member nations of OECD (the Organization for Economic Cooperation and Development). Then investigate and describe technological production in six of the member nations.

4 Match the invention, the inventor, and the year for the following major technological inventions of the past three centuries:

INVENTIONS	INVENTORS	YEAR
adding machine	Alexander Graham Bell	1764
airplane	Harvard University	1769
assembly line	David Bushnell	1774
atom bomb	James Watt	1776
internal combustion engine	Orville and Wilbur Wright	1835
compact disc	Jack Kilby and Robert Noyce	1850
computer	Charles Babbage	1876
digital computer	William Burroughs	1879
facsimile machine (fax)	Henry Ford	1884
incandescent light bulb	Wilhelm von Röntgen	1885
microchip	Gottlieb Daimler	1895
radio	James Harrison and Alexander Twining	1901
refrigerator	Arthur Korn	1903
spinning jenny	Frank Whittle	1907
steam engine	Philips and Sony	1914
submarine	John Logie Baird	1926
telegraph (electric)	George Louis Lesage	1937
telephone	James Hargreaves	1939
television	Thomas Alva Edison	1944
transistor	Otto Frisch, Niels Bohr, Rudolf Peierls	1948
turbo jet	Guglielmo Marconi	1959
x-ray	William Shockley, John Bardeen, Walter Brattain	1979

Figure 14.3 Inventions of the past three centuries

5 The economic policies of some Communist governments have promoted development at great costs to the environment. Disasters such as the nuclear meltdown at Chernobyl, the desertification of the area surrounding the Aral Sea, and oil pipeline ruptures in the fragile taiga of Siberia are all examples of an economic system out of touch with the environment. Environments in free-enterprise economies have not fared much better. In Hong Kong, over 1.6 million cubic metres of waste flow into Victoria Harbour every day. The water quality in this city-state is fifty-two times worse than minimum European safety standards. In Taiwan, cancer rates have doubled over the last thirty years and asthma cases among children under ten have increased 400 per cent because of air pollution.

Design a research project to examine and compare the effects of free-market economies and centrally planned economies on their environments. Your design proposal should elaborate the steps you would take to develop a well-informed position on the issue. Outline the following in your proposal:
a) identification of the issue
b) alternative positions on the issue, with a brief justification for each
c) a list of questions that need to be answered to develop your own position on the issue
d) important evidence you will need to gather—examples and case studies to support alternative positions—and sources you will consult
e) analysis of the strengths and weaknesses of the alternative positions
f) synthesis of your position
Your proposal should be no longer than two pages.

6 What economic advice would you offer to each of the following about (i) increasing its national wealth; (ii) ensuring sustainable development of its resource base?
a) an emerging industrial nation
b) a subsistence-based economy
c) a diversified economy with a negative trade balance
d) an oil-rich developing country

IDENTIFYING ECONOMIC ISSUES

Economic issues have cultural, environmental, political, and resource implications. The contrasting paradigms of different cultures convey various attitudes towards earning a living. The growth of the labour union movement in the nineteenth and twentieth centuries in western Europe and North America limited the demands that employers could make on employees. Legislation was enacted to protect workers from dangerous working conditions, unreasonably long workdays, and frequent unemployment. Improvements such as better wages, hours, and safety standards resulted.

Today, in developing countries, working conditions reflect these same ills associated with industrialization. Workers, desperately in need of enough money for daily survival, are often exploited for the benefit of wealthy factory owners. Illegal sweatshops operate, with little or no attention paid to the health or safety of thousands of underpaid labourers.

An environmental point of view that relates

Figure 14.4 Strikers in New York City, 1913

to economic issues is gaining strength. Environmental concerns are frequently at odds with economic development. Some developers argue that it costs too much to protect the natural environment. Environmental protection is an added financial burden for any industry, where the objective is to maximize profits and minimize expenses. It is very costly to produce goods cleanly, but a new understanding is essential if long-term economic development is to continue. As William Clark, president of the International Institute for Environment and Development has stated: "The key to the future is . . . sustainable development. . . . To be sustainable, development cannot ignore long-term costs for short-term gains. Concern for the environment is not a luxury that only rich nations can afford. If some development project is damaging forests or soil or water or clean air, then it is not true development" (*GAIA: An Atlas of Planet Management* [New York: Anchor Books, Doubleday, 1984], 232). In fact, the roots of

environmentalism can be found in the reaction to the rise of industrialization. The original protest movements against circumstances in the mills and factories of eighteenth- and nineteenth-century Britain not only denounced the terrible working conditions, long hours, and poor wages, but also the destruction of nature.

Politics and economics often go hand in hand. Economic prosperity can mean support and re-election for governments in power; economic failure can lead to dissatisfaction among the population, strikes, demonstrations, and even overthrow of government. It has been suggested that the United States was acting as much in its own economic self-interest as in the interests of Kuwait when it spearheaded the United Nations military coalition against Iraq in the Gulf War of 1991. If oil production had been curtailed in the Persian Gulf, world oil prices would have soared, American industrial production would have become more costly, and consumer gasoline prices would have risen. The increased energy costs would have triggered an economic recession in the developed world. The United States could not afford the disruption to its economy. In developing countries as well as in the developed world, policy makers must make decisions based on the maintenance and enhancement of their national economic interests.

Resource development is essential to economic growth. Oil exploration and industrial development off the coast of Canada's Atlantic provinces may offer a return to economic prosperity for a region reeling after the loss of its main resource—cod fish. Resource development can also bring affluence to depressed economic regions outside the economic mainstream. The exploitation of natural resources in such widely disparate regions as parts of the Amazon, northern Labrador, coastal British Columbia, and the North Sea off the coast of Scotland has resulted in economic growth and a source of income for the people dependent on the resources of these marginal areas.

Whatever the economic issue, it is closely linked to other focuses—of culture, environment, politics, and resource development.

EMPLOYMENT EQUITY AS AN ECONOMIC ISSUE

Economic issues reflect cultural values and attitudes. Employment equity is a case in point. In Western societies it is considered unfair when women with the same experience and qualifications are paid less for the same type of work that men do. Yet economic inequality between the sexes remains an issue. As illustrated in Figure 14.6, rarely in either developed or developing countries are women paid the same wages as men for equal work in manufacturing.

In developing countries, women are exploited by industrialists to an extent that goes beyond economic inequality as wage earners. The emerging industrial economies of China, Indonesia, Taiwan, and Thailand are supported to a large extent by a female labour force that earns low wages. The *Investor's Guide* in Taiwan advertises the availability of low-paid female workers. According to the publication,

women's wages are 10 to 20 per cent below those of men, and the female workers form an easily manipulated workforce that accepts their work without any complaints. Young teenage

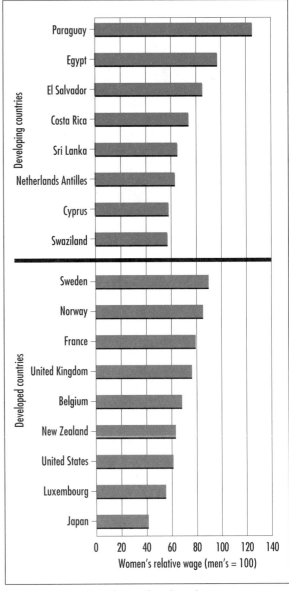

Figure 14.6 Unequal pay for equal work—relative wages in manufacturing, women/men

Figure 14.5 A child worker in a garment factory in Bangkok, Thailand

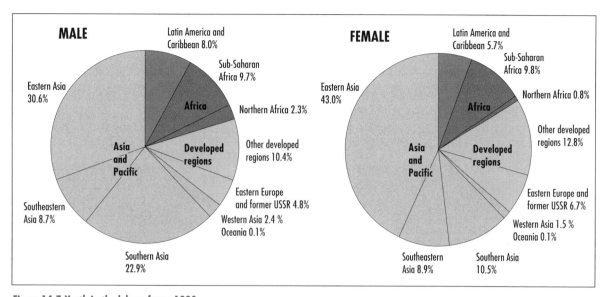

Figure 14.7 Youth in the labour force, 1990

girls work in sweatshops making any number of consumer goods throughout the Export Processing Zones of China, Indonesia, Taiwan, Hong Kong, and Thailand. In southern China young women often live in dormitories within a factory. They are not allowed to leave the factory compound, cannot associate with men, and are actually locked in their rooms at night. In 1993, nearly 2500 workers across Asia were burned to death in these fire traps, because of unsafe housing conditions.

Child labour is another issue of inequitable employment practices. In North American society today, it is illegal for children to work in factories. Nevertheless, child labour was a fact of life in Canada's early development and even today children as young as ten are expected to help out with farm chores in many rural districts. In several parts of Southeast Asia, Latin America, and Africa, boys and girls are counted on from an early age to contribute to family income. Figure 14.7 indicates that at the beginning of the present decade, the economies of many regions depended on substantial youth

participation in the labour force. In Asia and the Pacific, female participation is greater than that of males.

Attitudes towards children differ among societies. In some societies couples have many children to increase the number of wage earners in the family. However, children are often less able to concentrate on the boring piece work they are asked to perform for hours at a time. Ironically, developing countries where children are exploited usually have high rates of unemployment. The rapidly growing child population takes jobs away from adults. This results in lower family incomes and an even lower standard of living for an already impoverished economy.

There are many reasons that children are hired over adults. Children cannot fight for their rights. Trade unions are not formed where young children work, and employers are able to exploit them. Children are not allowed the same rights as adult members of the work force, so they work for lower wages and accept poor or even hazardous working conditions without question. In many countries, children are chosen to do specialized jobs—jobs that need nimble workers with small hands or bodies to work in spaces where adults would not fit. Rug weaving in southern Asia is a prime example. Thousands of girls in Pakistan, India, Afghanistan, and Iran spend weeks at a time weaving the beautiful oriental rugs sought after in North American homes. The glass industry in India employs many children to take molten glass on long rods from furnaces, where the temperature can be 700°C, to the adult glass blower. They have to run fast on floors strewn with glass shards and naked wires, so that the molten material does not solidify before it reaches the adult worker.

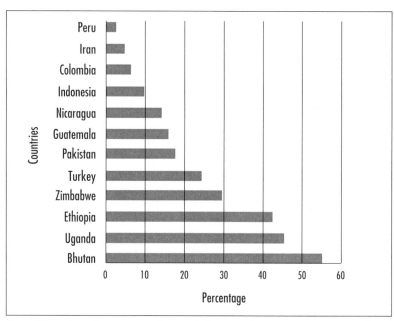

Figure 14.8 Estimated percentages of economically active children, ages 10 to 14, in selected countries, 1995

Study the graph. What generalizations can you make regarding child labour for each of the following: country, level of development, gross national product, birth rate? Why do you think some countries have more child labour than others?
Source: Statistics from ILO, 1996.

In 1975, the total world labour force of children under fifteen years of age was estimated at 54.7 million. The situation is improving as governments move to legislate against child labour and crack down on sweatshops in slum areas. It is projected that the number of child workers will drop to 37.3 million by the year 2000. Most regions are expected to show a decrease in this economic practice, except in east Asia where the "economic miracle" is progressing at the expense of child workers. The governments are more concerned with increasing industrial output than with child welfare. Furthermore, many people regard the opportunity for employment as an advantage rather than a problem.

Child labour poses many problems for both children and their countries. Many working children do not do very well at school and often drop out at a young age to pursue other jobs and agricultural activities. They never develop literacy or skills that would help them to get better jobs. Many are trapped in abusive relationships with employers. Moreover, children who work excessively long hours suffer from malnutrition, underdevelopment, and lack of care. Should economic development continue at the expense of children in developing countries?

GLOBAL RECESSION AS AN ECONOMIC ISSUE

A recession is defined as three consecutive quarters in a year where the economy does not grow. The money supply dries up, investment wanes, unemployment increases, and governments start to worry. Over the past several decades, a number of economic downturns have plagued the global economy but none have been

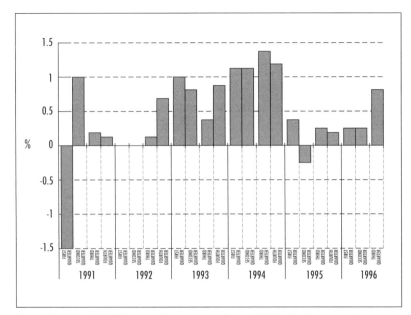

Figure 14.9 Canada's GDP quarterly percentage change (1996 prices)

A recession is defined as three consecutive quarters in which there is a decline in GDP.

Source: STATSCAN National Income and Expenditure Account (1993, 1994, 1996).

as serious as the Great Depression of the 1930s.

The Second World War spurred on economic growth, re-employment of the workforce, and increased industrialization in the 1940s. Participation in the war effort ultimately forced the economies of developed nations out of the period of economic collapse brought on by the 1929 stock market crash and the depression that followed. Manufacturing thrived as weapons, equipment, and supplies were produced for military forces. When the war ended, soldiers came home and a new wave of optimism and prosperity swept Western economies. Governments promised economic growth at home and vowed to help eradicate hunger and poverty in developing nations by funding megaprojects around the world.

Probably the main reason for the continued economic prosperity in the West resulted from legislation to liberalize banking practices. After the Great Depression, bank acts were revised to ensure there would be reserves available to pay off depositors if another depression occurred. In the economic optimism of the 1950s and 1960s, the Bank Act in Canada was relaxed. Similar legislation was passed throughout Western democracies. Banks were able to invest funds more aggressively than in the past. Also, consumer credit became readily available. Consumers, many of whom did without in the war years, wanted material goods and did not want to wait until they could afford them. Anxious to expand business with the increased funds at their disposal, banks provided almost unlimited credit. Consumer, corporate, and government debt skyrocketed during the twenty years after the war. The economy was stimulated and a stagnation such as the Great Depression after the First World War was averted in the US, Canada, and the other developed countries of the world. Of course, all this spending and increased borrowing put pressure on the money supply, so interest rates went up as did inflation. People were reluctant to save because today's dollar might be worth eighty cents in two years. Consumers of the 1970s and 1980s continued to borrow and spend lavishly.

By the early 1990s the bubble finally burst. People could no longer make purchases on future earnings potential. Inflation dropped, property values decreased, and many people found themselves unable to carry existing mortgages. With consumer spending at an all-time low and most people overextended financially, the economy entered a period of stagnation.

Figure 14.10 A resident of the Love Canal area, forced to leave his home because of chemicals seeping into the basement

Unemployment rose and the economy continued to spiral downwards. The post-war boom had finally ended.

POLLUTION AS AN ECONOMIC ISSUE

Economic growth often occurs at the expense of the environment. Because raw materials are taken from the environment and industrial waste is put into it, it stands to reason that economic activities affect the environment. While the first example (economic equity, page 218) is local in scale, and the second (global recession, page 220), continental, pollution has become an economic as well as an environmental issue, with global implications.

In the Niagara region of upstate New York, the residential area along the Love Canal became notorious in the 1970s and early 1980s.

Residents were frequently ill; congenital disabilities became endemic; and the cancer rate was many times the national average. In addition to the personal suffering, the economic costs in medical bills and absenteeism from work were substantial. The residents of this subdivision, built along an abandoned canal, did some investigating and discovered that the canal had been the dump site for a chemical company before the Second World War. When the land was filled in and houses were built on it, people forgot about its earlier use. After a much publicized investigation, the residents were evacuated and the area was sealed off. Today, twenty years later, the region has been declared safe for human habitation.

In Ukraine, the former Soviet Union was responsible for the worst nuclear accident ever to occur. In 1986 an explosion blew the roof off one nuclear power plant at Chernobyl. A meltdown of monstrous proportions was underway! Huge quantities of radioactive dust shot up into the atmosphere. Drifting westward with the prevailing winds, this dust caused widespread destruction. Four years after the accident, leukaemia and other forms of cancer continued to strike many children who had been exposed to the radiation. Sheep and dairy products as far away as western Great Britain could not be sold because they contained high doses of radioactivity. Even reindeer in Lapland were contaminated. It is impossible to know how many people's lives were curtailed by this disaster. All this took place because safety standards were relaxed to reduce operating costs.

Ten years later, the derelict power plant is a potential time bomb ready to explode again. Radioactive fuel bundles are still exposed within the plant. The radioactivity will take thousands of years to abate naturally. After the explosion, remote-control bulldozers and volunteer cement-truck drivers were used to seal the plant in a concrete sarcophagus. The sarcophagus was poorly and hastily constructed. Ten years of weathering have further eroded this structure. Scientists monitoring the plant have warned officials of the potential disaster, but Ukraine is a new nation trying to balance its economy. The government cannot afford the millions of dollars needed to repair this temporary structure. The economic costs to the people and the government are enormous. People are absent from work because of illness, and are not as productive as they might have been. The country's economic resources are being depleted as the government tries to tend to the disaster area.

Environmental controls are often limited or lacking altogether in order to maximize economic development. This may be a short-sighted view. As environments are destroyed their economic usefulness is diminished. Developed and developing nations alike must encourage and enhance models of economic growth that protect our natural environments now and into the future.

CONSOLIDATING AND EXTENDING IDEAS

1 Outline how specific economic issues relate to each of the following focuses:
   a) cultural issues
   b) environmental issues
   c) resource issues
   d) political issues

2 a) List the economic issues that are described in this chapter.
   b) For each issue describe contrasting opinions that people might have.
   c) Explain why people might have these attitudes.
   d) What is your opinion of each issue? Debate it in a group.

3 a) Review past issues (January 1995 to present) of all or some of the following periodicals to gather a list of articles dealing with economic issues:
   • *US News and World Report*
   • *World Press Review*
   • *Time*
   • *Scientific American*
   • *Maclean's*
   • *Current History: A Journal of Contemporary World Affairs*
   • *Canada and the World*
   b) In small groups, discuss the pervasiveness of economic concerns in our modern world.

4 Select an economic issue from this chapter or another economic issue of your choice.
   a) Collect articles about this issue from a variety of periodicals and newspapers. Consider using electronic information retrieval systems, which are available in some school libraries and most large reference libraries.
   b) Conduct an inquiry using the approach described in Chapter 4 on page 28.

CHAPTER 15

INDUSTRIAL DEVELOPMENT: DEFINING THE ISSUE

Industrial development is considered by many people to be the key to economic growth. In the nineteenth century, the Industrial Revolution set the course for economic prosperity in the Western world. Yet the situation today is quite different than it was 200 years ago. In today's complex economic climate, industrialization of developing countries presents challenges that are both domestic and international.

HISTORICAL PERSPECTIVES

THE FIRST INDUSTRIAL REVOLUTION

Experts on economic development maintain that the world has undergone three industrial revolutions: the first occurred primarily in Great Britain during the eighteenth and nineteenth centuries, and then spread to the nations of

Figure 15.1 Inventions from the three Industrial Revolutions

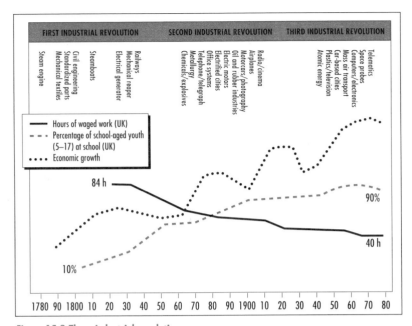

FIRST INDUSTRIAL REVOLUTION   SECOND INDUSTRIAL REVOLUTION   THIRD INDUSTRIAL REVOLUTION

Steam engine
Mechanical textiles
Standardized parts
Civil engineering
Steamboats
Electrical generator
Mechanical reaper
Railways
Chemicals/explosives
Metallurgy
Telephone/telegraph
Office systems
Electrified cities
Electric motors
Oil and rubber industries
Motorcars/photography
Airplanes
Radio/cinema
Atomic energy
Plastics/television
Car-based cities
Mass air transport
Computers/electronics
Space probes
Telematics

— Hours of waged work (UK)
--- Percentage of school-aged youth (5–17) at school (UK)
···· Economic growth

84 h

90%

40 h

10%

1780 90 1800 10 20 30 40 50 60 70 80 90 1900 10 20 30 40 50 60 70 80

Figure 15.2 Three industrial revolutions

Source: *Planet Under Stress: The Challenge of Global Change* by C. Mungall and D. J. McLaren. Copyright © The Royal Society of Canada. Reprinted by permission of Oxford University Press Canada.

western Europe; the second arose and evolved in the United States in the nineteenth and early twentieth centuries; the third revolution began very recently, in the 1950s, and is taking place throughout the nations of the developed world. Increasing economic growth during these three phases of industrial development has had an impact on populations; it has disrupted economies, resulting in social and political adjustments (see Figure 15.2).

The original **Industrial Revolution** began in the early 1700s with revolutionary changes in agriculture that greatly increased the amount and variety of food produced. Changing patterns of land ownership, new farming practices, and the invention of various types of farm machinery led to increased populations and rising demand for goods. Innovation and inventions that transformed the textile, iron, transportation, and communications industries began in Great Britain and spread throughout

Europe and North America during the eighteenth and nineteenth centuries. Initially, textile mills and factories had to be located near streams so that running water could turn mill stones and power industrial machinery. But with the invention of the steam engine, people were no longer dependent on energy produced by wind and water mills or human muscle. Wood and coal could be burned as fuel to heat water and change it into steam. The tremendous pressure created by the expansion of steam provided a new power source and energy for industrial processes.

At the same time, the introduction of larger blast furnaces revolutionized iron and steel production. The new furnaces used coal, instead of charcoal derived from trees, to smelt iron ore. This process vastly improved the quality of the finished product because the iron could be smelted at higher temperatures (allowing more impurities to be removed), and the steel hardened to a greater extent when it was forged. The steel industry expanded rapidly in regions such as northen England and the Scottish Lowlands where rich coal deposits could be exploited. From 1720 to 1839, pig iron production in Britain increased from 25 000 t annually to 1 347 000 t.

The first industrial revolution also brought sweeping changes to the textile industry. The invention of new machines, such as the flying shuttle, the spinning jenny, and the power loom, produced better-quality cloth at a fraction of the cost of homespun materials. As the domestic system was replaced by the factory system in the textile industry, cloth production dramatically increased. By 1835 Britain was manufacturing 60 per cent of the world's cotton goods. The socio-economic life in these emerging industrial economies was changed forever. People started migrating to the fast-growing industrial cities where they worked in textile mills and other factories. Entrepreneurs made great fortunes selling manufactured goods abroad and to the emerging consumer class. In time, social reforms followed, which benefited the workers who were emerging as a potent political force. Education, labour, pension, and other social reforms evolved in these early industrial nations. Education reforms required all children to attend school. Labour reforms set the number of hours a person could work each week, established minimum wages, and defined working conditions. Pension reforms obligated workers to set aside a portion of their wages for old age pensions.

THE SECOND INDUSTRIAL REVOLUTION

While Great Britain was the birthplace of the original Industrial Revolution, the second industrial revolution started in the United States with the invention of electrical power. This led to the creation of new communications technology such as the telegraph, the telephone, the camera, the radio, and the phonograph. These inventions facilitated communication across the United States and around the world. The telegraph and, later, the telephone linked isolated and distant communities. The camera replicated exact images that could be distributed in newspapers and magazines across the nation. The development of the radio further unified people in Canada and in the United States across a great continent. It could be argued that the information age we are experiencing today is simply an extension of this second industrial revolution.

The invention of the electric light-bulb and

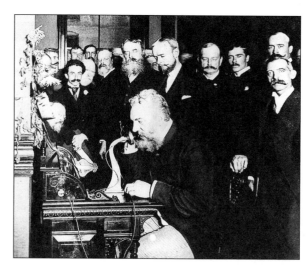

Figure 15.3 Alexander Graham Bell inaugurating the New York-Chicago telephone line

electrical engines revolutionized manufacturing. Industries no longer needed to locate near energy sources such as coal deposits because electrical power plants could now transmit energy across distances to other locations. Instead, industries began to locate close to raw materials and markets. The dispersion of industry outside of the established industrial centres resulted in the growth of new urban areas in North America.

Just as the steam locomotive was the main mode of travel by the mid- to late 1800s, the automobile became the common form of transportation in the 1900s. The invention of the internal combustion engine and refinements in steel production led to the establishment of the automobile industry, centred in Detroit, Michigan. It was now possible to transport products across land much more easily. Transportation was no longer primarily dependent on bands of steel rails needed to build railways. Automobiles and trucks could also be used. The versatility of the automobile allowed industry to expand throughout North America. Agriculture became mechanized as tractors took over much of the traditional work of people and animals. Increased agricultural and industrial production led to the rise of the consumer class. People wanted to own cars and many of the new consumer products that were flooding the stores. Automobile manufacturing offered new job opportunities to workers. In time, these job opportunities extended to entirely new industries, including car repair, road construction, insurance agencies, motels, drive-in restaurants, and many others.

THE THIRD INDUSTRIAL REVOLUTION

The third industrial revolution centred on the expansion of the computer into virtually every part of the economy in developed nations. Computers in their simplest form were actually used as early as the eighteenth century in the textile mills of France. By the 1950s they had become important as a means to collect and analyse data. It was not until the 1960s and 1970s that computers started to transform the way people thought about doing business. The information age had begun!

In the communications industry, computers allowed people to process words more efficiently than typewriters. Satellites with integrated circuits were developed to relay signals around the world in a matter of minutes. When John F. Kennedy was assassinated in 1963, news agencies flashed the tragic event to broadcasters in time for the evening news. In the print media, newspapers, magazines, and books no longer had to be manually typeset by printers; using specialized software, computers and printers were now able to produce all manner of high-quality printed materials. Television broadcasting has expanded to include hundreds of channels, and satellite dishes allow people everywhere to see their favourite programs or sporting events. Telephones have become more versatile. In the past, they had to be connected physically by wires; today, microwaves permit people to use cellular phones, which are wireless.

Tremendous growth in communications technology is changing the way that people work. The use of E-mail, the Internet, fax machines, and personal computers makes it possible for people to have access to virtually any information they need within seconds. The

Figure 15.4 Canada's "Silicon Valley" in Kanata, Ontario

nature of the workplace is also changing, as employees working from their homes are linked to other workers and clients through communications technology. Ever more sophisticated technology encourages innovative approaches to the world of work. As an example, the combination of cellular phones and laptop computers with fax capabilities has allowed people to have mobile offices.

The computer industry started in "Silicon Valley" in central California. With a location close to the American aerospace industry, a well-educated population, and an excellent climate, this region grew to be the research and development centre of the world in the 1980s. Now other regions have developed their own computer industries centred close to high-tech industries and a well-educated labour force. Computer designers increased the speed and capability of computers by transistorizing components and developing micro-electronics so that electronic pulses could travel through circuitry very quickly. Computers have became so adept at calculating massive amounts of numbers, they have revolutionized bookkeeping, accounting, statistical analysis, and all other

An interesting trend is evolving in Japan. Noted for its early use of robotics in the manufacture of automobiles, Japan developed a reputation for quality and uniformity in its cars. Today, however, many Japanese manufacturers are "retiring" their robots in favour of human workers! They have found that these miraculous machines lack the versatility of their human counterparts. Whenever a manufacturing plant has to retool, the robots have to be reprogrammed or replaced. Human workers are able to change how they perform a job more quickly and at a lower cost than machines. Also, humans have been found to be just as accurate as robots. We have come full circle in this paradigm—back to the concept of valuing human workers over machines.

forms of data manipulation.

Today, industrialization centres on the production of computers, communications equipment, satellites, robotics, and laser equipment. Unlike the first two revolutions, the third industrial revolution is not labour intensive. Advanced industrial machines and sophisticated robots perform the work of several people. As a result, the proportion of the labour force working in primary and secondary industries has declined in most developed nations, while industrial output has increased. Thus, it has become very difficult for developing nations to compete with the highly mechanized industrial production that characterizes most developed nations. Furthermore, this type of mechanized industrial production does not suit the majority of populations in developing nations. These countries have huge numbers of workers who have not had the opportunity for training and for whom employment must be found.

INDUSTRIALIZATION IN THE FUTURE

In the eighteenth and nineteenth centuries economic growth through industrialization was much simpler than it is today. At that time, an abundant supply of workers was

Figure 15.5 Truck assembly at a General Motors plant

needed to operate the simple machinery that launched the Industrial Revolution. In Great Britain, the thousands of people who flocked to the cities could readily find work. The growth of a middle class was largely responsible for much of the economic prosperity. People built up personal wealth, which enabled them to purchase consumer goods, thereby stimulating the economy still further.

Present-day technology is much more sophisticated. Machines are more advanced and require fewer workers to operate them. Modern industrial processes cannot provide jobs for the escalating number of workers in the developing world. Therefore, developing nations must employ different industrial development strategies from those that were followed by Western nations during the Industrial Revolution.

When the first industrial revolution began, there was much less competition because very few countries were at that stage of development. Today, the global marketplace is fiercely competitive. Manufacturers must offer the best value by producing high-quality goods at low prices. Economic planners need to find an economic niche for their countries in world markets. The country that produces the least expensive or best item can dominate a particular industry. For example, Singapore has become a world leader in the manufacture of personal computers.

Most industrialization of developing countries came only after the Second World War, or even later. By the 1960s, as the prices of primary goods (raw materials, cash crops) had fallen relative to the prices of imported manufactured goods, decision makers in developing countries tackled their trade deficits with their developed-nation trading partners. Policies were adopted that encouraged manufacturing. Instead of importing foreign commodities such as textiles, steel, foodstuffs, plastics, and electrical goods, efforts were made to manufacture these products at home. Industrialization/import substitution strategy became conventional wisdom. Unfortunately, reliance on imported fuel, technology, and parts; a preference for establishing assembly plants rather than production plants; and dependence on financial support from foreign corporations detracted from most benefits of industrialization.

The security of the world, whether it be food, financial, or military, depends on the creation of strong and stable economies. Many of the problems arising from industrialization, development, and global trade can be solved with the assistance of international organizations. One group most active in formulating common economic goals is the OECD, also known as the "G-24." Headquartered in Paris, it provides a forum for discussion and coordination of economic policy. Periodic economic meetings, such as the "G-7" summits, have also encouraged international economic cooperation by addressing global economic reform issues.

Strategies for economic decision making within developing nations themselves must be revolutionized. The decisions of the élite who emerged in most developing countries after their independence were conditioned by and linked to the interests of major foreign business groups. Regional policies of developing nations must be created to address issues that are in the public interest. Economic policies cannot continue to reflect the economic priorities of those concerned with international trade (multinational corporations, national entrepreneurial groups) at the expense of the local populations.

There is a basic assumption on the part of many economists and politicians that economies must grow. Yet continual economic growth is a relatively new phenomenon. For centuries prior to the original Industrial Revolution, there was little or no economic growth. Stability was the order of the day; people were not looking for ever-increasing economic gains. Pessimists would say these economies were stagnant. Optimists, on the other hand, would say they were stable. Can one achieve peace of mind in the economic rivalry common in diversified and industrial societies? Harmony has not been a feature of

Figure 15.6 A TV satellite dish in a village in western India

The Group of Seven, more commonly known as the G-7, is the economic organization of the world's seven largest free-market economies. Meeting periodically to discuss common goals and issues and to coordinate economic policy, the member nations of the G-7 are Britain, Canada, France, Germany, Italy, Japan, and the United States. The European Union is represented at summit meetings of the G-7 by an additional delegate.

the past 200 years! War, economic uncertainty, and social turmoil are all features of the modern age in which competition is valued over harmony. Is unlimited economic growth really what we want or is it even achievable? The more an economy grows, the more natural resources are exploited, the more pollution is created, and the more people want. There is a limit to the resources the planet can provide. Does the paradigm need to change so that planners throughout the world value stability rather than growth?

CONSOLIDATING AND EXTENDING IDEAS

1 Research one invention that was instrumental in launching the original Industrial Revolution. Explain what the invention did, how it affected industrial production, and how it influenced patterns of human geography. Present your findings to the class.

2 Explain why industrial development is more difficult today than it was 200 years ago.

3 a) Define the concept of economic growth. Why is this paradigm usually disruptive to environmental systems?
   b) What alternatives are there to economic development? Explain your answer.
   c) Debate in a group whether economic growth is more important than economic stability.

4 Study Figure 15.7.
   a) Prepare a multiple-line graph to show the data in the chart. What patterns do you observe?
   b) Which markets are expanding? Which markets are shrinking? Which markets do you think are most favourable for economic development? Explain.
   c) How do these statistics oversimplify the issue of economic development?
   d) Do you believe these trends will continue at the present rate or will events dictate changes? Explain your predictions and illustrate them by extending the line graph into the future with dotted lines.

YEAR	EUROPE	NORTH AMERICA	SOUTH AMERICA	ASIA	AFRICA
1750	20	-	2	65	13
1800	21	1	2	66	10
1850	23	2	3	63	8
1900	26	5	4	57	8
1950	23	7	7	55	8
2000	15	5	9	59	12

Figure 15.7 Distribution of world population (%)—selected continents

COUNTRY PROFILE: NAURU

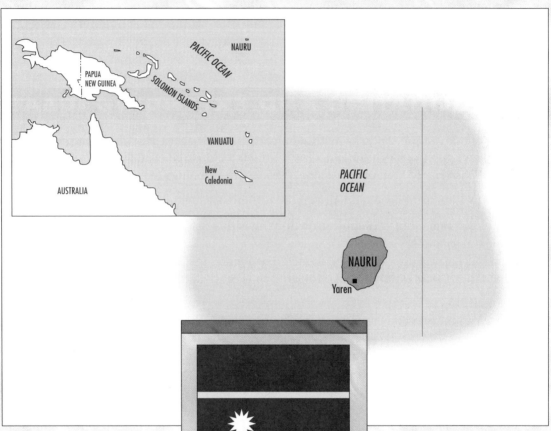

NAURU—FACTS AND FIGURES

GNP per capita	$30 000
Population	10 149
Birth rate*	n/a
Death rate*	n/a
Infant mortality**	41
Female literacy	99%
Defence expenditures	–

*per 1000 population
**per 1000 births

Figure 15.8

Source: Statistics (1990–95) from *World Resources, 1996–97.*

You might think that the wealthiest country in the world was one of the Mideast oil nations, or Switzerland, or the United States. But this is not the case. The richest people in the world live on the tiny island of Nauru in the South Pacific. Life on this island seems idyllic. Every Nauruan citizen automatically receives about $30 000 each year. Education and medical services are free and there are no taxes. Unfortunately, the present source of this island's wealth does not seem to have a bright future!

The wealth stems from phosphate deposits that are mined to make fertilizer. These deposits are the richest in the world and are derived from the accumu-

Figure 15.9 Phosphate mining in Nauru

lation of thousands of years of bird dung! However, the island is only 21 km^2, so the deposit will be depleted in ten to twenty years. The island has no other resources—not even freshwater, wildlife, or farmland. Everything has to be imported. The government is taking action now to ensure the country's future. Soil is being imported to cover the rocks in the hope that food crops eventually can be planted. Many other types of investments are being made—from hotels in mainland Europe to theatrical productions in London, England. The government is counting on these investments to support the population when the phosphate deposits are exhausted.

CHAPTER 16

INDUSTRIALIZATION: PATTERNS OF GEOGRAPHY

PATTERNS OF PHYSICAL GEOGRAPHY

Why do some regions develop industrial economies while others do not? At one extreme of this spectrum are **environmental determinists**, who believe that environmental factors such as climate and relief are responsible for economic development. At the other extreme, **environmental possibilists** are convinced that human ingenuity can overcome any environmental obstacle to development. Perhaps the reality lies somewhere in between these two points of view.

CLIMATE

Climate is one of the most important factors influencing economic development. Before the Industrial Revolution, societies were predominantly agricultural. They were able to thrive in temperate climates that provided moderate temperatures and adequate rainfall. The heat and humidity of the tropics and the bitter cold of the polar regions did not allow agriculture to flourish. Thus, societies expanded and prospered in temperate regions. Eventually they transformed from agricultural to industrial societies. As industries grew to support the agricul-

GENERAL MANUFACTORY OF THE BRUSH ELECTRIC COMPANY, CLEVELAND, OHIO.—INTERIOR OF MAIN MACHINE ROOM.

Figure 16.1 An early factory in the northeastern United States

tural base, the economy expanded, cities developed, and service industries emerged. Today, most of the world's diversified economies are found in temperate lands. They evolved from agricultural economies to industrial societies to their present diversified forms.

The view expressed above is very deterministic. The possibilist would argue that anything is possible even in the most severe climates. There are, of course, many anomalies that support the possibilist point of view. Some of the most successful economies in the world are located in tropical or subtropical regions. Hong Kong, a thriving industrial centre, experiences oppressively hot summer temperatures accompanied by heavy monsoon rains. Singapore, another industrial success story, is located close to the equator and experiences extreme heat and humidity. There are a variety of reasons that Singapore and Hong Kong have developed industrial economies, while other tropical regions have not. The presence of large markets and the proximity of trade routes are two major factors. Also significant in the cases of these city-states are political stability and a highly motivated entrepreneurial population with access to the

Even within countries, it is usually the most temperate regions that are the most highly developed. In Canada, 60 per cent of all industry is located within 200 km of our southern border. Agriculture can thrive in the warmer climates of southern Canada, whereas it is impossible in much of the north. Of course, other factors have also motivated most Canadians to live in the south; for example, the presence of good soils and the proximity of the large American market.

capital required to develop enterprises. In southeastern Brazil, the industrial triangle of Rio de Janeiro, São Paulo, and Belo Horizonte lies on the edge of a modified tropical rainforest climate, yet this region accounts for 80 per cent of Brazil's industry. It had greater potential for economic growth than other regions in the Southern Hemisphere's largest country for a variety of geographic and historic reasons. The region has fertile soils, which supported a rich plantation economy in past centuries. The discovery of mineral resources in the **hinterland** further enhanced the economic growth of the area. In addition, Rio has one of the best harbours in the world and was the first area settled in southern Brazil.

Figure 16.2 Agricultural workers in Nepal

Figure 16.3 An asbestos factory in São Paulo

LOCATION

Location often plays an even greater role than climate in the economic success of a nation. Hong Kong is a natural barren **archipelago** with what many people would consider to be an inhospitable climate. But it is strategically located off the coast of China in the path of the trade routes between this densely populated country and the West. As a result, it became an important British trading post in the nineteenth century. Over time, it developed its own industrial base and emerged as a thriving economic community. Similarly, Singapore developed because of its strategic location along the important trade routes on the Strait of Malacca. Every ship sailing to and from Asia had to pass its harbours. It was only natural that the city-state would become an important post along this route. The environmental possibilist might argue that the people of these two **entrepôts** made the most

In Africa most large cities are found along the coastal plains. Climatic factors make coastal cities more temperate. Most of these cities developed as **transshipment** ports for natural resources from the hinterland destined for export to Europe and North America.

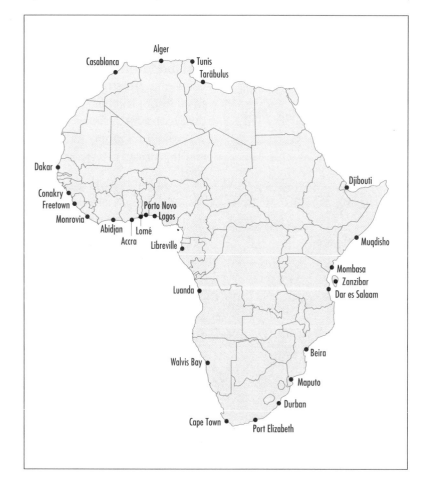

Figure 16.4 Major cities along the coast of Africa

of their opportunities despite the adversities presented by physical geography.

Brazil's industrial triangle did not develop from either its proximity to large markets or important trade routes. With one of the finest natural harbours in the world, it is little wonder that Rio de Janeiro emerged as the most important port in South America. Furthermore, Rio, São Paulo, and Belo Horizonte all surround a rich hinterland that supplies agricultural products such as beef, coffee, and sugar cane; gold, iron, manganese, bauxite, nickel, and zinc; and tropical resources such as rubber, hardwoods, and nuts. These three cities developed the industries needed to process Brazil's many natural resources. Numerous other cities around the world have prospered because of their adjacent location to natural resources.

LANDFORMS

Landforms are another physical factor that influence industrial development. Most industrial regions are located in valleys or on plains. The relatively flat terrain facilitates the construction of processing plants, transportation routes, and other aspects of the industrial infrastructure. By contrast, it is difficult to construct industrial complexes where the terrain is high in relief and elevation. The determinist points to many of the world's great industrial regions and argues the importance of the landscape in their success. Consider the Great Lakes-St. Lawrence Low-

The Great Lakes-St. Lawrence Lowlands in North America, the Po Valley in Italy, the Ganges Plain in India, and the Ruhr Valley in Germany are all thriving industrial centres located in areas of low relief.

Figure 16.5 Italy's Po Valley encompasses manufacturing centres such as Turin, Milan, and Brescia.

lands, one of the top ten most productive regions in the world. It is situated on a relatively flat lowland region with good access to transportation routes that run through broad river valleys.

Mountains and deeply eroded highlands are ill suited to industrial growth. Agriculture is also difficult in most regions of high relief, although possibilists point out some exceptions. People have modified mountain terrain by building terraces in many mountainous agricultural regions. These step-like structures provide a flat surface for growing crops. In Southeast Asia and the Andean countries of South America, terraces have created agricultural economic bases in what would otherwise be marginal land. From the agricultural base afforded by some of these mountain regions, industries such as coffee and tea processing have developed. South American cities have developed in the high plateaus between mountains. La Paz, the capital of Bolivia, Caracas, the capital of Venezuela, and Bogotá, the capital of Colombia, are all major cities located at high altitudes. These centres have become the industrial heartlands of their respective countries.

RESOURCES

Another element of physical geography that affects industrial development is proximity of resources. These may be natural resources such as water, coal, iron, or gold, or agricultural resources provided by rich farmland. Water is essential in most industrial processes for cleaning and cooling, and as an energy source. More water is used than iron in the manufacture of steel. In the production of aluminum, water power is often the main locational factor. Arvida, located near Chicoutimi in Quebec's Saguenay Valley, became the site of a huge aluminum plant because it was close to huge hydroelectric power plants.

As explained in Chapter 15, coal fuelled industrial development in Great Britain, the United States, and other Western nations in the eighteenth and nineteenth centuries. It provided energy to power steam engines and was an essential ingredient in the production of iron and steel. Today, developed countries generally prefer cleaner-burning fuels to coal, but this fossil fuel is still important to developing nations. Coal-producing areas of Columbia, Venezuela, Chile, Peru, Brazil, South Africa, Botswana, Zambia, Zimbabwe, China, India, and South Korea enable manufacturing and industrial development. In China, for example, huge coal deposits are permitting the country to build hundreds of coal-fired generating stations and steel foundries. While environmentalists oppose this industrialization because it causes massive pollution, China is merely following the path of earlier industrialization by Western countries.

Food and other agricultural products can also spur economic growth. Food-processing industries such as milling, brewing, and meat packing often develop in agricultural regions. Food-processing is a natural extension for food and other agricultural products in farm belts. Because more labour goes into manufactured goods than into raw materials, economic growth expands as raw materials are increasingly processed. Take the production of sugar as an example. The raw product has little value. But if it is refined, it is worth a great deal more and adds to a region's prosperity. If the refined sugar is processed even more, the amount of **value-added** will even be greater. For this reason, many Caribbean countries prefer to export rum (a liquor made from sugar) rather than the raw sugar itself.

OVERCOMING THE LIMITS OF PHYSICAL GEOGRAPHY

Modern technology has made it possible to modify the environment and thereby reduce the influence of physical geography on economic development. Air conditioning has enabled modern cities in even the hottest climates to flourish. Office towers are appearing throughout the Middle East, where abundant oil supplies permit air conditioning in one of the world's hottest regions. Modern hospitals, universities, and international corporations have been established in the desert regions of the Arabian peninsula. Inhospitable arctic regions are also being developed wherever natural resources are found. Oil-drilling platforms populate the stormy North Sea. Siberian mining towns and the huge nickel mining developments in northern Labrador exist because of advances in environment modification. What other examples can you think of where people have modified a hostile environment to allow economic development?

Figure 16.6 A North Sea oil exploration platform in Falmouth Harbour, England

CONSOLIDATING AND EXTENDING IDEAS

1 a) Identify the environmental factors that (i) encourage, and (ii) limit industrial development.
  b) Give examples in which industrial development occurred despite environmental limitations. What made this development possible?

2 Study Figures 16.7 (a) and (b).

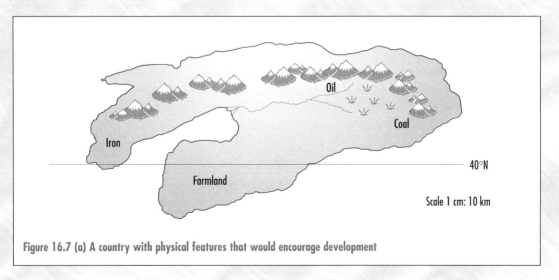

Figure 16.7 (a) A country with physical features that would encourage development

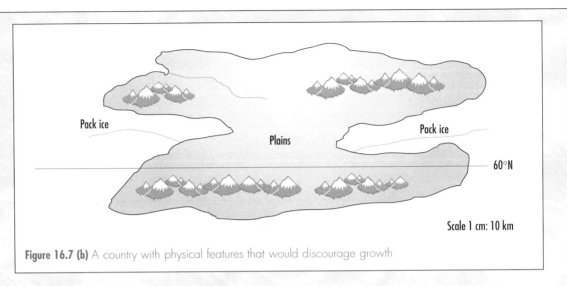

Figure 16.7 (b) A country with physical features that would discourage growth

a) In your opinion, which country has the greatest potential for industrial development based solely on environmental factors?
b) For the country you chose in part (a), find the best location for industrial development. Write an explanation for your choice.
c) Outline the way you would expect development to proceed based on the decisions you made in part (b).
d) How could environmental possibilism enable the country you did *not* choose in part (a) to develop an industrial economy?

3 a) Explain what an environmental determinist is and what an environmental possibilist is.
 b) Which view do you hold? Explain.

4 a) Using only an atlas, identify a developing country that you believe has the essential elements for industrial expansion.
 b) Draw a sketch map of the nation showing the area of development, and make an overhead of your map.
 c) Outline how each of the following factors of physical geography is favourable to development: climate, location, landforms, resources.

PATTERNS OF HUMAN GEOGRAPHY

Industrialization depends on more than just physical geography. There are many factors that either encourage or limit industrial development in various nations throughout the world. Culture, labour force and markets, transporta-tion, economic orientation, and political ideologies together determine the extent to which a country industrializes.

CULTURE

The term *culture* describes a fairly complex concept: the way of life of a specific group of people. Our understanding of culture includes ideas about how human beings interact with each other—socially, politically, and economically. A culture is grounded in its heritage, its traditions, and its norms of behaviour. It is typified by its language(s), symbols, and arts. A culture's attitudes towards education, religion, the role of the family, and the roles of women and men are reflected by its social institutions. Through its political institutions, a culture defines how it will be governed and how power will be distributed within the society.

Of course, there is great diversity between cultures in our world. Two different cultures confronting the same physical environment will deal differently with the opportunities and limitations of the surroundings. One culture may resist industrialization and economic development because they are not in the best interests of the community. Another culture may seize technology and science to speed up resource extraction and economic growth.

Industrial growth in a country often occurs when companies from a well-established industrial nation expand to foreign markets. This growth is often fostered when the attitudes and beliefs of the two nations are similar. Thailand, for example, has become an economic satellite of Japan. The dominant religion in these two countries is Buddhism and the people share similar attitudes towards work. It made a lot of sense for Japanese industrialists to expand into Thailand, and many of the successes the companies had in Japan have been emulated in this developing country. The concept of branch plants opening in countries with similar cultural values is not recent. Canada is a case in point. When the United States wanted to expand its auto industry into foreign markets, Canada was a natural candidate. The people speak the same language, have fairly similar tastes and attitudes, and are well educated and hard working.

LABOUR AND MARKETS

After the Second World War, the United States provided financial assistance to Japan so that the war-torn nation could rebuild. At first, the country produced inexpensive consumer goods: toys, textiles, and other low-technology items for the US market. In the 1950s, the term "made in Japan" was synonymous (to North Americans) with inexpensive and not necessarily well-made products. The Japanese were successful in those early years because they had a large, low-paid labour force as well as plentiful financial aid from the United States. As the nation rebuilt and became increasingly industrialized, Japanese industry expanded its steel production and shipbuilding, and more valuable consumer goods manufacturing. During the 1970s, Japanese wrist-watches, transistor radios, and cameras were traded around the world. By the 1980s Japanese products were highly regarded. High-value consumer goods such as automobiles, computers, and other electronic goods started flooding Western markets.

The availability of labour and access to markets are interconnected. Workers often provide the market for the goods they produce. Today, Japanese consumers purchase many of the consumer goods their nation produces. So industry continues to grow even though exports have stabilized as Western nations restrict Japanese exports in an attempt to balance trade inequities.

Financial aid and assistance from the US, the stability of Japanese governments after the Second World War, and trade barriers and import restrictions all contributed to Japan's economic expansion. Japan's labour force was part of a "company culture" that stressed teamwork and commitment beyond the workday and the work site. Employees typically worked six days a week and many were employed by their companies for life. Business organization, industrial flexibility, and government funding of research and development supported the rebuilding of Japan's industrial base for three decades.

The 1990s, however, have brought tough economic times to Japan. Between 1992 and 1995, Japan's GDP grew at an average of 6 per cent a year. Unemployment is rising and only 20 per cent of the workforce still enjoys life-long employment benefits programs. Multinationals are no longer investing in Japan; instead, they are turning to China, Southeast Asia, Europe, and the US.

It has been suggested that the very qualities of the Japanese labour force that encouraged economic success—obedience and selflessness—stand in the way of the country's future development. Ichiro Ozawa, leader of the main political opposition party, suggests that the

"Shuffling, stooped, mostly male, a ragged group of homeless people waits patiently for a cup of hot soup, their main meal for the day. The food line, organized by Shinjuku Renraku Kai, a citizens group aiding Japan's homeless, has grown steadily longer over the past two years. . . .

Many are older people who have lost their construction jobs and company-provided bed space. Some have been abandoned by children, a once-unthinkable turn of events in a society that has traditionally venerated the elderly.

(*Maclean's*, 8 January 1996: 21.)

"Japanese lack self-reliance and a sense of the individual without which there is no democracy or creativity . . . [What is needed] is close to a revolution. What's demanded is a change of Japanese consciousness . . . " (*Time*, 22 April 1996: 34) Devotion to employer and sacrifice for the greater good are becoming labour practices of the past as Japan undergoes it most profound restructuring since the Second World War.

TRANSPORTATION

The transportation infrastructure often determines the level of industrial success in a nation. This infrastructure includes highways, railway lines, airports, and seaports, as well as storage and warehouse facilities. Countries with good transportation infrastructures are often able to industrialize more easily

Figure 16.8 The Belém link of the Trans-Amazonian Highway in Brazil

than countries without these facilities because they enable resources to be moved efficiently from place to place. Raw materials from the hinterland can be transported to processing facilities in other parts of the country or to coastal ports for shipment to market destinations around the world.

CAPITAL AND MONETARY POLICIES

Industrialization also requires financial resources. Capital must be available in order to develop industries, and local financial institutions must be able to provide credit for day-to-day operations, otherwise a country's industrial development is seriously hampered. For this reason, many countries welcome foreign investment. Foreign investors earn profits from the new industries while the developing country creates employment, expands its manufacturing base, and increases its export income. Investors can participate in the economic growth of a developing country in two ways. First, they can lend money, which will be repaid with interest. This transaction benefits both lender and borrower. The borrower benefits from earning interest on the money invested. Many mutual funds are now investing heavily in the developing nations of east Asia. Economic growth is so much greater than in developed countries that the rate of investment return is significantly better. Borrow-

ers gain because they secure the capital needed for industrial expansion—capital that they may not be able to raise locally. However, investors may lose their investments if the venture fails and goes bankrupt.

A second alternative is for investors to provide capital in return for a share of the profits. Investors may still lose their capital if the venture is unsuccessful. But if it succeeds, they stand to make a great deal of money. Often a combination debt/equity arrangement is made. A foreign company or government puts up capital to initiate a project, and provides training and management expertise. In return, it receives a portion of the profits and repayment of the loan, or other concessions.

In addition to having investment capital, a country must have a stable monetary policy. This includes a stable inflation rate and currency and at least one local stock exchange to ensure that publicly owned shares and commodities can be traded within the country.

Many economists feel it is also important for governments to support and encourage economic expansion. This can be accomplished in a number of ways. Subsidies and tax relief are commonly used by many countries. Provincial and municipal governments in Canada have eliminated property taxes or even given companies capital funding so that they would build in a particular

> In the 1980s, rampant inflation in Argentina made investment in the country a risky proposition. The inflation rate was over 1000 per cent a year! A million dollars had the purchasing power of only $1000 a year later, and $1 the following year. Not surprisingly, Argentina experienced no economic growth during this period. Consumers in the country would spend money as soon as they earned it and not save. There was no point. This attitude had a devastating effect on Argentine money markets. It was not possible for companies to borrow money even over the short term for commercial operations.

Figure 16.9 Assembly-line workers at the Chrysler plant in Mexico City

region. Many of the huge auto plants in southern Ontario were built because of government incentives. Another tactic is for governments to pass legislation that favours industry. Many Latin American countries, for example, have less stringent pollution controls than the United States. Manufacturing companies can produce goods less expensively than in a developed country. Production costs are reduced because of lenient environmental standards and low labour costs. Brazil and Mexico both have used this policy to attract business. As a result, the air pollution in Mexico City and São Paulo is among the worst in the world.

Other legislation regarding labour regulations also

Political uncertainty in developed countries can also cause investors to be concerned about the future of their investments. In Canada, the Canadian dollar weakened on the international money markets when the issue of Quebec's sovereignty was put to a public referendum in 1995. The issue affected investors' attitudes towards Canada. They did not want to buy Canadian currency because the economic health of the nation seemed precarious at the time.

induces industry to move into developing countries. If the minimum wage is low, if labour unions are prohibited, and if regulations do not prohibit exploitation of labourers, industry will be profitable in developing countries.

POLITICS

Political stability is another important element in economic development. War, civil unrest, and political uncertainty are all deterrents to foreign investment. When a country's political future is uncertain, investors do not have confidence in its economy and are unwilling to risk their money. Political stability offers a greater guarantee that investments will return a profit in the future.

Countries that share a common heritage or have similar political views are usually more likely to cooperate in economic development. Canada derived most of its early economic support from Great Britain. When the "home country" wanted to expand its economic empire, it naturally chose its colonies for expansion. Thus, Canada benefited along with Australia, New Zealand, South Africa, and many other former colonies. Of course Britain benefited also from the raw materials the former colonies could provide; so much so that many consider Britain to have gained more from the exercise than any of its former colonies.

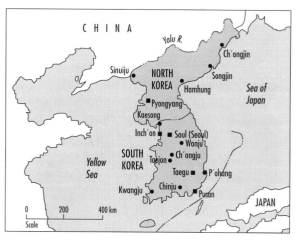

Figure 16.10 North and South Korea

CASE STUDY

South Korea

Following the Russo-Japanese War of the early 1900s, Korea was annexed by Japan in 1910. When Japan surrendered at the end of the Second World War, the northern half of the country was occupied by the Soviets and the southern half by the Americans. After unsuccessful attempts to reach an agreement with the Soviets regarding unification, the United States turned the problem over to the United Nations. The Soviets' refusal of access to their occupation zone led to the creation in 1948 of the Republic of Korea in the south and the Democratic People's Republic of Korea in the north. Two years later Northern forces invaded the South, resulting in the Korean War. Sixteen member nations of the UN supported the Republic of Korea, with the US contributing the bulk of the foreign forces. China committed itself to supporting the North, but lost heavily in its efforts to dislodge the UN forces. An armistice was signed on 27 July 1953. Korea remained divided.

In the 1960s bullock carts could still be seen on the streets of Seoul, the capital of South Korea. At that time, 70 per cent of the population were farmers. Today, most people work in manufacturing and 65 per cent of the population lives in the cities, especially Seoul, where one Korean in four resides. Industrialization began in the early 1960s with small factories that produced textiles and shoes using untrained labour. As the steel industry developed, manufacturing became more diversified (for example, shipbuilding and automobile manufacturing) and workers more highly trained to meet the needs of their burgeoning industrial economy. By the late 1980s, South Korea was pro-

Figure 16.11 Workers assembling circuit boards for telephones at an electronics plant in South Korea

ducing electronics and computers. Today, South Korea has an economic growth rate of 12 per cent per year—one of the highest in the world—and a $10 billion trade surplus.

South Korea has little advantage in terms of physical geography. It is located in a small peninsula jutting into the Sea of Japan. Winters are cold and wet. Summers are pleasant. There are few natural resources and most of the peninsula is mountainous. Located on the Pacific Rim, the country lies between China, Japan, and the eastern border of Russia.

South Korea's success has been determined more by human geography than anything else. When General Park Chung Hee gained power in 1961, he employed the services of American economists to help rebuild the war-torn country. The first thing he did was to provide an economic and monetary infrastructure that would encourage foreign investment. Park raised interest rates to very high levels and foreign money started pouring in. The entire nation became a supply depot for the American troops based in Vietnam. The Vietnam War lasted more than a decade and South Korean heavy industry grew as a result. The Korean people contributed to the capital available for investment. Koreans traditionally save about one-third of their income. This is a necessity in a country that does not have old age pensions and many other social benefits.

Several huge conglomerates evolved partly because of government support, American contracts, and foreign capital investment. Kim Woo Chong, head of the $7 billion Daewoo Corporation, credits the country's success with its ability to copy and modify designs made in other countries and, most important, to manufacture them at a lower cost than anywhere else. While wages are low and there are few social-welfare benefits, people in South Korea enjoy a high standard of living.

Politically, the country is unusual because it has been governed by a succession of dictators intent on warding off Communism. Democratic movements lead by student activists have been suppressed. Fearful of an invasion from China through North Korea, the country is in a perpetual state of alert. Located 40 km from the demilitarized zone between the rival nations, Seoul is ready for an invasion at any time. Billboards hide radar installations, northern walls in many of the city's buildings are reinforced against shell bombardment, most of the coast is fenced to discourage amphibious landings, and patrols check everything. Still, the country prospers.

CONSOLIDATING AND EXTENDING IDEAS

1 Summarize how human factors aid in industrial development.

2 Continue the case study you started on page 240 (activity 4). Research the human geography of South Korea and determine how it would influence industrialization.

3 a) Outline the factors that have led to the economic development of South Korea.
   b) What factors did the South Koreans have to overcome in order to grow economically?

4 In two paragraphs, reflect on the economic development of South Korea. In your first paragraph, suggest how this nation might have progressed if Communism had become the prevailing political belief and, as a result, all economic decisions were based on central planning. In your second paragraph, describe how capitalism encouraged industrialization in South Korea.

5 Prepare an organizer or a chart summarizing the advantages and disadvantages of foreign ownership of industry in developing countries.

6 Debt/equity swaps began in Chile in 1985. Fifteen highly indebted countries, mainly in Latin America, have followed suit to reduce billions of dollars of debt to foreign banks. When considering a debt/equity swap, governments of indebted countries must decide whether their primary goal is to increase foreign investment or to reduce debt.

   Suggest how an indebted government might (i) increase foreign investment, and (ii) reduce debt to foreign banks. Consider currency exchange, interest rates, and foreign ownership of enterprises.

CHAPTER 17

INDUSTRIALIZATION IN THE DEVELOPING WORLD

Political, social, and economic problems often impede a country's efforts at industrial development. War, civil unrest, famine, or other crises can prevent a country from establishing a development strategy. While another country may be relatively stable, it too may not be ready to embark on the path to industrialization. According to American economist Walter Rostow, this is because all countries must pass through several stages of economic growth to reach full industrialization.

STAGES OF ECONOMIC GROWTH

In 1960, Rostow introduced a theoretical model to explain how countries industrialize. Rostow's model illustrates, from an economic and historic point of view, the stages of economic development as they occurred in such countries as Britain, the United States, Canada, and Japan.

STAGE 1: TRADITIONAL SOCIETY

Traditional society consists of three classes: rulers and leaders at the top of the social structure, a small group of artisans in the middle, and the vast majority of the population at the lowest end of the social and economic scale. The majority of the people are peasant farmers. They earn a subsistence living from the land. Their way of life is stable, and they are unable to improve their status. Economic production per capita is low and is not increasing.

> When Italian merchants began to trade with their Chinese and Arab counterparts in the thirteenth and fourteenth centuries, they discovered new ideas about art, technology, and lifestyle. This triggered the cultural and economic reawakening in Europe known as the Renaissance.
>
> The 1949 Communist revolution in China marked the progression to Stage 2 in a much more violent way.
> In order to gain a greater measure of power and prosperity, the peasants had to turn to revolution and the overthrow of the government.

STAGE 2: PRECONDITIONS FOR TAKEOFF

A paradigm shift in society is triggered when the populace comes to embrace new ideas and inventions, and to demand social change. Three essential developments take place during this stage. People begin to accumulate wealth as they adapt to these new ideas. Banks, stock exchanges, commodities markets, and other economic infrastructures emerge. Improved socio-economic conditions lead to greater stability in government—people do not revolt when they are happy! The establishment of a stable government encourages economic growth. Demographic patterns change. Population growth increases as death rates decline, and birth rates remain high. The percentage of workers engaged in agricul-

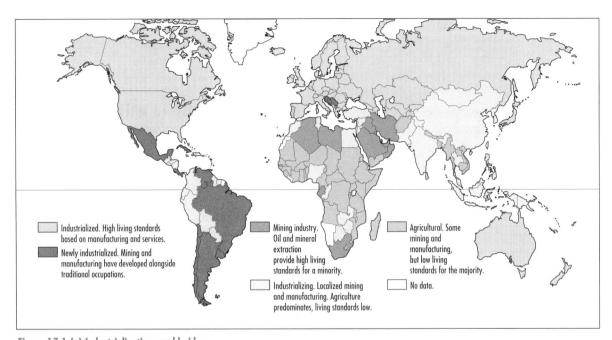

Figure 17.1 (a) Industrialization worldwide

Source: *The Third World* by Rex Beddis (Oxford: Oxford University Press, 1995). Reprinted by permission of Oxford University Press.

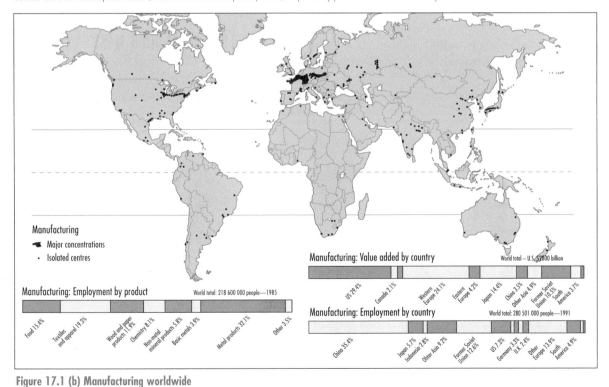

Figure 17.1 (b) Manufacturing worldwide

Source: *Goode's World Atlas* 19th ed., © 1997 by Rand McNally. Reprinted by permission of Rand McNally.

ture declines as more people migrate to the rapidly growing cities to earn a living in the new workshops. But the farming of cash crops remains an important activity as it is essential in supporting the growing urban population. This stage of the model presupposes that there are abundant energy resources.

STAGE 3: ECONOMIC TAKEOFF

As affluence increases, so too does literacy because people have more time to learn new things. They are not preoccupied with making enough money to get the food they need. They have money to spend on their children's education. Birth rates begin to decline and population growth slows down. A better-educated labour force and greater individual wealth spark an increase in manufacturing in order to meet the needs of this affluent new market. Essential consumer goods such as textiles, clothing, and iron and steel implements are produced for mass consumption. The development of improved agricultural machinery aids food production for a growing urban population. Transportation,

Figure 17.2 Industry in Hong Kong generates nearly 24 per cent of the GDP.

energy production, and economic infrastructure expand to support the growing economy. As a result, per capita standard of living increases, as measured by annual GDP and GNP.

STAGE 4: THE DRIVE TO MATURITY

During this stage, the dominant economic forces are heavy industries requiring greater capital investment and a better-educated workforce. The manufacturing of automobiles, electronics, and durable goods leads industrial growth. Light manufacturing and primary industries continue to play a significant role, but they are less dominant in this diversified economy based on primary, secondary, and tertiary industries. Universal health care, greater educational opportunities, and social security improve the standard of living. Per capita GDP is high and international trade in food and manufactured goods expands. Increased migration from rural areas leads to rapid growth of urban populations.

STAGE 5: THE AGE OF HIGH MASS CONSUMPTION

At the fifth stage the economy has shifted from a manufacturing base to a service-oriented one. Tertiary industries designed to meet consumer needs and to service the automated manufacturing sector are now dominant. Society in general is affluent, with a very high per capita GDP, and leisure is now an important aspect of society. International trade is very significant in this mainly urban society. The development of quaternary industries becomes the main area of economic growth. Society begins to question its own **materialism**. There is growing concern about the quality of life. The effects of industrial society on the environment are being seriously evaluated for the first time. Pollution, the destruction of ecosystems, and the depletion of natural resources emerge as important social concerns. Canada and other developed nations are currently at Stage 5 of Rostow's model.

CONSOLIDATING AND EXTENDING IDEAS

1 a) Using historic and geographic factors, evaluate the strengths and weaknesses of Rostow's model.
   b) What are the limitations of his model?

2 Generalizations are statements that describe relationships between two or more concepts. They allow us to discuss patterns and trends, and are used to illustrate connection, dependence, and interdependence. To describe such relationships we use phrases such as "is influenced by," "is associated with," "causes changes in," "is linked to," and others.

   Refer to the *Canadian Oxford School Atlas*, 6th ed., pages 118 to 143. Write generalizations that describe relationships between different concepts. For example, what relationship(s) can you detect between climate and economic development? Physical geography and environmental damage? Population and agriculture? Quality of life and energy?

3 Study the statistics in Figure 17.3. For each country, identify its stage in Rostow's model of industrial development. Explain your answers.

COUNTRY	BIRTH RATE (per 1000)	DEATH RATE (per 1000)	GDP PER CAPITA (US$)	% EMPLOYED IN AGRI-CULTURE	% EMPLOYED IN MANU-FACTURING	% EMPLOYED IN SERVICES	FEMALE LITERACY
A	45	42	120	91	3	6	4
B	48	22	177	70	14	16	11
C	16	6	6 462	18	27	56	94
D	20	9	5 800	13	34	53	95
E	39	14	201	57	10	34	22
F	14	8	21 561	3	19	77	99
G	23	7	2 677	29	16	55	80

Figure 17.3 Selected statistics showing stages of economic growth

4 Select one continent.
   a) Use the statistics in the Appendices to find representative countries on your chosen continent of each of Rostow's five stages.
   b) Does one level of development stand out for your continent? Describe the pattern.
   c) Form a small group with students who studied different continents and develop a general description of global development patterns.
   d) How could this type of generalization be a distortion of reality?

DEVELOPMENT STRATEGIES

Economic planners must first determine what they hope to achieve through industrialization. Is their goal to strengthen the economy by reducing foreign debt and balancing trade? Is their primary concern lowering unemployment and improving the standard of living? Any development strategy must take into account the characteristics of the country and the reasons for industrialization.

DISPERSION OR AGGLOMERATION?

Should industrial development be dispersed throughout the country, or should it be agglomerated, or concentrated, in one or two regions? **Dispersion** reduces regional disparity by providing jobs and income throughout the

> If you travel from Windsor, Ontario, to Charlottetown, Prince Edward Island, you can see a pattern of agglomeration. There are vast areas with little or no industrial development, interrupted by islands of industry in places such as London, the Golden Horseshoe of southern Ontario, Montreal, Quebec City, and Moncton.

country. It also eliminates overcrowding and urban sprawl by allowing industries to locate outside major population centres. This strategy requires government intervention. For example, in the 1980s authorities in Indonesia moved people in Java from their densely populated island to sparsely populated Borneo.

Agglomeration, on the other hand, is often related to the natural order of things. Industries tend to concentrate in certain regions because of geographic advantages such as climate, location factors, and resources. New development gravitates towards these established industrial areas because of their proximity to markets, labour, and transportation networks.

In time, however, agglomeration begins to lose its appeal. Overcrowding, pollution, crime, and the high cost of living make new industrial development in established centres more of a liability than an asset. Businesses may then decide to locate in less developed areas. Political considerations also favour dispersion. Encouraging industrial development in economically depressed regions is good for the economy. Development reduces the number of people who are unemployed or collecting welfare. It stimulates the local economy and allows for greater utilization of a country's resources. For this reason, governments often

Figure 17.4 San Diego's geographic advantages have attracted diversified industries such as aerospace and electronics equipment manufacturing, and fish canning. Its deep-water harbour is the base for a major commercial fishing fleet.

undertake projects in peripheral areas instead of in the economic heartland.

In developing countries, agglomeration is the primary development strategy because major cities have established transportation and economic infrastructures. This often leads to a dual economy. People living in the industrial heartland have a higher standard of living than those living in the countryside. But it also creates the problem of overcrowding. Today, huge cities of 10 to 25 million people are emerging in many developing nations. Overcrowding affects the quality of life. Pollution rises and municipal infrastructures collapse under the pressure. Traffic congestion makes travel difficult, and municipal water may be in short supply or of poor quality. Sanitation may decline as the sewage facilities become overtaxed. Increased population may

Rio de Janeiro is a favourite tourist destination in Brazil, with its beautiful beaches, exciting night-life, and sophisticated shops. But there is a side of the city that many tourists do not see. In some areas of the city living conditions are unenviable, many children are homeless and live in the streets, and crime is rampant. In fact, roving death squads have attacked homeless children. The government implemented a policy of dispersion in the 1960s when it moved the capital from Rio to Brasilia, but the policy has been largely unsuccessful. Residents do not want to leave the beautiful but problem-ridden former capital.

also result in inadequate police and fire protection, and strained educational and health facilities.

THE GROWTH POLE THEORY

The **growth pole theory** combines the advantages of both agglomeration and dispersion. New industry is developed in an economically depressed region that possesses certain geographic advantages but has not yet developed. The expectation is that the new industry will spur greater economic development by attracting other industries to the region. Population growth will result as people move into the area to work in the new industries. Economic development will lead to increased affluence, with a higher standard of living, higher rates of literacy, and a better quality of life. The new markets that are created

In the next century many of the world's largest cities will be in developing countries. Figure 17.5 indicates some of the fastest-growing cities.

	POPULATION 1990	PROJECTED POPULATION 2000
Tokyo	25 013 000	27 956 000
São Paulo	18 119 000	22 558 000
Bombay	12 223 000	18 142 000
Shanghai	13 447 000	17 407 000
Mexico City	15 085 000	16 190 000
Lagos	7 742 000	13 480 000

Figure 17.5 Some of the world's fastest-growing cities

will encourage service industries to locate in the area. In the end, the region prospers. Some developing countries have applied this theory to disperse economic development around the country without resorting to centrally planned economies.

The growth pole approach has been used by many countries with varying levels of success. Mexico used it to develop coastal Yucatán Peninsula in the region we know as Cancun. This coastal tropical rainforest had a sparse population descended from the ancient Mayans. Subsistence agriculture was the prevalent economic activity. The region lacked natural resources and was an economic backwater. But it did have certain geographic advantages. It lies adjacent to the Caribbean Islands—a tourist haven that attracts millions of vacationers each year. It has beautiful beaches and a hot, sunny climate in winter. The Mexican government provided tax incentives and the infrastructure for large American hotels to build in the area. The result has been the development of one of the most prosperous tourist regions in the Western Hemisphere. Many Mexicans are now employed in the resorts that have grown up on the edge of the jungle. The economic growth has spurred infrastructure jobs such as construction, police and medical services, hotel support services, and the food and beverage service.

In Brazil the same approach was used in the development of the Carajás region. Enormous iron deposits discovered in the 1980s led to the establishment of a huge iron mine, smelters, and related support industries. The intent was to create a strip of development along the railway line that links the mining region to its Amazon port, Carajás. The project has been enormously successful from an economic point of view. People have moved to the region and are employed in relatively well-paying jobs. And the companies that developed the region have made great profits. However, from an environmental point

of view the project has been a disaster. To facilitate mining operations, the tropical rainforest that once surrounded the area has been totally destroyed. The Tucuruí River was dammed to provide hydroelectricity. Native peoples were displaced when their land was flooded by the dams. They were unable to fish in the lakes that were formed because pollution resulted when the trees slowly rotted in the water. The construction of a railway line to Carajás has led to urban industrial sprawl through the jungle. Small workshops and local industries that depend on the rainforest have evolved. Environmentalists around the world have protested the destruction of this region.

The growth pole approach has some real advantages. Nevertheless, developers must realize that development in marginal areas often causes disruption to ecosystems. The approach requires government initiative and tremendous capital funding. While countries that do not have the money cannot initiate this approach, international lenders such as the World Bank and multinational corporations are often anxious to get involved in a potentially lucrative investment.

IMPORT SUBSTITUTION

A major economic problem faced by many developing countries is a negative balance of trade, or a **trade deficit**. The primary exports of most developing nations are low-value raw materials or cash crops. Yet most of their imports are high-priced manufactured goods. Consequently, more money leaves the country as import payments than enters it as export income.

To offset this trade imbalance, some developing nations turn to establishing their own manufacturing base in order to produce the products they need locally. This is called **import substitution**. This practice eliminates the high cost of importing manufactured goods, and also strengthens the local economy because more money remains within the country.

Today, several developing nations practise import substitution. India has developed its own transportation industries rather than rely on the products of British, American, and Japanese multinational corporations. Everything from bicycles to locomotives is produced in this rapidly growing industrial nation. Brazil has used its tremendous agri-

Canada practised import substitution in the early twentieth century. Steel plants were established in Hamilton and Sault Ste. Marie, Ontario, even though the costs of producing steel there were higher than in the United States. Protective trade barriers and government subsidies allowed these industries to remain competitive and play an important part in Canada's industrial development.

cultural potential to reduce its dependency on imported oil. Sugar cane is grown on huge plantations. This common crop can be converted into alcohol, which can be used as a substitute for gasoline.

The drawback to import substitution is that production costs are higher, quality is often lower, and future growth is limited because the industry is designed solely for the domestic market. To protect the industries, artificial trade barriers must be erected. These barriers often make

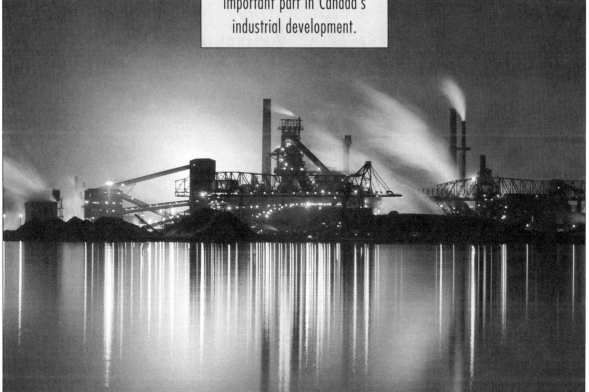

Figure 17.6: Stelco steel plant in Hamilton, Ontario

Figure 17.7 Aswan High Dam in Upper Egypt
Lake Nasser, the reservoir created when the dam was built, provides 25 per cent of Egypt's hydroelectric power, but also catches most of the silt that had formerly replenished the Nile Delta.

locally produced materials much more expensive than they would be if products were able to flow from one country to another without restrictions. Often, it is low-wage earners who suffer when trade barriers are erected because they have to pay inflated prices for a product that could be imported more cheaply. Studies conducted by the World Bank show that export-based economies grow more quickly than those that practise import substitution. Still, for some developing nations import substitution has been a successful path to industrial development. It certainly helped Canada to industrialize in the early twentieth century. Located next to the world's largest economic entity—the United States—Canada could never have competed with its neighbour. The American steel mills with their huge economies of scale could consistently outbid Canadian companies for big contracts. Similar situations exist in other countries such as India and Brazil, where it is necessary to protect budding industry from bigger, more powerful, enterprises.

MEGAPROJECTS AND GRASS-ROOTS DEVELOPMENT

Should development centre on megaprojects such as massive hydroelectric power plants and industrial complexes? Or is development money better spent on small projects that benefit people living in local villages?

Immediately following the Second World War, megaprojects were considered the best means of achieving industrial development. Millions of dollars were spent on these massive enterprises. In theory, it was believed that these funds were well invested because in the long run local citizens would benefit from a more highly developed infrastructure. Increased opportunities through the establishment of industries would trickle down to everyone in society. In reality, this was seldom the case. People living in local villages rarely derived any benefit from the huge development projects. The sponsors of the projects generally gained considerable prestige from their accomplishments.

The foreign investors usually made a generous return on their money. Members of the country's ruling élite were often the greatest beneficiaries because they owned companies that supplied construction materials or they gained a controlling interest in the completed venture. The only people who did not benefit were those who needed help the most.

Today, development strategies take a different approach. **Grass-roots development** is considered a more effective way of improving living conditions among people in remote villages. Small-scale initiatives in the local community encourage agricultural development and the establishment of **cottage industries**—small-scale operations in which people do simple manufacturing tasks at home. Local incentives increase prosperity and contribute to growth in the national economy. This type of development is especially appropriate in societies that are in the first or second stage of Rostow's model.

Theories of industrial development correspond to those of foreign aid. The foreign-aid concept of "trickle-down" holds that funding should go to the government and large corporations. It will then trickle down through the economy and help people at all levels of society. In actual fact, little financial aid actually reaches the bottom levels of society. Many NGOs today subscribe to the notion that aid should go directly to people in the villages and "bubble up" through the economy. This approach has done much to alleviate suffering and instil dignity in millions who have not benefited from aid in the past.

CASE STUDY

The Native Peoples of Peru

Descended from the ancient Inca, the Quechuan-speaking peoples of the Andes have long subsisted by growing grain on their terraced fields and raising goats, sheep, and llamas. The women are noted weavers who make beautiful handicrafts from the wool their animals produce. Ponchos, ski hats, and shawls feature traditional patterns and intricate woven designs. However, the villages where these people live are so remote that few outsiders have access to these goods.

In recent years women have organized themselves into cooperatives. Each month one or two llamas loaded with woven crafts head down the mountain to market towns such as Chivay. From here they are sent to port cities such as Arequipa. The goods are exported to cities throughout the world. The women work together to market their crafts so that they can sell directly to retailers. Because many women work together they can provide the quantities needed by foreign suppliers. The money earned from selling the handicrafts buys plastic film, which is used to build greenhouses so that the people can grow green vegetables and improve their diets. Corrugated sheeting is also brought up the mountain to fortify thatched roofs. In addition, specialty foods such as tea and sugar are purchased. The experience of these Native peoples is being replicated around the world in places such as Nepal, Mozambique, and Ethiopia.

CONSOLIDATING AND EXTENDING IDEAS

1 Many nations of the world are struggling to deal with debt. Some are in a better position to deal with their debts, while other nations may never be able to pay off their debts. Using only the information in Figure 17.8, decide which countries are in better shape than others with respect to repaying their debts. In a paragraph, summarize your conclusions and explain your thinking.

EXPORTERS OF NONFUEL PRIMARY PRODUCTS

Afghanistan	Madagascar
Albania	Mali
Argentina	Mauritania
Bolivia	Myanmar
Burundi	Nicaragua
Côte d'Ivoire	Niger
Cuba	Peru
Democratic	Rwanda
Republic of Congo	São Tomé & Principe
Equatorial Guinea	Somalia
Ghana	Sudan
Guinea-Bissau	Tanzania
Guyana	Uganda
Honduras	Vietnam
Liberia	

EXPORTERS OF FUEL (mainly oil)

Algeria
Angola
Congo
Iraq
Nigeria

EXPORTERS OF SERVICES

Cambodia	Jordan
Egypt	Panama
Jamaica	

EXPORTERS OF MANUFACTURED PRODUCTS

Bulgaria	Poland

DIVERSIFIED EXPORTERS

Brazil	Ecuador	Mozambique
Cameroon	Kenya	Sierra Leone
Central African	Laos	Syria
Republic	Mexico	

Figure 17.8 Severely indebted countries by export categories, 1992

Source: Copyright © World Eagle 1995. Reprinted with permission from World Eagle, 111 King St., Littleton, MA 01460 USA. 1-800-854-8273. All rights reserved.

2 Read the article "The Deadly Ways of the World Bank" by Canadian geneticist and broadcaster David Suzuki, on page 259.
   a) Does Suzuki express a bias? Explain your answer.
   b) What is the World Bank's point of view about the development of Papua New Guinea's forest resources?
   c) Do you think the local people are benefiting from development, or are they being exploited? Explain your answer.
   d) What are the development strategies in Papua New Guinea? What alternatives does Suzuki propose?

e) Reread the article. Reflect on the development of forest resources in Papua New Guinea. Using a mind map/concept web format, address one or more of the following issues as it relates to the Papua New Guinea scenario.
   i)   Should developed countries set the standards for quality of life?
   ii) To what extent should environmental concerns restrict economic growth?
   iii) Should economic growth and development be a primary goal for all nations?

   The issue could be placed in the centre of your visual, with related topics, ideas, and examples radiating from the centre.

THE DEADLY WAYS OF THE WORLD BANK

Port Moresby, Papua New Guinea—When this nation gained independence in 1975, it was encouraged to accept development loans from the World Bank, invite foreign investment in mines and logging, and generate revenue from coffee and oil palm plantations.

After the Arab oil embargo in the 1970s and the collapse of world coffee prices in the '80s, the World Bank forced the country to accept a structural adjustment program. Bank advisers were sent to oversee cuts in social services but the debt has grown to US $1000 per capita and now consumes a quarter of the national budget on interest alone.

Mining and logging generate enormous sums (US $10 billion and $20 billion a year), but most of the money flows directly or indirectly out of the country. Now the World Bank is pushing a land mobilization program that will allow people to use the land as collateral for loans.

Money is a new and alien concept to many of these tribal people and even where they still live traditionally, clan leaders can be bought off for a fraction of the value of the trees and minerals ultimately extracted. Denuded land and polluted rivers are often the legacy left after the companies move on.

It is criminal. People from foreign-owned companies lie, bribe, and cheat landowners to gain access to resources for a pittance.

I met people who had sold logging rights for 500 kina (Canadian $500) per family when a single tree fetches far more than 500 kina. Most skilled logging and mining jobs go to outsiders. Most politicians turn a blind eye to what is going on and are among the country's wealthiest people.

In a speech in Port Moresby, I suggested that by any objective assessment of natural resources, the country must rate among the wealthiest nations on earth. If Papua New Guinea could push the World Bank and transnational companies off their backs, I suggested, then they could define development that makes sense in their culture and chart their own course into the future.

The next day, I was challenged by Ajay Chhibber, World Bank division chief for Indonesia and Pacific Islands. He told me I was misinformed and demanded an opportunity to set me straight. When I entered my room the phone was ringing. It was Pirooz Hamidian, senior country economist, East Asia Department, World Bank.

I met them at 6:30 a.m. and Chhibber

began by saying: "You and I are well off. We don't have the right to deny the poor people of the world an opportunity to improve their lives." I agreed. But who is really "poor" and how do we define "improvement," I asked. What are the fundamental assumptions underlying the kind of economics the World Bank is forcing onto the so-called "developing" world?

I suggested that global economics overvalues human capital while externalizing natural capital. That's why forests and rivers, for example, that have provided a people with a living for millennia, only have economic value when humans "develop" them.

There are alternatives to the World Bank's ideas. All nations share the same biosphere. Papua New Guinea's forests remove greenhouse gases we produce disproportionately, so it's in *our* interests to pay the country to maintain its forests.

We can encourage forest protection by ensuring the country gets full value for wood cut sustainably. And we can inform people in the South by our experience that a high-consumption, profit-driven way of life has enormous social and ecological costs.

To this Chhibber retorted: "People are better off now than they have ever been. There's more food than ever before in history, and it's because of economic development."

But, I said, leading scientists of the world are warning about ecological catastrophe. Chhibber replied: "Scientists said twenty years ago there would be a major famine and large numbers of people would die. It never happened. They often exaggerate."

His facile dismissal of scientific expertise and some 10 million deaths by starvation annually reveals why the World Bank mentality is so deadly.

David Suzuki. From the *Toronto Star*, 12 August 1995.

THE DECISION-MAKING PROCESS

People make decisions every day, from what clothes we should wear to how we should research and present our geography assignment. Simple decisions are often made on a whim. We wear certain clothes on a particular day simply because we *feel* like wearing them. This approach is fine for decisions that do not have long-term implications. Other decisions require more careful thought. In deciding which courses to take at school, for example, we need to consider our interests and abilities because this decision will affect our academic future as well as our career.

How should we evaluate choices and make reasoned decisions? In any decision-making process, we must first establish our alternatives. Suppose you want to buy a car. The first step is to decide what kind of car you would like. Once you have considered the alternatives, you have to choose the type of car that fits your needs. Then you have to determine the criteria that will help you reach a decision about which make of car to buy. When buying a car, most people consider cost, performance, reliability, fuel economy, design, and so on.

Each criterion can be rated in an organizer or a chart similar to Figure 17.9. The alternatives are listed across the top; the criteria are listed down the side. Each alternative should be rated from 1 (the least preferred) to 6 (the most preferred). For example, let's say that cost is a major factor since you have only limited funds. The subcompact car is the least expensive, so it would rate a 6; the luxury car is the most expensive, so it would rate a 1. You evaluate each car until all criteria have been rated, then you add up the figures for each alternative. The car with the highest total is the one that best meets your needs.

Sometimes one criterion may be more important than another. If this is the case, a **weighting factor** should be included to give emphasis to the more important criteria. For example, you may consider performance a more important consideration than cost. Therefore, you might add a weighting factor of 5 to performance, which means that you would multiply the scores for this category by 5. Add up the figures for each alternative. The car with the highest total is the one that best meets your needs. Usually, the weighting criteria are based on a scale of 1 to 10, and more than one criterion can have the same weighting, as shown in Figure 17.9. In this example, the best choice is the subcompact car, followed by the truck/sports utility vehicle, the classic car, the luxury sedan, the van, and the sports car. Of course, the best choices will differ according to the weighting and criteria that are selected.

CRITERIA/ ALTERNATIVE	WEIGHTING	SPORTS CAR		CLASSIC		SUB-COMPACT		TRUCK OR SPORTS UTILITY		VAN		LUXURY SEDAN	
		Rank	Weighted	Rank	Weighted	Rank	Weighted	Rank	Weighted	Rank	Weighted	Rank	Weighted
Cost	10x	2	20	5	50	6	60	4	40	3	30	1	10
Reliability	8x	2	16	1	8	3	24	5	40	4	32	6	48
Performance	4x	6	24	3	12	1	4	4	16	2	8	5	20
Fuel economy	8x	1	8	5	40	6	48	3	24	4	32	2	16
Image	2x	5	10	4	8	1	2	3	6	2	4	6	12
Ease of operation	4x	5	20	4	16	6	24	2	8	1	4	3	12
Usefulness	4x	1	4	3	12	2	8	6	24	5	20	4	16
TOTAL		22	102	27	146	25	170	27	158	21	130	27	134
			6th		3d		1st		2d		5th		4th

Figure 17.9 Decision-making organizer

CONSOLIDATING AND EXTENDING IDEAS

1 a) Use Figure 17.9 to decide which vehicle would be best for you.
  b) Complete the organizer for (i) a family of four, (ii) a single person, (iii) a retired couple.

2 Prepare a new decision-making organizer or chart to determine your choice for one of the following: (i) a summer vacation, (ii) a university, (iii) a career. Present your findings to your group or your class.

3 Discuss how a decision-making process might help in reaching decisions about industrial-development strategies. Consider a list of criteria that could be used. These might include
  • criteria that relate to your values and attitudes;
  • criteria that should solve the problem;
  • criteria that help to preserve the environment;
  • criteria that benefit all members of society.

CASE STUDIES

Industrial Development in India

Over the past thirty years, India has embarked on many development schemes. As you read about some of these projects, consider which ones were the most appropriate for India at the time the development occurred. Use Figure 17.15 on page 269 to evaluate each project.

BACKGROUND

India is a fascinating country of many contrasts. With a population of 846.2 million, it is the second-most populated country in the world, after China. The population of India is expected to surpass that of its giant neighbour in the next century. By the year 2025, it is projected to have a population of 1.4 billion people. Yet parts of India are barren wastelands where few people live. The extreme northern limits of the country include the mighty Himalayas, the tallest mountains on earth. Covered with snow year round and devoid of significant vegetation in its upper reaches, this mountainous northern section is virtually unpopulated. In the northwest state of Rājasthan, desert and semi-desert provide such a small resource base that population density is very sparse. Farther south on the west coast, the rugged Western Ghats create a rain-shadow that makes the southern interior a difficult land to farm, though it supports a moderately dense rural population. Nevertheless, much of India is extremely fertile. The Indo-Gangetic Plain in north-central India, the delta lands of West Bengal in the east, and Gujarat in the west provide rich alluvial sediments as well as abundant rains

Figure 17.10 The many faces of India: *top left*, villages in the foothills of the Himalayas; *top right*, the Maruti car plant in northern India; *bottom left*, overcrowded conditions in east Delhi; *bottom centre*, highrises in Bombay; *bottom right*, traffic congestion in Delhi

in the summer monsoon. These fertile lands are where most of the people live.

India was a colony of the British Empire from the late eighteenth century until 1947, when Mahatma Gandhi, considered to be one of the greatest patriots of modern times, led the people in a revolt of peaceful non-cooperation. Although the country was exploited by the forces of imperialism during the colonial period, the British established transportation and economic infrastructures that

benefit India today. Since independence, industrial expansion has been facilitated by the excellent transportation infrastructure. India is a land in which incredible poverty exists amidst rapid economic growth. So, while some areas are experiencing economic growth, other regions are suffering from the poverty that has plagued much of India for centuries. The country's sizeable population makes it the seventh largest world market. India has implemented a program of import substitution to encourage local industry. With such a large domestic market, industry can flourish through sales to Indians alone. Competition with other countries for foreign markets is not as important as it would be for less-populated countries. An emerging industrial power, India is still an agrarian society with most people engaged in subsistence agriculture. The nation is poised for swift economic growth in the twenty-first century.

Prior to independence, India was primarily an exporter of agricultural products to Great Britain. Since then, industrial development has been an economic priority. To stimulate this development, economic strategists applied the growth pole theory to establish four industrial centres near existing cities. One was established in the north near Delhi, the second in the west between Ahmadabad and Bombay, the third in the south around Bangalore and Madras, and the fourth in the Damodar Valley upriver from Calcutta. These industrial centres were supposed to stimulate the economy, encourage growth, and provide economic infrastructures. Later, small-scale projects were initiated in the villages to distribute more widely the advantages of industrialization and to reduce urban migration and regional disparity. It was also hoped that the greater affluence and higher literacy rates that would result from industrialization and better educational opportunities would help to control population growth.

THE DAMODAR VALLEY SCHEME*

The hinterland upriver from Calcutta has many advantages for heavy industry. Like Britain and the United States, it has substan-

Figure 17.11 Steel industry in the Damodar Valley

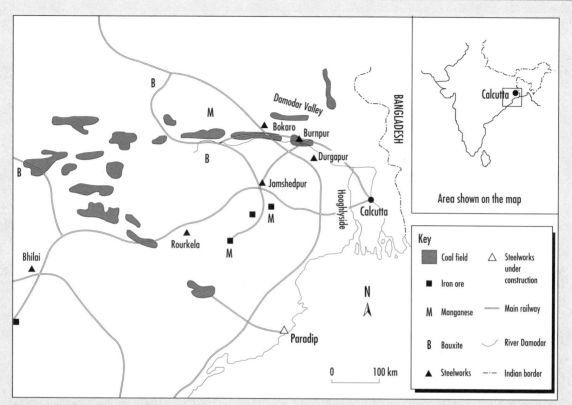

Figure 17.12 The industrial northeast of India

tial coal reserves, iron ore, and other industrial minerals. With a good railway system, financial markets in Calcutta, and an extensive labour pool, the region seemed a natural choice for development. However, its physical geography had limited development in the past. Monsoon rains and unsustainable agricultural practices caused soil erosion and a decline in agricultural output. As a result, the region was economically depressed and many rural peasants migrated to Calcutta annually in search of work.

The first three five-year plans that India implemented in the 1950s and 1960s (from 1950 to 1955; from 1956 to 1960; and from 1960 to 1965) were devoted to the development of the Damodar Valley. The first stage of the plan was to dam the Damodar River. This reduced flooding during the monsoons and provided water for irrigation during the dry winters. More important, the dam was the first step towards the creation of a hydro-electric generating station.

The second stage of the plan was to establish mines for the extraction of coal, iron, and other industrial minerals. Since the deposits were large and close to the surface, low-cost open-pit mines could be operated. The existing railway lines provided a means of transporting the raw materials to smelters and steelworks.

The third stage of the plan was to

develop iron foundries and steel plants. But the cost of constructing these facilities proved to be much more than planners had originally estimated. As a result, the foundries and plants were never fully developed and the project was a failure. Planners had hoped the project would increase employment and stimulate heavy industry. The project never created a **multiplier effect**. Prosperity did not trickle down through society as a whole.

Several other factors contributed to the failure of the project. Funds slated for the scheme had to be redirected to importing food, following several devastating famines. An ongoing border dispute between India and Pakistan meant that further funds were siphoned off to pay for military confrontations with India's neighbour. While there was a large workforce available, most of the people lacked the education and training needed to perform specific industrial tasks. In the end, the only people who benefited from the scheme were the investors and the engineers who planned and managed the project. It was finally completed in the early 1970s with funds from foreign investors. Although the scheme was widely criticized by political opposition parties and low-paid working people, it provided the heavy industry needed to expand the economy. Today, India is one of the major steel exporters in Asia. Many national industries owned by wealthy industrialists use Damodar steel to produce their manufactured goods.

THE TELCO TRUCK WORKS*

Though not funded by the government of India, Telco truck works is a local, privately funded industry developed with considerable government encouragement and financial sup-

port in the form of tax concessions and favourable labour laws. Started by the Tata family on a site near the city of Poona southeast of Bombay, the truck works is an Indian operation. Most of the equipment used in manufacturing the trucks is designed and made in India. The plant utilizes an assembly line system that is labour intensive but requires limited training. The trucks produced are smaller than North American trucks, and are extremely sturdy. This makes them well suited to the road conditions in India. In addition, they are more economical to operate and require only basic tools and equipment for repairs. They sell well in India and throughout Southeast Asia as they are appropriate to the level of development in these markets.

The truck works currently employs 13 000 workers and has plans for expansion to increase production. Telco has had a positive influence in stimulating the local economy. The company provides loans to its workers to buy land and build homes. It also loans money to small business ventures that benefit the truck works and the local economy in general. For example, one workshop that produces wooden seats for the trucks employs 350 people. Employee-run training workshops for new workers are funded by the company. The rural economy also benefits from development in this industrial centre. Farmers are able to sell their produce to the more affluent city dwellers. Many factory workers send money to their families, which helps to develop the rural infrastructure.

Many people gain from this enterprise. As in any capitalist venture, the owners of the company, in this case the Tata family, benefit from its success. The factory workers have steady work and some degree of security, as well as the opportunity to obtain low-interest

Figure 17.13 The TISCO steel mill, the largest sector in the Tata empire

THE HERO BICYCLE COMPANY*

The Hero bicycle company is the largest producer of bicycles in the world outside of the United States. Located in the town of Ludhiana in the Punjab, the factory produces 5000 bicycles a day. It actually makes only the frame, crank, and wheels for bicycles. The other parts—seats, pedals, handlebars, and so on—are made in small, privately owned workshops in the town. Many of these subsidiary plants are run by former employees of the bicycle works, who used their earnings from the plant to start their own businesses.

Hero employs over 5000 workers and is growing rapidly. It plans to diversify its operation by expanding into the production of mopeds. These mopeds are actually bicycles with two-cycle engines attached. The company has also signed a contract with Honda to produce motorcycles. The manufacturing of four-wheel vehicles is another possibility for the future but substantial capital investment will be needed.

The success of the Hero factory has been based on a number of factors. The manufacturing of bicycles in India is labour intensive,

loans. Agricultural workers on the rural outskirts profit from increased sales of farm products and from improvements they are able to make to their farms with money from relatives working in the factories. The government receives tax revenues from the sale of the vehicles abroad.

Figure 17.14 A bicycle mechanic in Bombay

unlike North American assembly plants where automation is commonplace. The enormous low-paid labour force enables the company to produce the bicycles inexpensively, with a minimum of capital invested in costly machinery. The bicycles are not assembled at the plant; instead, the parts are sent as kits to retailers, who assemble them in their shops rather than pay a higher price for assembled bicycles. A large number of children and adult workers with less training are employed in the smaller works that produce items for the Hero factory. While child labour is banned under Indian law, small workshops are rarely investigated so the practice goes undetected.

*Adapted from N. Law and D. Smith, *Decision-Making Geography*, 2d ed. (London: Stanley Thornes Ltd., 1991), 107–108, 112. © Stanley Thornes (Publishers) Ltd. Reproduced with permission.

OPPORTUNITY OR EXPLOITATION?

Some people might argue that the local workers in these companies are being exploited because they earn a fraction of what North American workers are paid. Working conditions are poor—long hours are spent in badly ventilated conditions. Moreover, there are no unions to protect the workers' interests. However, in a country where unemployment and underemployment are endemic, the employees themselves may view their jobs from a different perspective. Factory work may be considered a good job opportunity. It is better paying than employment in other workshops or than agricultural labour. Faced with high unemployment and few job opportunities, most workers hope to keep the jobs they have. This situation applies to most manufacturing in India and, indeed, across south Asia.

CONSOLIDATING AND EXTENDING IDEAS

1 a) Compare the three industrial development projects in India in a decision-making organizer or chart similar to Figure 17.15.

   b) In small groups debate the merits of each development project.

CRITERIA/ALTERNATIVE	WEIGHT	DAMODAR		TELCO TRUCK WORKS		HERO BIKE CO.	
		R	W	R	W	R	W
Profitability							
Employment growth							
Working conditions							
Marketability of the product							
Suitability to worker training							
Infrastructure growth							
TOTAL							

R: Rank   W: Weighted score

Figure 17.15 Case studies: decision making

2 For this role-play assignment work in groups of five. Your group will include a company president, a factory worker, a government agent, an unemployed worker, and a minister of finance.

   You are hoping to convince the World Bank to finance expansion of current operations in your industrial development project. Collaborate with your group members to develop persuasive arguments that will be presented to a committee of the World Bank.

   a) Each group member should be assigned one of the above roles.

   b) Brainstorm the details of your industrial development project and reasons for additional financial support.

   c) Do some research. What are the aspects of this industry that will help you to be persuasive about expansion?

d) Develop each role-player's position.
e) Rehearse your presentation.
f) Perform the role-play for the class.

3 Study Figure 17.16. It compares two development models discussed in this chapter.
a) Prepare a decision-making organizer or chart to decide which choice would be preferable.
b) You are a member of the federal cabinet of India. Your main responsibility is economic development. You believe that textile manufacturing is an important area of economic growth, but you must decide which economic strategy would be more effective. Should you support the growth pole theory and construct a new factory in an undeveloped part of the country near Nagpur? Or should you encourage grass-roots development and support the growth of small cooperatives around the country?
c) Write a press release announcing your decision.

	FACTORY MODEL	VILLAGE DEVELOPMENT MODEL
Organization	• modern factory • highly automated, using Indian-made equipment • 400 employees • high wages • no child workers	• cooperative system • spinning, dyeing, weaving done by hand in village setting • 400 workers • low wages but a share of the profits • children work alongside parents
Raw materials	• uses chemically produced polyester from imported oil stocks	• uses locally grown cotton and commercial dyes
Finances	• company profits ploughed back in to increase production	• profits shared equally by all members of the cooperative
Benefits	• health services available at factory	• doctor available • village school
Product	• polyester fashion goods	• good-quality cotton goods
Market	• huge middle-income market in developed countries	• domestic market • low-cost clothing

Figure 17.16 Two development models

Source: Adapted and reproduced with permission from *Decision-Making Geography* by N. Law and D. Smith. © Stanley Thornes (Publishers) Ltd.

4 There is great debate over which development strategy is most effective. Some people believe that agglomeration is the most desirable form, while others are convinced that import substitution, dispersion, and grass roots are more useful. Still others maintain that megaprojects and the growth pole model are most beneficial.

Which is the most effective means of achieving industrial development?

In this position paper you must defend a point of view.
Your final paper should include:
a) an introductory paragraph that identifies the issue and alternative positions (points of view).
b) a paragraph that presents one position on the issue as indicated above.
c) a paragraph that presents the other position on the issue as indicated above. Each paragraph (b and c) should argue the advantages of the position, and support the point of view with specific facts and examples.
d) a concluding paragraph that defends *your* position on the issue.

PART 5

ENVIRONMENTAL ISSUES: EARTH, AIR, AND WATER

CHAPTER 18

OUR CHANGING EARTH

ECONOMIC DEVELOPMENT AND ENVIRONMENTAL RESPONSIBILITY

In 1948, the British futurist and novelist Aldous Huxley said of modern civilization that "industrialism is the systematic exploitation of wasting assets. In all too many cases, the thing we call progress is merely an acceleration in the rate of that exploitation. Such prosperity as we have known up to the present is the consequence of rapidly spending the planet's irreplaceable capital.

"Sooner or later [humankind] will be forced by the pressures of circumstances to take concerted action against its own destruction and suicidal tendencies. The longer such action is

Figure 18.1 *Left,* the *Exxon Valdez* spilled 41.6 million litres of oil into Prince William Sound, Alaska; *top right,* vehicles are a major source of pollutants; *bottom right,* air pollution by industry has caused countless deaths and millions of hours of lost production.

postponed, the worse it will be for all concerned … Treat nature aggressively, with greed and violence and incomprehension: wounded Nature will turn and destroy you …" (Aldous Huxley, "The Double Crisis," in *Themes and Variations* [London: Chatto and Windus, 1950]).

Although Huxley expressed his views fifty years ago, they are even more valid today. Evidence of environmental destruction caused by human activity surrounds us. Pollution clogs rivers, lakes, oceans, and skies. Acid rain threatens the survival of lakes and forests. Greenhouse gases are altering the climate at an alarming rate. The **ozone** layer that protects us from the sun's harmful rays is thinning. Wilderness habitats are being destroyed and countless animal species are becoming extinct.

In 1983, the World Commission on Environment and Development was established by the United Nations "to formulate a global agenda for change." The commission's mandate was to explore environmental and development problems facing the world and to formulate realistic solutions. Its report, published as *Our Common Future* in 1987, served notice that it was critical that economic interests be tempered by environmental concerns and that governments and citizens take responsibility for the damage being inflicted on the environment. The report introduced the concept of sustainable development—that is, economic development must meet "the needs of the present without compromising the ability of future generations to meet their own needs." It called upon affluent societies to reduce their resource consumption and to practise a more sustainable lifestyle. It also recommended that developing countries control their population growth in order to foster economic development in harmony with environmental systems.

> *Our Common Future* is widely known as the Bruntland Report, after its chair, Gro Harlem Bruntland, former prime minister of Norway. With a central message of sustainable development, the report emphasizes the importance of integrating environmental concerns into policy making in all economic sectors.

THE ROOTS OF THE PROBLEM

Inevitably, development projects affect the environment. Whenever natural resources are removed from the environment or industrial by-products are added to natural systems, there is bound to be disruption. The challenge is to remove the resources with as little damage as possible and to dispose of waste products in a responsible manner. While there are companies with strong reclamation policies—Cardinal River Coals in Alberta, for instance—unfortunately, not enough industrial producers accept their responsibilities in cleaning up the environment after extraction of mineral resources.

MINING

Mining is one way in which development disrupts the environment. Although relatively little of the earth's surface is affected by mining, it is incredibly destructive to the natural environment. Mining operations strip away natural vegetation, rock, and soil in order to get to the valuable ore deposits. When the ore is smelted, the waste, or slag, is discarded around the mine. Smelting produces acid deposition, which pollutes the air and poisons the water supply. Often, ecosystems are destroyed for kilometres downwind and downstream of mining sites.

Today, mining companies in developed

Figure 18.2 Mining and smelting operations have disrupted the environment around Sudbury, Ontario.

countries must restore the land to its former condition before the mine was built. When an underground mine is closed down, all evidence of its existence must be removed. Officials from ministries or departments of natural resources monitor all aspects of mining operations including mine closures. Shaft entrances have to be sealed. Material that was removed from tunnels and shafts (overburden) must be covered with soil and reforested. In the case of **open-pit mines**, the giant hole left by the operation has to be filled in or at least reforested. In populated areas, these open-pit quarries are often used as landfill sites. Soil is mixed with the garbage so that it can decompose, and the whole

At Inco Limited in Sudbury, Ontario, seedlings are produced in underground greenhouses that have been set up in abandoned mine shafts. They are then transplanted throughout the city in areas where acid rain and other forms of pollution have destroyed the natural vegetation.

area is reforested once the quarry is filled in. Of course, this practice leads to other environmental concerns. In most diversified economies pollution controls around mines are stringent. Not only is solid waste covered and reforested, but liquid and gaseous waste are also treated. Groundwater is tested to ensure that poisons do not enter the environment. Smelting plants must remove pollutants before water is pumped into local streams. Air quality is also measured. Scrubbers and other devices remove ash and chemical contaminants before the gases are released into the atmosphere. Some chemicals such as sulphur are made into by-products, which are then sold to other industries.

In developing countries where government regulations are less stringent, restoration is often non-existent. Since the primary goal is generally to increase production, costly anti-pollution controls are not a priority. When mines are worked out they are abandoned, causing a threat to public safety and the environment. There are usually few restrictions on the pollutants that mining companies dump into the environment. Acid rain, ash, heavy metals, and industrial poisons fill the air and water. The lack of controls on environmental protection makes it less costly to extract minerals in many developing countries than it is in Canada and other developed countries. Lower extraction and labour costs allow many foreign operations to be more profitable than domestic producers. Although Canada still has enormous mining potential, many Canadian mining companies are concentrating their efforts in developing nations. While the companies can maximize their profits in these countries, they often inflict severe environmental damage that would not be tolerated in North America. We need to ensure that mining companies everywhere accept the responsibility of practising sustainable methods of development that do not permanently destroy natural systems.

Taking Action, a UN report released in April 1996, estimates that between 150 and 200 species of life become extinct every twenty-four hours! Blaming this mass annihilation on humankind's unsustainable methods of production and consumption, the report details a long list of serious and growing environmental problems: acid rain threatening forests, water, and wildlife globally; increased air pollution levels worldwide; ever-decreasing amounts of available freshwater; and annual expansion of the ozone hole over the Antarctic.

AGRICULTURE

Many modern farms have become factories. They take from the land and disrupt natural systems. This process starts the moment the land is cleared. Trees and other plants are removed. Animals indigenous to the area must move or die. As wilderness areas continue to shrink—a result of expanding farmland—there is less and less space for wild animals. We are currently experiencing a period of mass extinction similar to the one that occurred 65 million years ago. That extinction, which saw the death of the dinosaurs, resulted from a natural disaster. The current period of mass extinction is a result of human dominance of the world's ecosystems. Habitat destruction is the main cause of species extinction today. In 1996, World Wildlife Canada measured habitat destruction in Canada at 97 ha per hour. As a result, more than 275 species of mammals, birds, fish, molluscs, reptiles, amphibians, plants, and lichens were classified as vulnerable, threatened, or **extirpated** (a species no longer existing in the wild in Canada).

In developing countries, threats to wild animals and plants include the increasing disappearance of natural habitats, toxic chemical pollution, and unsustainable use of natural resources. The crush of too many people living on too little land has resulted in marginal areas being turned into farmland. Semi-arid regions are irrigated, mountain slopes are terraced, rainforests are cleared, and wetlands are drained so that more land can be farmed. The net result on ecosystems is disastrous. Native plants are removed and the ani-

Figure 18.3 Irrigated fields in Texas near the Rio Grande

income may find it difficult to consider the rights of wild animals.

Agricultural production is responsible for other harmful environmental effects. After the land is cleared, the soil is ploughed or cultivated so that farmers can sow seeds into the ground. Much of the soil's fertility is lost as the soil is ploughed. Moisture evaporates from the soil and the forces of erosion are able to remove topsoil that may have taken nature a thousand years or more to produce. To compensate for the lost moisture and fertility, the land is often irrigated and artificial fertilizers are applied. While this ensures good crop growth, these additives can also harm the environment. When water is added, so too are dissolved chemicals such as sodium and potassium compounds. Salts build up in irrigated soils to the point where many irrigated regions have become wastelands where nothing grows. This is particularly true in dry areas such as Rājasthan in northwest India and the Tigris-Euphrates Valley in Iraq, where high rates of evaporation and extensive irrigation over thousands of years have left the land contaminated with salt. The addition of fertilizers affects local ecosystems whenever it rains. Phosphorous, nitrates, and other chemical additives end up in local streams and wetlands. They stimulate algae growth to such an extent that these plants remove much of the oxygen from the water. Other plants and animals that depend on the oxygen die out, leaving the water a sterile place suitable for only primitive algae growth.

As the crop grows, pesticides are added to ensure that weeds, fungi, and insect pests do not

mals that depend on them for survival are also doomed. Entire food chains collapse when a wilderness area is converted to farmland. Even wilderness regions surrounding agricultural land suffer from agricultural practices. Wild animals often enter farmland in search of food. Farmers in Africa are sometimes terrorized by foraging elephants and lions. In Canada, coyotes have become a real menace. Yet the animals have every right to search for food wherever they can find it, especially since their natural habitats have been destroyed. Nevertheless, farmers who depend on the land for food and

destroy the crop before it is harvested. As with fertilizers, these chemicals usually end up in the water table. Plants and animals are killed off by these environmental poisons. In time, the few remaining wilderness areas in rural communities are destroyed by agricultural practices.

MANUFACTURING AND SERVICE INDUSTRIES

It is not only primary industries that damage the environment. Large factories create air pollution, water pollution, and toxic waste products. These factories use raw materials transported from farms, wood lots, and mines by trucks, ships, or trains. The act of transporting the materials creates carbon monoxide, nitrous oxides, and carbon dioxide as hydrocarbons are burned. The greenhouse gases generated by the internal combustion motors needed to transport goods are creating climate changes never before experienced by modern peoples. Inside the factories, industrial processes generate waste products in enormous quantities. These pollutants must be disposed of carefully, with as little harm as possible to the environment. For example, when paper is made chlorine compounds are used to bleach the paper. Chlorine-based waste

The Holland Marsh north of Toronto provides many of the vegetables for the huge market in southern Ontario. The area was originally marshland, created from glacial melt water after the last ice age. Insects, frogs, fish, marsh birds, and mammals such as foxes, coyote, and deer all populated this wilderness. The land was drained in order to create farmland. Today, the marshland has been replaced with a modified agricultural environment able to sustain millions of people. But there have been environmental costs. Fertilizers, herbicides, and pesticides end up in the Holland River and eventually in Lake Simcoe, where algae growth is a problem. The algae thrive on the fertilizers and consume so much oxygen that other forms of aquatic life cannot survive.

effluents enter the water table and damage ecosystems. Chemicals also pollute the air; solid ash and dust can be seen, but the most serious pollutants are usually invisible. Often, it is only when the incidence of cancer, respiratory diseases, and skin rashes rises that action is taken. In the diversified economies of the world, where the level of education is high and people are politically active, the situation has improved somewhat over the past twenty to thirty years. But in many developing nations where production is a priority, environmental protection is not yet a significant consideration.

Even tertiary and quaternary industries have an impact on the environment. Every time a natural resource is used, environmental systems are disrupted. Cars burn gasoline as workers commute to work. The resultant emissions cause global warming, tropospheric ozone, and other pollutants. Air conditioning and heating draw on electricity, creating demands for power. Fossil fuels are burned or hydroelectric dams are built to generate the power. We all have an impact on the environment. It is our responsibility to become aware of the repercussions our activities in order to preserve the planet for future generations.

CASE STUDY

Industrialization in the former Soviet Union

The former Soviet Union is one of the most polluted places on earth. Following the Second World War, the Communist government made industrialization a top priority. During the Cold War (1950–1990), the Warsaw Bloc, made up of the USSR and Eastern European countries it took over after the Second World War, was pitted against the growing industrial and military might of the United States and Western Europe (NATO). The Soviet government feared that the US would invade the

USSR to stop Communism from spreading. Similarly, the NATO countries were afraid that the USSR would continue to extend its empire by invading Western Europe. As a result, industrial production was given precedence in both blocs so that military machinery and weapons could be stockpiled for a third world war, which many people believed was inevitable.

By the 1970s, pollution controls in the NATO countries forced industry to clean up

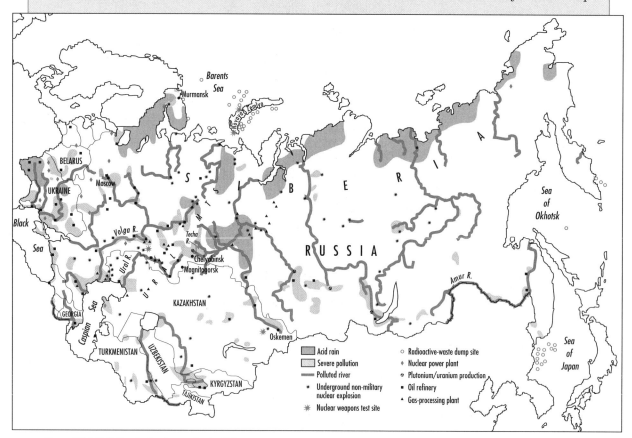

Figure 18.4 Pollution in the former Soviet Union

Source: *National Geographic*, August 1994: 80–81.

Figure 18.5 A mining operation in Vorkuta in northeast Russia

some of its practices. These same controls were not enforced in the USSR because the government did not consider them to be of prime concern. So environmental damage continued to escalate as industrial production grew. Billions of dollars were invested in industrialization projects, with little consideration for the environment or the health of the population.

The Soviet government set out to develop heavy industry. Iron and steel were needed to make tanks, submarines, and airplanes. Uranium was required for nuclear reactors to power submarines, while other military vehicles needed petroleum. Mining expanded, the petrochemical industry boomed, iron and steel plants grew, and huge industrial complexes appeared seemingly overnight. Magnitogorsk, a city of 435 000, developed along the Ural River beside the world's largest steel mill. The mill produced 12 million tonnes of steel a year, using antiquated equipment con-

fiscated from Germany during the 1945 invasion. Sulphur, carbon monoxide, nitrous oxides, fly ash, and other solid wastes poured out of the many smokestacks, which lacked gas and dust filters. Over the years the surrounding area became a poisoned wasteland, but it was the Soviet citizens who suffered the most. Today, a third of the people in the city suffer from chronic bronchitis, asthma, and other respiratory diseases.

In the Ural Mountains along the Techa River lies the town of Chelyabinsk. Between 1948 and 1951, the Techa River served as a sewer for a plutonium plant located upstream. In 1957 an explosion dumped 120 million curies of radiation into a local aquifer. The radiation worked its way through the water system into the Techa River. The people of Chelyabinsk drank the river water, and today they suffer from a variety of diseases, including several forms of cancer and disorders of the nervous system.

At the mouth of the Volga River around the city of Astrakhan, there were rich deposits of hydrocarbons containing hydrogen sulphide, which was needed for industrialization. In the 1960s, the Soviet government built a sulphur plant at the site, but did not build a power plant to generate the additional electricity needed to process the sulphur. As a result of the increased demand for electricity, there were many power failures. During these failures poisonous hydrogen sulphide escaped into the atmosphere. This created acid precipitation downwind from the industrial complex, and led to a variety of health problems, such as rashes, lung diseases, and an increase in stillborn babies. In order to resolve the situation, in the late 1960s the government detonated fifteen underground atomic bombs to create huge storage tanks for the poisonous gas. Thirteen of these caverns collapsed and were unusable. The poisonous gases continued to contaminate the area. Another environmental problem emerged—the nuclear explosions had left the soil radioactive. The residents were subjected not only to chemical pollution, but also radioactive pollution. Today, scientists are still monitoring the radioactivity leaking from the ground above the collapsed caverns.

In Murmansk, the Russian port on the Barents Sea, radioactive decay jeopardizes the marine life of the Arctic Ocean. For decades radioactive waste was routinely dumped into the Arctic off Novaya Zemlya. Fourteen nuclear reactors (four containing exceedingly dangerous fuel rods), seventeen ships loaded with spent fuel rods from nuclear power plants, and thousands of barrels of solid nuclear waste were secretly dumped into the ocean from the 1960s up until the collapse of the Soviet Empire. This occurred in spite of the fact that the Soviet Union had signed an international treaty in 1976 banning such dumping. The consequences for marine food chains in the high Arctic have been devastating. Plants and animals have died of nuclear contamination. Those that survive are often deformed or have various forms of cancer. Although the full extent of the problem is still unclear, Norwegian scientists speculate that everything from plankton to whales is tainted with radioactivity.

Even Moscow, the Russian capital, is afflicted with pollution problems. Sewage sits in fields on the outskirts of the city. Sewage-treatment facilities are unable to eradicate the heavy metals, dioxins, PCBs, and toxins in the waste. It has been estimated that 15 t of heavy metals enter the sewers in Moscow each day. There are no pollution controls or filters in the factories. As a result, industrial waste passes directly into the sewers. Eventually, it finds its way into the water resources. In some of the most seriously affected communities children are born with congenital disabilities. The city's mortality rate is twice that of the average Canadian city.

East of Moscow in the Republic of Kazakhstan, a lead-and-zinc smelter in Oskemen has poisoned many of the residents in this city of over 100 000 people. The smelter was built using reclaimed German machinery after the Second World War. The life expectancy rate for males is only fifty-five years. Fifty-eight per cent of all children suffer from abnormalities to their immune systems. Lead and zinc emissions, laced with touches of cadmium and arsenic, have found their way into everything from the air to garden vegetables to breast milk.

CONSOLIDATING AND EXTENDING IDEAS

1 The quotation from Aldous Huxley at the beginning of this chapter calls in question several assumptions that people in developed countries have made about "progress."
   a) Reread the quotation.
   b) Form groups of four to evaluate concepts in Huxley's statements. What differing perspectives can be used to understand "industrialism," "wasting assets," "progress," "prosperity," and "humankind"?
   c) To what extent should environmental concerns restrict economic growth? Pair off within your groups to debate this issue. Debaters should develop arguments to advance opposing perspectives: environmental concerns should restrict economic growth *or* environmental concerns should not restrict economic growth.
   d) Using debate format, the original groups of four students present their arguments to the entire class.
   e) Now write a position paper to respond to the issue:

   To what extent should environmental concerns restrict economic growth?

   In this position paper you must defend a point of view.
   Your final paper should include:
   i)   an introductory paragraph that identifies the issue and alternative positions (points of view).
   ii) a paragraph that presents one position on the issue as indicated in part (c).
   iii) a paragraph that presents the other position on the issue as indicated in part (c).
        Each paragraph (ii and iii) should argue the advantages of the position, and support the point of view with specific facts and examples.
   iv) a concluding paragraph that defends *your* position on the issue.

2 Prepare a cause-and-effect organizer or chart that details industrialization in the former Soviet Union. Across the top of your chart use the headings Human Effects and Environmental Effects; along the side of the chart, Location and Industrial Practices.

3 Create a collage of visuals that illustrates the advantages/disadvantages of several of the following alternatives. As you reflect on the pros and cons, consider environmental responsibility in the context of standard of living/quality of life.
   a) heating your home with (i) a wood stove, (ii) solar heat, (iii) natural gas, (iv) an oil furnace
   b) travelling to school (i) alone in a car, (ii) on foot or by bike or roller blades, (iii) in a car pool, (iv) on the school bus
   c) shaving with (i) an electric razor, (ii) a safety razor, (iii) a disposable razor
   d) buying beverages in (i) cans, (ii) plastic bottles, (iii) drinking boxes
   e) eating (i) prepackaged single-serving dinners, (ii) meals prepared at home using fresh ingredients, (iii) in a restaurant, (iv) food that has been delivered from a fast-food outlet

4 Write a paragraph to explain
  a) what you think the cartoonist's message is;
  b) human activities that "Nature" would include in a global clean-up (refer to the section "The Roots of the Problem" on pages 275 to 279).

Source: *Calgary Herald*, 6 January 1997.

IDENTIFYING ENVIRONMENTAL ISSUES

Environmental issues involve cultural, political, economic, and resource factors. As we discovered in the previous case study, politics and economics in the Soviet Union played an important role in steering a course of industrialization that created some of the worst pollution problems in the world.

Culture may influence environmental attitudes. Many traditional societies are more in tune with nature's systems than urban societies are. People who derive their living from hunting, fishing, trapping, and subsistence farming often have a deeper understanding of how their environment and its ecosystems operate. Native peoples in Canada and other countries often protest development projects that interfere

Source: *Toronto Star*, 12 August 1995. ©Tribune Media Services, Inc. All rights reserved. Reprinted with permission.

Figure 18.6 Milton Born With A Tooth (*front, right*) and other Lonefighters protesting the Oldman Project

with natural systems and their traditional ways of life. In the early 1990s, during the controversy surrounding the building of the Oldman River Dam and Reservoir in southern Alberta, a group of Peigan Indians, known as the Lonefighters Society, attempted to halt the construction project. Led by Milton Born With a Tooth, the activists opposed the building of the dam because it would flood ancestral lands and sacred Native burial sites. The society's action against the $353 million construction project included appeals to government and the media, attempted diversion of the water in the Oldman River around a government-owned weir, and an armed stand-off between the Lonefighters and the RCMP. Successful negotiations and the arrest of their leader ended the group's protest, and the Oldman River Dam and Reservoir were completed in the spring of 1991.

Logging, hydroelectric dams, and overfishing are only a few of the issues that directly affect the way of life of many aboriginal peoples. Their attitudes are often echoed by other concerned citizens who want to see the natural environment protected and preserved.

One project that attracted much public attention in Canada was the Kemano 2 Power Diversion Project. In 1952, Alcan Aluminum built the Kemano 1 Dam on the Nechako River. This dam provided the hydroelectric power needed in southern British Columbia at that time, but reduced by about one-third the water flowing through the river. In the early 1990s the company wanted to build another hydroelectric dam on the Nechako. Water from the reservoir created behind the dam would turn turbines and generate power for the growing population on the west coast. However, the new dam would completely block 88 per cent of the water flowing down the river, allowing it to flow downstream only after it had passed through the power plant. About 20 per cent of BC's commercial salmon swim up the Nechako River to spawn. If they could not get to their spawning grounds, they would be unable to reproduce and the fishery would be destroyed. Although Alcan decided it would build artificial spawning beds, salmon are very particular about their spawning grounds, and biologists predicted that the beds would not succeed. Native peoples such as the Cheslatta Carrier Nation depend on the salmon fishery for food and income. Commercial fishers and environmentalists were outraged that the Kemano 2 Project had the go-ahead from British Columbia's government. After many protests, the government cancelled the project in 1995 even though $535 million had already been spent on the dam.

The controversy is not over yet. The Cheslatta Carrier Nation is still concerned about the Kemano 1 Dam. After this dam was built, the Cheslatta were relocated to the village of Grassy Plains because their traditional lands were now prone to flooding. Their homes were burned and the ancient burial grounds of fifty ancestors had to be abandoned. Over the past fifty years the land has flooded eighty times. The flooding could be relieved if a new dam were built to restore water levels to their pre-1952 levels. Alcan had planned to build the new dam if the Kemano 2 Project proceeded. Now these plans have been suspended because of the huge loss that Alcan sustained. Similar land disputes have occurred elsewhere in Canada and around the world.

Of course, there are also economic and resource factors that influence the way in which people regard the environment. Often, residents of communities that depend on natural resources have different attitudes towards development than residents of cities. Consider a mining town where the only industry is the mine. New government regulations that require the mine to act in a more environmentally responsible manner could reduce profitability and may prompt the company to close the mine because it no longer generates enough income. The miners would likely have to move to another mining town or be unemployed. Do you think the families of these miners would be as concerned about the environment as other people?

In developing countries, there is often conflict between people who want to preserve the

> On 1 April 1999 the map of Canada will change. The new territory of Nunavut (Inuktitut for "our land") will be created, which includes the central and eastern parts of the Northwest Territories—an area that the Inuit have occupied for at least a thousand years. In the past, First Nations peoples and Europeans had different paradigms concerning land ownership. Native peoples believed that humans belonged to the land, which should be respected if it was to provide for their needs forever. Many Europeans felt that people could own land and capitalize on its resources. Soon, the Inuit will be able to control development in Nunavut and ensure that the land is preserved forever.
>
> See also pages 379 to 380.

environment and those who want to earn a living from exploiting the resources. When a peasant farmer's family is starving, the preservation of nature cannot be the family's primary concern. Peasants living in Ethiopia and other parts of the Sahel understand how important trees are to the preservation of their environment, but when there is no other fuel to cook their food, they must chop branches off the trees and cut down the vegetation that has kept the desert at bay for centuries. Meanwhile, development agencies supported by environmentalists strive to reforest the semi-arid slopes.

Conservationists condemn the depletion of the gene pool—the result of deliberate weeding out of variety by multinational agribusiness—and the extinction of plant and animal species—the result of over-hunting and destruction of natural habitats.

Coral reefs are declining because of pollution from oil spills. Soil sediment is being washed into the seas. Power stations, sewage pollution, and runoff of agricultural chemicals result in discharges of heated water. If we want to maintain coastal fisheries and avoid severe coastal erosion, we must address and reverse these practices.

There will always be tension between the advocates of economic development and those

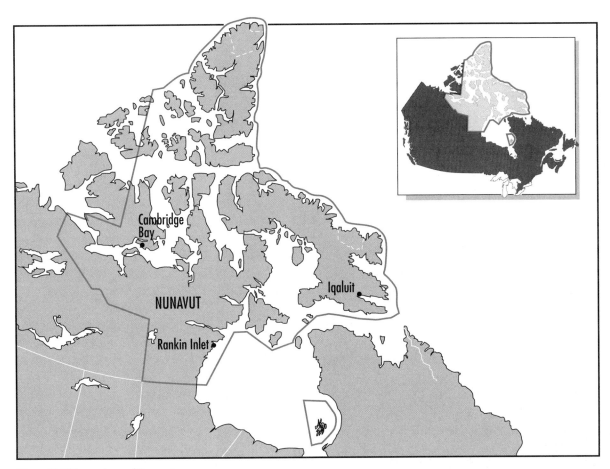

Figure 18.7 The territory of Nunavut

of conservation. Many of the issues surrounding our global environment *are* issues because we have not paid attention to the potential repercussions of economic development. Historically, decisions about land use have been made by landowners focused on immediate gains and seeking the highest financial returns. We have come to realize only recently that land-use policies are necessary to protect the public interest.

DEFORESTATION

The destruction of the tropical rainforests is one of the most widely publicized environmental issues in the world today. Television and

film documentaries, as well as newspaper and magazine articles, have called attention to the plight of the earth's most fascinating biome. We know that rainforests are the lungs of the planet. They purify the air, absorb carbon dioxide, and produce oxygen as a by-product of photosynthesis. When the forest cover is removed, there are fewer plants releasing oxygen and using carbon dioxide to produce carbohydrates. The rainforests of the world are major sinks—naturally occurring storage reservoirs for natural substances—in which substantial quantities of carbon are stored. If the rainforests are cut down, the carbon will be free to enter the atmosphere as CO_2. The

Figure 18.8 Stripping bark from Pacific yew trees for taxol extraction

rainforest plants are also effective medicines: quinine is used in the cure of malaria; the Pacific, or western, yew is a source of taxol, one of the most potent anti-cancer drugs known; andiroba and copaiba trees yield potent medicinal oils. Traditional medicine, based largely on tropical plants, nurtures 80 per cent of the world's population. Rainforest plants provide essential ingredients for pharmaceuticals—an industry worth tens of billions of dollars annually.

Many developing countries look to the tropical rainforests to improve their economies. Large populations in coastal and interior plains are putting so much pressure on local resources that planners are encouraging people to move into the heavily forested interiors of these tropical regions. In central Africa, people are moving up the volcanic mountains into the domain of the mountain gorillas, popularized by the movie *Gorillas in the Mist*. Filipinos are also moving into the mountainous interior of their islands nation, and in Indonesia similar migrations

rainforests are also the habitat of thousands of plant and animal species, many of which have yet to be identified. The rosy periwinkle, a plant found in the rainforests of Madagascar off the southeast African coast, is known to moderate the effects of leukaemia. Other

are taking place. But the most publicized development, or exploitation, of the rainforest has been in Brazil. While the following case study looks at the situation in Brazil, many other rainforest regions in the world are undergoing similar development.

CASE STUDY

Developing the Brazilian Rainforest

Much of the vast and rapidly growing population of Brazil is located in the economically depressed northeast of the country. The *nordestinos* have not had the opportunity of formal education and live in difficult circumstances. In order to survive, they move to wherever they can find jobs.

Brazil is burdened with an enormous foreign debt. These loans were used in the 1960s and 1970s to build the modern country of Brazil, but economic growth did not match political aspirations, so much of the debt is unpaid. The government struggles to pay the interest each year. To service the debt and to improve the lives of millions of people without incurring further debt, the government's policy has been to develop the resources of the tropical rainforest.

In Brazil (and in other rainforest countries), there are five ways in which the government seeks to develop the rainforest: subsistence farming, lumbering, industrial development, cattle ranching, and extractive reserves.

SUBSISTENCE FARMING

Peasant farmers displaced from the land because of unsustainable practices or overpopulation have been given the opportunity to start a new life in the Amazon jungle. In the early 1970s, the government introduced a plan to give farmers their own farmland on large tracts in the rainforest. Roads were built deep into the western Amazon. People moved into the region hoping to build a new future by making the land productive. Theo-

retically, it appeared to be a sound idea. This approach had been successful in North America when European settlers transformed the prairies into one of the world's most productive grain-growing regions.

The problem with the Brazilian resettlement program was that the developers did not fully understand the complexities of the rainforest ecosystem. Unlike other ecosystems, the nutrients necessary for plants to grow are not contained in the topsoil; they are deep in the contact zone where the soil meets bedrock and in the living plants and animals themselves. Soil profiles are extremely deep, having been created over thousands or even millions of years. Nutrients are transferred to the soil as the bedrock is broken down through chemical processes. Rainforest plants have extremely long root systems that penetrate to the bottom of soil horizons to obtain necessary nutrients from this contact zone. These nutrients allow the plants to flourish even though the topsoils are infertile compared to soils in temperate zones. When a plant or animal dies, its remains quickly decompose and the nutrients are recycled by other life forms. If the forest cover is removed, the surface nutrients are also removed.

Domesticated plants sown by farmers do not have the root systems to reach the necessary nutrients in the soil-bedrock contact zone as much as 20 m below the earth's surface. Once the trees are cut down and the brush is burned, the soil is exposed to the heavy rainfall common to the region. Leaching occurs when water-soluble nutrients are absorbed by the rainwater and washed away or deposited deep in the soil. The land is also susceptible to erosion as the water washes away huge amounts of soil and creates gullies on even the gentlest slopes. Contrary to what many people believe, there is a short dry (relatively) season in the region of the Amazon that has been developed. Plants can be sown during this season without the danger of seedlings being washed away. However, the harvest often occurs when rainfall is at its heaviest, so the drying of grains and other crops for storage poses a serious problem. During the first couple of years of the program, adequate yields were harvested because the plants derived nutrients from the ashes of the burned brush. By the third and fourth years, the soil had become so infertile that nothing of value would grow. The settlers moved farther up the highway, clearing new land and beginning the cycle all over again. Today, few of these subsistence farm communities remain in the western Amazon. Most of the people have returned to the cities of the southeast. The jungle has regained the deserted plots.

LUMBERING

The rainforests have long been considered an inexhaustible source of wood. Many rainforest trees have unique characteristics that make them useful as furniture and construction material. Logging is common in all of the world's rainforests, especially in areas that are accessible by road or river. There are particular problems associated with harvesting trees in the tropics. Marketable species are usually spread over an immense region and are not concentrated in stands as they are in Canada. On average only about 5 per

cent of the vegetation in the jungle is marketable. In the past, **selective cutting** was practised—only the commercially valuable trees were cut down. However, in order to gain access to the trees, much of the other vegetation was destroyed, leaving the soil exposed to the physical elements. Food chains collapsed as fragile, highly evolved species of plants and animals were eliminated. The erosion that resulted from the denudation of the forest cover caused the rivers and streams to become clogged with silt, disrupting aquatic ecosystems many kilometres downstream. At one time logging was an extractive operation; reforestation seldom happened. The jungle was so irreparably damaged that recovery of the marketable species was in doubt. Although the disruption to natural systems is still cause for concern by environmental groups and aboriginal peoples, methods in the Amazon are less destructive today. Modern equipment has made it possible to use non-commercial wood for pulp or charcoal.

In the mid-1960s an attempt was made at reforestation in the eastern Amazon. Along the Jari tributary a single species of tree was planted over an area the size of Prince Edward Island. But various diseases destroyed the entire plantation. The introduction of a **monoculture** in this genetically diverse ecosystem was bound to fail. Plants of the same species have a better chance of surviving if they are spread out. This prevents pests that live off one particular plant from spreading over a wide area.

Today, the new Jari Project has been hailed as a great success. Scientific and technical knowledge have improved, resulting in sustained yields of reforested pulp-wood being produced each year. Other advances

Figure 18.9 A section of the Jari Project

have reduced the susceptibility of monocultures to infestations. New blight-resistant eucalyptus trees have been developed through selective breeding and cloning. These new trees grow so fast that they can be harvested every six to eight years. This project may be an economic success, but what about the natural ecosystems? There is no question that the region has been so disrupted, it will never return to its former state of wilderness. Still, much has been done to reduce the impact of development. Rows of native forest at least 400 m wide have been left between monoculture plantings. These ensure the disbursement of selective blights and diseases and allow natural ecosystems some opportunity for survival. The area is no longer sprayed because the pests quickly mutate and become resistant to the pesticides. Eight genetic reserves totalling 20 000 ha shelter 530 different types of indigenous trees. Ecologically sensitive areas on steep slopes, around springs, and along rivers are also protected.

INDUSTRIAL DEVELOPMENT

Soil and trees are not the only resources in the Amazon. The region contains huge mineral deposits of gold, diamonds, and industrial minerals. It also holds a large proportion of the world's freshwater supply. The only country with more freshwater is Canada. The Amazon's water resources have the capacity to provide hydroelectric energy for the growing cities of Brazil and Argentina.

One massive development project in Brazil is the Grande Carajás Project. This open-pit mine produces over 50 million tonnes of iron ore each year from the world's largest iron ore deposit, estimated in excess of 17 billion tonnes of high-grade ore. Financed by the World Bank and US Steel, the development has generated much debate. Preservationists condemn the environmental damage. Natural vegetation had to be cleared to gain access to the mine area. A railway line linking the mine to port facilities 1000 km away disrupted the jungle ecosystem and intruded on its 4500 residents, many of whom had not been in contact with the outside world. A hydroelectric plant was built on the Tucuruí River. The dam created a reservoir over 2000 km^2 in area. The oxygen in the water was depleted by rotting vegetation, and a huge dead lake replaced the rainforest. (The oxygen levels in the lake have now recovered and an aquatic ecosystem has evolved.) In order to produce the charcoal required for smelting the iron ore, trees were chopped down and burned. Those who oppose the project argue that the damage to the ecosystem is not worth the profits made by the developers.

CATTLE RANCHING

Once a region in the Amazon has been logged and mined, the land is often only suitable for grazing. As with other agricultural activities, the fertility of the soil decreases each year because erosion washes away essential nutrients. So the amount of land needed to sustain an animal increases each year. The land becomes completely exhausted and cannot regenerate itself because the cattle eat anything that grows. In spite of the environmental consequences cattle ranching continues in the Amazon. Most of the beef from these cattle is destined for the export market. Much of it is transported to North America where it is used to make hamburgers in several fast-food chains.

EXTRACTIVE RESERVES

The extractive reserve is the only modern development strategy that seeks to preserve the rainforest in its entirety while still using its valuable resources. The idea is based on a practice that developed in the nineteenth century. Rubber tappers made small fortunes collecting latex from the rubber trees indigenous to the Amazon Basin. They collected a small amount of sap every day much like Canadians collect sap to make maple syrup. This is a sustainable practice since the trees do not die.

Most of the demand for natural latex has been replaced by synthetic rubber, which is made from oil. To compensate for their reduced incomes, rubber tappers in the Amazon have extended their activities to the collection of other valuable forest products. Brazil nuts, mangoes, bananas, cassava, taro,

and other tropical fruits and vegetables are collected for sale in local markets and for export to other regions. Marketable crops have been planted along forest trails to help make the forest more profitable. Because the forest cover is not disturbed, this type of extensive **horticulture** does not deplete nutrient levels in soils the way intensive agriculture does.

The extractive reserve is the legacy of the rubber trapper Chico Mendes. He became a martyr to the cause when he was killed by cattle ranchers in 1988. After his

Figure 18.10 A rubber tapper in Brazil's rainforest

death the Brazilian government gave its support to the establishment of extractive reserves in the Amazon. Up until then, it had supported only large-scale projects such as Jari and the Grande Carajás schemes. By the mid-1990s, four extractive reserves had been established in the Amazon. The largest is 6 million hectares. No mining, forestry, or other development is permitted in these areas. They are reserved strictly for small-scale horticulture and the gathering of forest products. With the support of the World Bank, deforested areas of the rainforest are recovering. It is hoped that, eventually, they will also be developed as forest reserves.

While Canadians often protest deforestation in other countries, sometimes they are unaware of local deforestation. The Oak Ridges Moraine and the Niagara Escarpment are two beautiful forested areas of southern Ontario. Many people live in new subdivisions on the moraine, but do not realize that a woodland area has been razed to allow for development. The forests not only provide a pleasant landscape adjacent to urban Ontario, but they also act like a sponge, collecting and purifying rainwater in aquifers. Much of the drinking water in rural Ontario and small towns north of the Golden Horseshoe comes from these aquifers.

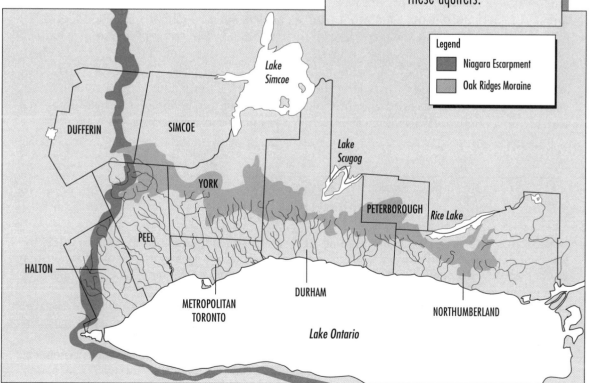

Figure 18.11 Oak Ridges Moraine and Niagara Escarpment

CASE STUDY

Biodiversity in Vietnam

In the fall of 1994, a Canadian-Russian research group (funded by the Royal Ontario Museum and organized through the Academy of Sciences' zoological institute in St. Petersburg) was granted access to the unexplored forests of Vietnam. Located in isolated pockets in the interior of the country and along the steep mountains on the border of Laos and Vietnam, these jungle forests are home to species of creatures never seen or identified by the scientific world.

Because of their remote location and limited use by indigenous populations, the forests of Vietnam did not gain worldwide attention until 1992, with the discovery of a new species of bovid, locally known as the *sao la*. While the number of wilderness areas yet to be discovered is dwindling each year, Vietnam offers decades of potential work to taxonomists. Most of this nation's plants have not been identified, and the tracts of primary forest are a refuge for rare species of frogs, snakes, and insects of great diversity.

After decades of isolation, cultural and industrial developments now threaten Vietnamese forests and indigenous species of wildlife. Since 1943, the nation's original forest cover has been reduced from 44 per cent to 20 per cent of its territory. Because the Vietnamese are primarily an agrarian population, heavy deforestation has increased to allow for the creation of farmland. Poaching and trafficking in critically endangered species, such as gibbons, clouded leopards, pangolins, Malayansun bears, snakes, lizards, turtles, and birds, have reached crisis proportions. Biological diversity, which is of fundamental importance to all ecosystems and economies, is lost forever as a result of these practices.

ENVIRONMENTAL WARFARE

Actions of war are responsible for some of the world's worst environmental disasters. Deliberate environmental warfare is waged against the enemy's environment to prevent access to resources, subdue the civilian population, punish insurgents, and thwart the use of guerrilla tactics. "Scorched-earth" strategies were used by Britain against Mau Mau rebels in Kenya in the 1950s; by France against the Algerians from 1949 to 1962; by the Soviet Union against the Afghans in the 1980s; and by Iraq against Kurd nationalists in the mid-1990s.

The term **ecocide** (the killing of ecosystems) describes the tactics used by Iraqi forces as they left Kuwait at the end of the Persian Gulf War of 1991. Oil wells, pipelines, and storage facilities, as well as loaded tankers, were dynamited to cause severe pollution of Kuwait's air, sea, and groundwater.

The Vietnam War (1960 to 1973) is another example of ecocide. To counter guerrilla warfare and civilian opposition, the United States military used daily bombing runs over South Vietnam to create a landscape pitted with 250 million bomb craters. Herbicides (plant-killing chemicals) such as Agent Orange and Agent Blue were sprayed on forests and crop lands to defoliate the trees and rice paddies, which provided cover for the North Vietnamese.

Adapted excerpt from R. Stefoff, *Environmental Disasters* (Broomall, PA: Chelsea House, 1994). Reprinted by permission of Chelsea House Publishers.

DEGRADATION OF THE EARTH

It has been estimated that 11 per cent of the earth's total vegetated surface has already suffered moderate to extreme soil degradation. Causes include deforestation, overgrazing, and unsound agricultural practices such as monoculture, shorter fallow periods in shifting cultivation, cultivation of hillsides without adequate erosion-control measures, and over-irrigation. In turn, soil loss and degradation result in reduced agricultural productivity and place increased stress on still-arable lands and agricultural technology.

Cities and industries generate tremendous amounts of solid waste, some of which takes the form of chemical toxic waste. Many household products can be hazardous if released into the environment. Over time, small quantities of these substances can accumulate to toxic levels and contaminate the soil. When illegally dumped or leached over time into the earth's surface, these toxins create wastelands.

Urban populations in OECD countries are serviced by municipal waste-collection agencies that participate in massive recycling and incineration programs. These agencies use **landfill** sites to dispose of rubbish locally. Waste products from nuclear power stations, sewage, and industrial waste are dumped into the world's oceans. Some cities and companies have responded to strict disposal laws at home by shipping toxic waste overseas to be burned, stored in tanks, or buried in landfills. This process has become known as "Third World dumping" because many of the recipient nations are developing countries.

DUMPING WESTERN WASTE

As the continents of Africa and South America are no longer willing to take toxic wastes from Western industries, dumping has shifted to the countries of eastern Europe and the former Soviet Union, where regulations are few and governments are eager for cash. Recently, 180 t of acidic wastes from the German state of Saxony waited at a Hungarian train station for further transport, while Albania received a shipment of 455 t of pesticide by-products from Berlin and Schwerin, Germany. Environmental groups have documented 30 million tonnes of Western waste destined for dumping in Russia since 1991. The countries of origin included the United States, Belgium, Spain, and Switzerland.

Christiane Grefe. Adapted from *Wochenpost* (Berlin), (*World Press Review*, March 1994): 44.

Related to this disposal of toxic wastes is the practice of manufacturing and selling products known to be toxic. Some American and European companies continue to make and sell DDT (an insecticide) abroad, though it has been banned in developed nations since the 1970s. Millions of kilograms of fertilizers and pesticides, which are either banned or restricted in developed countries, are exported annually to developing nations. These products are used to boost agricultural productivity in the short term, but in the long term they bring about environmental disaster—human and livestock poisoning; pollution of the air, water, and soil;

and food contamination.

World Resources 1996-97 estimates that 20 to 50 per cent of the solid waste generated in cities of the developing world remains uncollected. This uncollected trash is the usual cause of blocked urban drainage channels in Asian cities; it also increases the risks of flooding and carrier diseases. Municipal solid waste sites in developing countries often handle both household and industrial wastes, including hazardous wastes. Because of the lack of appropriate disposal legislation and limited enforcement of existing regulations, illegal dumping of toxic and hazardous wastes is common.

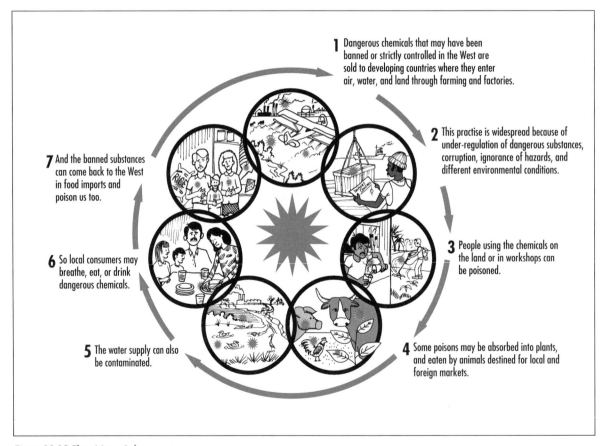

1 Dangerous chemicals that may have been banned or strictly controlled in the West are sold to developing countries where they enter air, water, and land through farming and factories.

2 This practise is widespread because of under-regulation of dangerous substances, corruption, ignorance of hazards, and different environmental conditions.

3 People using the chemicals on the land or in workshops can be poisoned.

4 Some poisons may be absorbed into plants, and eaten by animals destined for local and foreign markets.

5 The water supply can also be contaminated.

6 So local consumers may breathe, eat, or drink dangerous chemicals.

7 And the banned substances can come back to the West in food imports and poison us too.

Figure 18.12 The vicious circle

In a world of growing food interdependence, we cannot export our hazards and then forget them.

Source: *New Internationalist*, November 1993. Copyright ©New Internationalist. Reprinted by permission of Guardian News Service Limited.

CONSOLIDATING AND EXTENDING IDEAS

1 Outline how specific environmental issues relate to each of the following:
   a) cultural issues
   b) economic issues
   c) resource issues
   d) political issues

2 a) Assume one of the following roles: First Nations Chief, developer, finance minister in a government, tourist agent, president of a conservation club, president of a local association of hunters and fishers, or commercial salmon fisher.
   b) Prepare an argument in support of or against the Kemano 2 project.
   c) Hold a formal debate in class about the establishment of the Kemano 2 power plant.

3 Refer to the case study "Developing the Brazilian Rainforest" on page 289.
   a) Prepare point-form notes that outline the advantages and disadvantages of the different development strategies discussed in the study.
   b) Develop a glossary of terms specific to the rainforest. Include definitions for the following:
   • canopy
   • ecosystem
   • emergent layer
   • floor
   • rainforest
   • reforestation
   • subsistence farming
   • wilderness area
   • wildlife sanctuary
   c) Map the earth's green belt by locating and identifying the rainforests of South America, the Caribbean, Mexico, Central America, North America, Australia and the Pacific Islands, Africa, south Asia, and Southeast Asia.

4 Refer to Figure 10.4 on page 130 to remind yourself of the "future wheel" format.
   Create a future wheel that predicts the outcomes of successful lobbying/protests over environmental protection and conservation. Consider the efforts of a variety of special interest groups in your predictions.

5 Chapter 1 explained that paradigms are rules and conditions that we use to understand those things we perceive. In several brief paragraphs, speculate about the nature of paradigms held by each of the following individuals towards environmental issues:
   • a provincial minister of the environment
   • a rancher in southern Alberta
   • an environmental activist
   • a Native Canadian

6 **Biodiversity** is one area of concern for environmentalists today. The loss of species is occurring at potentially disastrous rates. Collect pamphlets/newsletters of global organizations that work to protect biodiversity on our planet (for example, World Wildlife Fund, Canadian Wildlife Federation, Conservation Foundation, Canadian Nature Federation, Greenpeace, National Audubon Society, Sierra Club). Create a classroom bulletin board to display your materials, and add to your display throughout the course.

7 a) Working in pairs, spend an afternoon cleaning up a local park or the shores of a nearby waterway. As one student picks up debris, the other records the items, noting whether they are made of glass, plastic, paper, metal, or organic matter. Later, in class, compile the results and use pictographs to show the amount recovered of each type of trash.

  b) Conduct a similar study in the school cafeteria.

  c) Spearhead a campaign to encourage litter-free lunches.

CHAPTER 19

OUR CHANGING ATMOSPHERE

DEFINING THE ISSUE

Every human being requires 2 kg of freshwater, 1 kg of nutritious food, and 16 kg of air each day. We are sustained on the surface of a planet that is surrounded by a precious 10 km layer of air and water—the atmosphere. It provides all life on earth with a steady supply of oxygen and, in synchronization with the oceans, brings us the climate and weather that are responsible for our planet's cycles of life. The atmosphere supplies carbon dioxide, which is essential for photosynthesis. It helps to stabilize the differences in temperature between equatorial and polar regions by trapping solar energy in the form of heat, thereby preventing it from escaping into space (this is commonly known as the **greenhouse effect**).

Two environmental issues that have gained public attention in recent years are **global warming** and **ozone depletion**. The delicate ozone layer protects life on the planet by shielding it from ultraviolet radiation. This layer is threatened by synthetic substances called **CFCs (chlorofluorocarbons)**, which are used by industry (see page 309). CFCs rise into the upper atmosphere and cause thinning of the earth's ozone layer. Global warming occurs when the delicate equilibrium between energy entering and leaving the atmosphere is disrupted. It is the result of the build-up of greenhouse gases generated by human activity. Climate models predict a rise of 1.5 to 4.5°C over the next century. Although this may seem insignificant, it is the largest temperature increase in recorded history, and is almost equal to the temperature increase from the height of the last ice age to the present.

Many scientists predict that the growing amount of **greenhouse gases** in the atmosphere will cause world temperatures to increase. While it is impossible to predict exactly how this increase will influence weather patterns and long-term climate change, one thing is certain—disruptions to established atmospheric patterns will be so great that there will be major changes in the way people live. Ozone depletion could also bring about

Figure 19.1 Scientists predict that the incidence of violent storms will increase as a result of global warming.

enormous disruptions to our ways of life. The increased **ultraviolet radiation** reaching the earth's surface is expected to result in a dramatic rise in the incidence of skin cancer. It also has the potential to impede crop yields severely, and may cause a breakdown in marine ecosystems as microscopic plants and animals succumb to harmful **UV rays**. While global warming and ozone depletion are often linked, they are in fact quite distinct. They have different causes and different solutions. Both exist because of the world's increasing dependency on fossil fuels and other industrial chemicals.

THE HUMAN AND ENVIRONMENTAL COSTS OF GLOBAL WARMING

Even a relatively small increase in global temperatures could radically alter natural and human systems. Sea level would rise as continental ice-caps melt. People living along low-lying coastal plains, in river estuaries and deltas, and on coral atolls would be forced to migrate to higher ground. Island nations such as Vanuatu in the South Pacific would disappear altogether. Coun-

tries such as Bangladesh and American states such as Mississippi and Louisiana would lose so much of their coastlines to flooding that enormous resettlement programs would be necessary.

The damage would not be confined to coastal regions. It is predicted that continental interiors would become drier because the higher temperatures would accelerate the rate of evapotranspiration. The midwestern United States, Ukraine, central Europe, and the pampas of South America contain some of the richest farmland in the world. These areas could become deserts if rainfall decreases and temperatures rise. This situation alone could lead to hunger and famine on a scale unknown in history.

More violent storms are also predicted as a result of global warming. The incidence of severe weather in the form of hurricanes, tornadoes, and blizzards is more common now than in the past. The insurance industry has seen claims related to weather damage increase from US $16 billion during the 1980s to US $48 billion during the first half of the 1990s. The potential sixfold increase projected for the entire decade

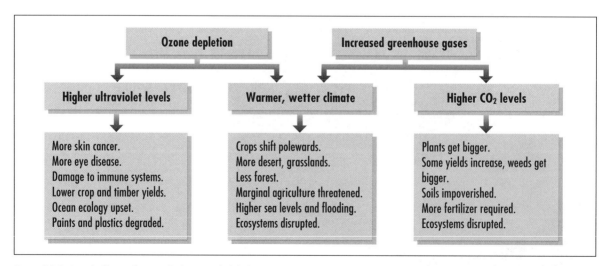

Figure 19.2 Social effects of ozone depletion and global warming

Source: *Understanding Our Environment* by S. Dunlop and M. Jackson. Copyright ©Oxford University Press Canada 1997. Reprinted by permission of Oxford University Press Canada.

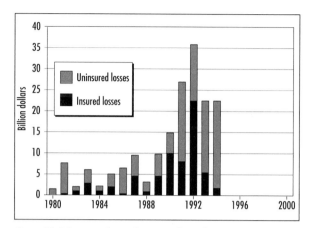

Figure 19.3 Economic losses from weather-related natural disasters worldwide, 1980–1994

Source: *Vital Signs 1996: The Trends That Are Shaping Our Future* by L. Brown, C. Flavin, and H. Kane. Copyright ©1996 by Worldwatch Institute. Reprinted by permission of W.W. Norton & Company, Inc.

is indicative of how much our climate is changing. A hurricane with winds of 270 km/h hit Bangladesh in May 1991. Approximately 139 000 people died, 1 million homes were damaged or destroyed, and the cost was in excess of US $3 billion. In 1992, Hurricane Andrew cost an estimated US $25 billion—the most costly natural disaster in American history. In 1995 and 1996, the prevalence and ferocity of hurricanes and tornadoes continued to be well above previous levels.

Not all countries would be adversely affected by global warming. Northern regions such as Canada, Alaska, and Siberia may benefit from warmer temperatures. Tundra could melt, leaving soil profiles that could be made suitable

Figure 19.4 The Athabasca Glacier: *top,* **in 1917;** *bottom,* **in 1986.**
The glacier's rapid retreat can be observed by using the small hill at the front and left of both photos as a reference point.

for forestry or commercial farming. Countries such as Britain and Norway might experience a longer, warmer growing season, which would enable them to expand the amount and variety of crops they grow. In higher latitudes, cattle may be able to winter outside without the need for expensive barns and hay storage facilities. In Canada, it is likely that the tree line would be extended farther north and higher up mountain slopes, which would benefit the forestry industry. An extension of western Canada's growing season is also predicted. Global warming could lead to the greatest human migration ever known. In the last ice age people migrated from cold lands to more temperate regions farther south. The next great migration may be in the opposite direction! People living in the south may migrate north to cooler, moister climates.

Not only are the atmospheric changes we are experiencing today global and dramatic, but they may also be irreversible, at least over the next 2000 or 3000 years. In the geological time scale, things happen very slowly (as compared to our human time span where a lifetime may be only eighty years). Adjustments in climate patterns, cataclysmic events such as volcanic eruptions and meteorites, and periodic changes in the earth's position and orbit affect climate in ways that we are just beginning to understand. But global warming is not a natural event; it has been caused by human activities. Therefore, the solution to the problem lies with us and the ways in which we use natural resources.

THE HUMAN AND ENVIRONMENTAL COSTS OF OZONE DEPLETION

The sun creates most of the world's energy. It is ultimately responsible for life on earth. Yet not all of the sun's rays are beneficial to life. Ultraviolet radiation causes mutations in the genetic structures of plant and animal cells. In humans,

Figure 19.5 Sunbathers in the northeastern US

A 1 per cent thinning of the ozone layer is expected to increase the incidence of melanoma by 4 per cent. The American Environmental Protection Agency predicts an increase of 153 million skin cancer cases by the year 2075.

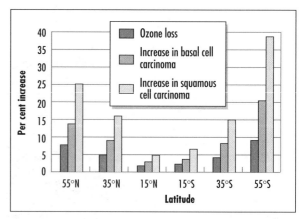

Figure 19.6 Increases in non-melanoma skin cancer related to latitude

While the connection between melanoma and UV radiation is not well understood, non-melanoma cancer is directly attributed to increased levels of UV-B exposure, as suggested by the graph.

these changes lead to skin cancer, disruptions to the immune system, and cataracts. A 1991 UNEP report predicted that a 10 per cent depletion in the ozone layer would cause a 26 per cent worldwide increase in **melanoma**, a skin cancer that is often non-fatal. Because UV radiation reduces plant growth, there is concern that the number of plant extinctions may grow, thereby reducing the earth's biodiversity. Crop yields are being affected as plants such as soybeans adjust to changing levels of ultraviolet radiation. Generally, a 1 per cent reduction in the ozone layer results in a 1 per cent drop in soybean yields

However, the most crucial issue is not human or crop abnormalities; it is the potential for a sudden change in ozone concentrations in the atmosphere. If this were to happen, would life be able to adapt? There have been prior episodes of mass extinction. Would the thinning of the ozone layer increase ultraviolet radiation to the point that terrestrial life could not survive? Since some UV rays can penetrate into clear water, algae, plankton, and fish larvae are at risk. Photosynthesis decreases because of damage to plant hormones and chlorophyll (plant cells essential for photosynthesis). Biologists have already detected changes in the food chains of southern oceans where **phytoplankton** surge to life each year following the end of the polar winter. When the sun appears after the six-month-long Antarctic night, microscopic plants start to grow. Phytoplankton is essential for marine ecosystems as it is the basis for most animal life. Nearly all marine animals either feed directly on phytoplankton or eat animals that depend on phytoplankton for nutrients. The quantity and diversity of these essential plants and animals have declined. But there is hope that these simple life forms will survive. Plankton is capable of rapidly adapting to its environment because it has such a short life span. Natural selection allows those mutations that are best suited to the environmental change to survive. Already, some zooplankton have developed dark pigmentation on their upper surfaces to shield against the sun's rays. Can other larger and more complex plants and animals adapt as quickly? Although we still do not know the extent to which ozone affects life, we do know that terrestrial life did not exist on earth before there was an ozone layer.

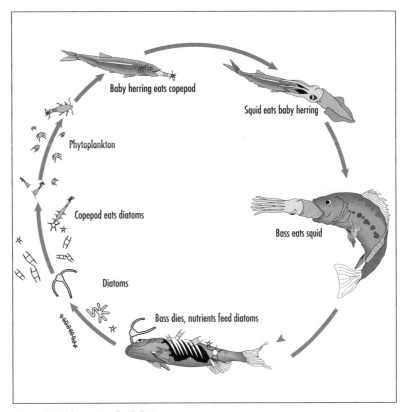

Figure 19.7 The marine food chain

STATLAB

1 The statistics in Figure 19.8 show global temperature change, using 1960 as a base year. Every other year is shown as either warmer or cooler than 1960.

   a) Prepare a line graph indicating temperature fluctuations from 1860 to 1990.

   b) Determine the increase in global temperatures by the year 2100 by extending the trend line from 1960 to the year 2100.

   c) In 1996 the Intergovernmental Panel on Climate Change predicted a 2°C rise in global temperatures by the year 2100. How close is your projection to the projection made by the IPCC?

YEAR	°C	YEAR	°C
1860	-0.8	1930	-0.2
1870	-0.5	1940	-0.1
1880	-0.4	1950	-0.1
1890	-0.3	1960	0.0
1900	-0.3	1970	0.2
1910	-0.3	1980	0.4
1920	-0.2	1990	0.5

Figure 19.8 Global temperature change

2 Study the maps and graphs in Figures 19.9. (a) to (d)

   a) Analyse each graph or map. What common trends do they show?

   b) Which parts of Canada seem to be experiencing the most change? Explain your answers.

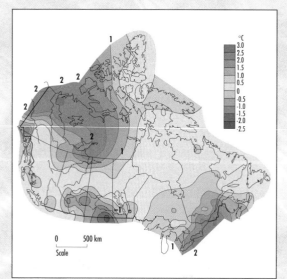

Figure 19.9 (a) Temperature departures from normal, winter 1997

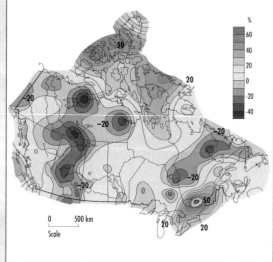

Figure 19.9 (b) Precipitation departures from normal, winter 1997

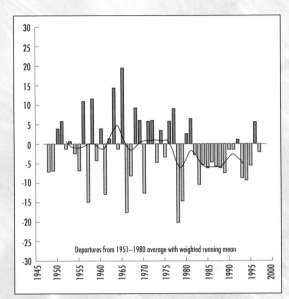

Figure 19.9 (c) Winter national precipitation departures with weighted running mean, 1948-1997

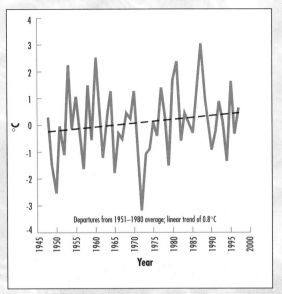

Figure 19.9 (d) Winter national temperature departures and long-term trend, 1948-1997

Source: Climate Research Branch—Atmospheric Environment Service, Environment Canada.

CONSOLIDATING AND EXTENDING IDEAS

1 Explain how global warming could make the lifestyle of your grandchildren significantly different from the one you enjoy today. Consider each of the following: clothing, leisure activities, housing, transportation, food, vacations.

2 For each of the following economic activities, determine if global warming has a positive or negative impact. Explain your answers.
   a) a ski resort in Quebec
   b) shrimp fishing in Thailand
   c) a theme park in southern Florida
   d) farming in northern Saskatchewan
   e) owning a home in Calgary
   f) snow removal in North Bay

3 Using the Internet or other electronic data retrieval systems, research the earth's ozone layer. Find out what has happened in the past decade and what is happening now. Develop visuals (maps, diagrams) to explain your findings.

PATTERNS OF HUMAN GEOGRAPHY

Over millions of years the atmosphere has evolved into its present form. The delicate balance that exists among natural systems is changing because of human geography. Our activities are releasing greenhouse gases into the atmosphere at an alarming rate. The burning of fossil fuels emits millions of tonnes of carbon into the atmosphere in the form of CO_2. Industrial processes discharge still more greenhouse gases. Even rice farming and dairy farming result in the creation of greenhouse gases.

CARBON DIOXIDE

Carbon dioxide levels have grown from about 280 parts per million (ppm) in 1850 to 350 ppm today—a staggering increase of 25 per cent. The rate continues to increase by 2 to 3 per cent every decade. The burning of fossil fuels such as coal, gasoline, and oil has added significantly to CO_2 levels. Since 1850, modern technological society and the processes of industrialization have consumed more and more fuels. It is hardly surprising that carbon dioxide levels in the atmosphere have become so dramatically high. Carbon dioxide is released when fossil fuels

Figure 19.10 *Top left,* preparing food in a microwave; *top right,* spraying air freshener; *bottom left,* rice farming; *right,* a forest fire. *Each of these situations contributes to the production of greenhouse gases.*

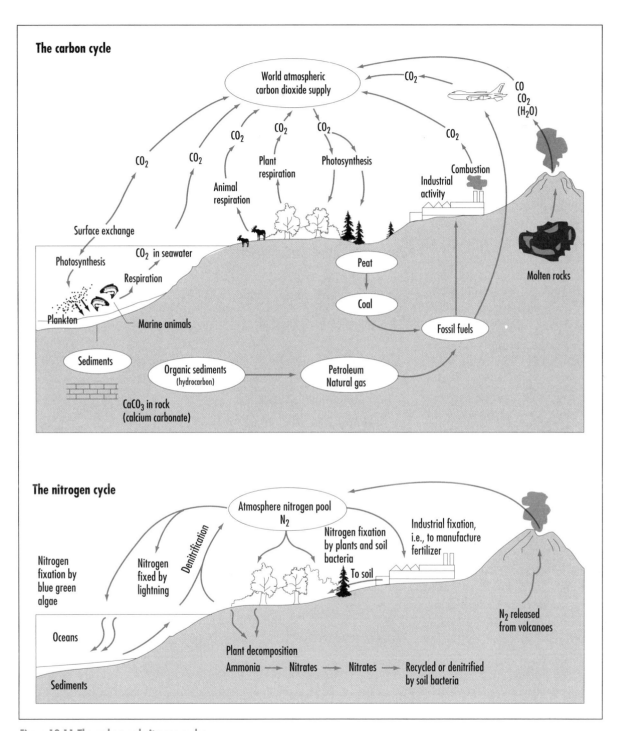

Figure 19.11 The carbon and nitrogen cycles

Source: (Carbon cycle) *People and the Environment* by Graves, Lidstone, et al. (Oxford: Heinemann Educational Books, 1987). Reprinted by permission of Heinemann Educational Publishers. (Nitrogen cycle) *Geography: A Study of Its Physical Elements* by Quentin Stanford. Copyright ©Oxford University Press Canada 1988. Reprinted by permission of Oxford University Press Canada.

(including oil, natural gas, and coal) are burned to produce electricity in power plants. These power plants service the demands for electricity by consumers and industry. The burning of forests around the world has also contributed to the increase in CO_2 levels. This problem is compounded by the fact that the number of trees has been drastically reduced, which means that less carbon dioxide is returned to the **biomass** through photosynthesis. Many developing countries rely on wood and animal manure for fuel. The burning of this biomass is another factor in the build-up of carbon dioxide. As the world's population continues to climb and as living standards rise globally, the amount of stored carbon entering the atmosphere will also rise.

NITROUS OXIDE

Nitrogen is the most common gas in the atmosphere. It becomes a gas when it combines with oxygen. This chemical reaction takes place when substances containing nitrogen are burned, or when food is digested. The burning of biomass or fossil fuels discharges nitrous oxide into the atmosphere. When nitrate deposits in the ground are mined to make artificial fertilizers, they mix with oxygen in the air to form nitrous oxide. With increased vehicular traffic, the burning of biomass, and the use of synthetic fertilizers, levels of nitrous oxide (N_2O) are increasing at about the same rate as

N_2O is also a pollutant responsible for acid rain. In the same way that sulphur from smelting plants mixes with rainwater to produce sulphuric acid, nitrous oxide produces nitrous and nitric acid when it reacts with rainwater in the atmosphere. Acid precipitation most often falls over busy cities where there is a great deal of traffic. As each car burns gasoline, nitrous oxide is emitted. Most North American vehicles now have catalytic converters that reduce nitrous oxide emissions.

carbon dioxide. At the present time, over 30 per cent of global nitrous oxide is generated by human activity.

METHANE

Like carbon dioxide and nitrous oxide, methane gas contributes to the earth's greenhouse effect. Methane is a naturally occurring gas in swamps and other areas where vegetation decomposes slowly in water. The world's wetlands release enormous amounts of this gas into the atmosphere. However, it is humans who are responsible for the large increases in this gas over the past 200 years. Today, there is twice as much methane in the atmosphere as there was before the Industrial Revolution, and its levels are rising at the rate of 1 per cent a year. Methane, or natural gas, is produced from food waste and lawn clippings that accumulate in landfill sites. As these materials rot, methane gas is created.

Methane is also produced when digestion occurs. As the number of domesticated farm animals (especially cows) climbs, the amount of methane generated also grows. This increase is faster than the rate of increase for carbon dioxide, but there is only 0.5 per cent as much methane in the atmosphere as there is carbon dioxide. However, it is a much more potent greenhouse gas than carbon dioxide. One molecule of methane absorbs twenty-nine times as much heat as one molecule of CO_2.

CHLOROFLUOROCARBONS

Chlorofluorocarbons are complex chemical compounds made up of chlorine, fluorine, and carbon. CFCs were invented in the mid-twentieth century for use in consumer products. They have been utilized as coolants in air conditioners and refrigeration systems, as solvents for cleaning computer components, and as propellants in aerosols such as deodorants and hairspray. Foams used in the manufacture of insulation, fast-food containers, and upholstery are blown into moulds by propellants. CFCs were considered to be a significant improvement over ammonia, the refrigerant they replaced. Ammonia has a strong odour, is poisonous, and is extremely difficult to handle. CFCs became popular because they were odourless, tasteless, and non-poisonous.

The problem with CFCs is that these compounds do not break down when they come in contact with other chemicals. They last for years because they are so stable. While this may be an advantage in a refrigerant, it is extremely harmful in the atmosphere. Unlike natural greenhouse gases, chlorofluorocarbons do not readily disperse. Instead, they continue to expand and to damage the atmosphere for

In parts of Southeast Asia and the United States, rice farming has steadily expanded in recent decades in response to growing demand and the Green Revolution. Methane gas is produced in rice paddies, in the same way as it is in natural wetlands.

decades before they finally decay. CFC levels are increasing in the atmosphere at a rate of 4 per cent a year. They can absorb up to 300 times the amount of heat that carbon dioxide can, which means they are a major factor in the greenhouse effect.

TROPOSPHERIC OZONE

Tropospheric ozone is a by-product of human activities in large urban centres. (Do not confuse this ozone with the natural ozone layer found in the stratosphere!) Low-level ozone is formed from chemical reactions between ground-level pollutants. About 75 per cent of all tropospheric ozone is created from internal combustion engines and industrial processes in which coal is burned. The high level of energy produced through combustion causes the same chemical reaction as that between UV radiation and oxygen in the stratosphere (see page 303.) One of the main components of smog, tropospheric ozone can damage crops and aggravate respiratory problems. Although tropospheric ozone is not a major contributor to global warming, it can have an effect on microclimates in urban centres. Consequently, as urbanization continues to spread, ozone pollution will steadily intensify.

CONSOLIDATING AND EXTENDING IDEAS

1 Study the flow charts in Figure 19.11.
 a) Write an explanation for each chart.
 b) Choose one instance where human activity has affected the cycle and explain its impact on the natural system.
 c) What human activities contribute to the increase in greenhouse gases?

2 Study the graph in Figure 19.12. How does it show that there is a relationship between temperatures and greenhouse gases?

3 Refer to page 128 in the *Canadian Oxford School Atlas*, 6th ed. (or to other materials that summarize atmospheric growth of greenhouse gases). Evaluate the impacts of carbon dioxide, methane, CFCs, nitrous oxide, ozone, and other trace gases by determining the following:
  a) Which have the highest annual rates of increase?
  b) Which have the longest lifetimes in the atmosphere?
  c) Which contribute most to global warming?

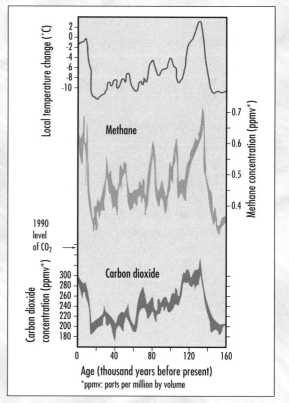

Figure 19.12 Comparison of local temperatures and atmospheric concentrations of methane and carbon dioxide in Antarctica over the past 160 000 years

Source: *Understanding Atmospheric Change*, State of the Environment Report No. 91–2, Environment Canada, 1991. Reproduced with the permission of the Minister of Public Works and Government Services, 1997.

STATLAB

1 Study the technology indicators on pages 450 to 457 in Appendix C.
  a) Prepare an organizer showing total methane emissions and CFC emissions by continent.
  b) Prepare an organizer showing the ten highest producing countries for CO_2 emissions per capita.
  c) Choose from either a) or b) and prepare an appropriate graph to illustrate your findings.

d) What conclusions can you make from this analysis?

e) Write a statement that outlines a plan to reduce greenhouse emissions in North America.

2 a) List the top ten contributors to global warming in order from greatest contributor to least. Explain how you came to your decision.

b) What generalizations can you make from this analysis?

3 a) Prepare a scattergraph comparing GDP and CO_2 production.

b) What patterns do you notice? What other variable would likely correlate with CO_2 production?

4 Prepare a graded shading map to illustrate the global distribution of one of the greenhouse gases.

5 Using information from this chapter and the statistics in Appendix C, determine the contribution of each of the greenhouse gases towards global warming.

6 Study Figure 19.13, which shows trends in greenhouse gas production.

a) Prepare a multiple-line graph showing the trend for each gas.

b) Analyse the graphs to determine which gases continue to be a problem and which seem to be less serious.

7 Using Figure 19.14, prepare a line graph showing global temperature change since 1950.

a) What trend is indicated?

b) What correlation appears to exist between these data and the data in activity 6?

YEAR	CARBON DIOXIDE (000 000 t)	CFCs (000 000 t)	NITROGEN (000 000 t)
1950	1620	0.042	7
1960	2543	0.150	12
1970	4006	0.640	18
1980	5172	0.880	22
1990	5943	0.820	26
1995 (estimated)	6056	0.300	27*

Figure 19.13 Trends in greenhouse gas production

Source: *Vital Signs* (New York: W. W. Norton & Co., 1995).

* (1993 data)

YEAR	TEMPERATURE (°C)
1950	14.86
1955	14.92
1960	14.98
1965	14.88
1970	15.02
1975	14.92
1980	15.18
1985	15.09
1990	15.38
1995	15.39

Figure 19.14 Global temperature change since 1950

Source: Goddard Institute for Space Studies, New York (January 1996).

GLOBAL SYSTEMS AFFECTING CLIMATE CHANGE

Natural systems are extremely complex—there are dynamic processes that work either to reduce or intensify the effects of climate change. Some natural processes accelerate the effects of the greenhouse but most work to moderate global climate change. It is likely that changes will occur gradually and then, at some point, will become rapid and irreversible. While it is possible that an increase in greenhouse gases would be offset by one or more of these processes, we must not be lulled into a false sense of security simply because the changes are proceeding so slowly. By studying the fossil record, geologists have learned that, in the past, minor temperature changes of 2 to 3°C have resulted in major changes to the biosphere.

VOLCANOES AND METEORS

Volcanic eruptions and meteors hitting the earth create stratospheric dust clouds that lower the amount of solar radiation entering the troposphere. It is possible that these cataclysmic explosions could counter the effects of global warming. Dust particles absorb and reflect incoming solar energy. This decreases the amount of solar radiation hitting the earth's surface. Recent evidence suggests that the dust that resulted when the Chicxulub meteor struck southern Mexico 65 million years ago plunged the planet into darkness. The meteor, with a diameter of 20 km, was travelling at speeds greater than 180 000 km/h. When it plunged to the earth, this massive extraterrestrial rock created a hole 40 km deep, right through the limestone sea floor and the crust to the semi-molten mantle of the earth. More than 3000 km³ of dust shot up into the stratosphere. There was no light for three months and most plant life died. More than 25 000 km³ of crushed and melted rock blew out of the crater, raining on the landscape across 5000 km. Because the vaporized bedrock was limestone, carbon dioxide levels jumped by 1000 per cent. Global temperatures increased

Figure 19.15 A volcanic eruption in the Democratic Republic of Congo

dramatically for a thousand years. Most of the terrestrial life that remained after the meteor struck was killed off by acid rain.

Less significant events, such as the June 1991 eruption of Mt. Pinatubo in the Philippines, modify the atmosphere for two or three years. When Mt. Pinatubo erupted, increased dust and sulphur dioxide in the atmosphere caused temperatures in the Northern Hemisphere to drop. The amount of solar radiation entering the troposphere was reduced because the dust particles reflected solar energy back into space. If the planet experiences a rash of similar volcanic activity, global warming could lessen over the short term. Periods of extensive volcanism are sporadic. There was considerable volcanic activity from 1750 to 1770 and from 1820 to 1835, but relatively little from 1857 to 1877 and from 1913 to 1962. Therefore, we cannot count on these random events to moderate global temperatures.

THE MODERATING INFLUENCE OF OCEANS

Large bodies of water moderate land temperatures. Water takes longer to heat up and cool down than land. It is heated to a greater depth because it is fluid and translucent. The vast amount of water that covers the planet moderates world temperatures considerably. It has been estimated that changes in air temperature take over 1000 years to be reflected in deep ocean temperatures. This in itself could slow the effect of increased greenhouse gases. If global warming were to occur, ocean levels would rise and there would be more surface water. This theory suggests that world temperatures may be moderated by the enormous energy sink the world's oceans provide. In addition, sea water absorbs carbon dioxide. If sea level rises, there would be more salt water to absorb the increased concentrations of CO_2 in the atmosphere. Thus, higher sea levels could help reduce the amount of carbon dioxide and thereby temper the effects of global warming.

Climatologists in New Zealand have predicted that a higher sea level and warmer temperatures could cause the massive Ross Ice Shelf on the icebound continent of Antarctica to break away and float north into the South Pacific. The degree of cooling in the southern mid-latitudes that the melting of this ice sheet would create could be great enough to lower global temperatures. It may also serve to reduce or eliminate the effects of global warming. In fact, some experts believe this melting ice could trigger another ice age!

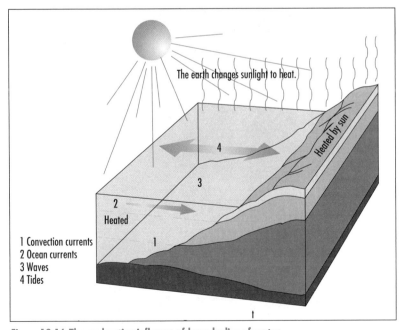

The earth changes sunlight to heat.

Heated by sun

4

3

2
Heated

1

1 Convection currents
2 Ocean currents
3 Waves
4 Tides

Figure 19.16 The moderating influence of large bodies of water

These forces help to mix sea water. It takes water longer to heat up and cool down because it is heated to a greater depth than land.

CLOUD FAMILIES

There are three categories of clouds: cirrus, stratus, and cumulus.
All other clouds are variations of these three types.
- Cirrus: high clouds made up of ice crystals that are so thin and feathery you can see right through them
- Stratus: low sheets of heavy, dark clouds
- Cumulus: fluffy white clouds that sail across the sky like large ships

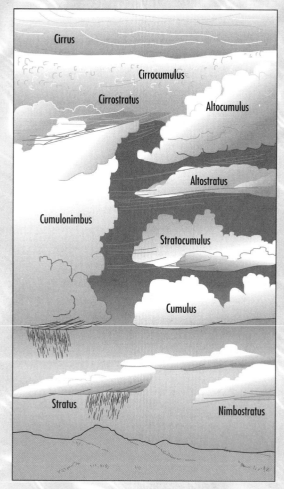

Cirrus

Cirrocumulus

Cirrostratus

Altocumulus

Altostratus

Cumulonimbus

Stratocumulus

Cumulus

Stratus

Nimbostratus

Figure 19.17 Types of clouds

Figure 19.18 *Top*, stratus clouds; *bottom*, cirrus with cumulus clouds

CLOUD COVER

Evaporation increases when temperatures rise, resulting in more cloud cover. The impact that clouds have on global temperatures actually depends on the type of cloud. Some clouds act like a blanket. They absorb heat and reradiate it to the earth's surface. Other clouds block the sunlight, preventing it from reaching the ground, thereby reducing temperatures. The stratus family of clouds tends to trap heat close to the earth's surface. Sunlight passes through the clouds but the water droplets in the clouds trap infrared radiation. On the other hand, cumulus and cirrus clouds are bright white, so the high **albedo** reflects solar radiation. These examples are over-simplifications of extremely complex patterns, which scientists do not yet fully understand.

VEGETATION

If carbon dioxide levels climbed dramatically and world temperatures increased, plant growth would be stimulated. This would have an effect on carbon dioxide levels. Biomass is one of the greatest CO_2 sinks the planet has. They hold vast amounts of carbon in the carbohydrates that make up their leaves, trunks, roots, and so on. The stimulated plant growth could help to stabilize carbon dioxide levels because plants would store more carbon in their structures.

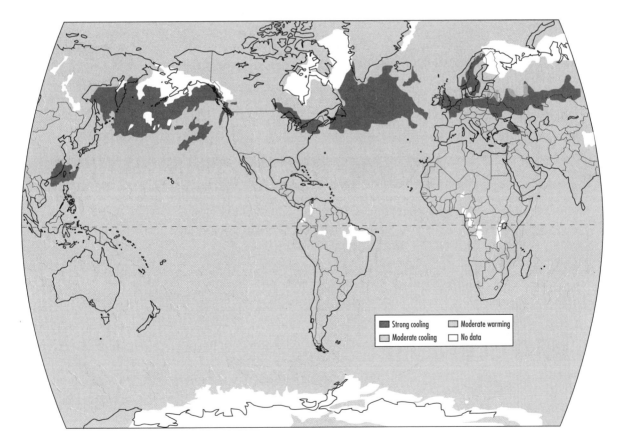

Figure 19.19 Net effect of clouds
Source: *Climate Change Digest 94–01*, Environment Canada.

CLIMATE MEASUREMENT AND PREDICTION

Some scientists believe that global warming is the result of technological improvements in measuring and monitoring equipment rather than of any real changes in the atmosphere. Our monitoring devices are much more sophisticated than they were even ten years ago. It is likely that measurements taken in the past were inaccurate compared with those we can obtain today. So if there is a noticeable change in recorded temperatures, does this reflect atmospheric change or simply better methods of measurement? Other researchers contend that the differences in measurement are negligible. We are just now beginning to be able to monitor natural systems accurately.

Another factor in global warming may be the influence of **microclimatology** on monitoring stations. Most weather stations are located at airports. Originally, these airports were situated in rural areas outside of the cities they served.

Over the past fifty years many urban centres have expanded, often to the point where the airports are now surrounded by urban development. Temperatures in urban centres are considerably warmer than those in outlying areas. Statistics for two airports in Edmonton illustrate this point. The international airport is located about 20 km from the city centre on flat farmland. The municipal airport is only 3 km from the central business district. Its annual temperature is 1.5°C warmer than that of the rural terminal and it has twenty-six more days of temperature over -2°C. Environment Canada reports that the winter minimum temperature in Montreal's city core is sometimes 5 to 8°C warmer than in the environs. Increases in temperature recorded at airport weather stations may reflect urban expansion rather than overall global warming. Local accumulations of greenhouse gases in these locations could skew data to a great extent. Consequently, comparisons to information that was collected when urban pollution was not a factor may be inaccurate.

CONSOLIDATING AND EXTENDING IDEAS

1 a) List the factors, past and present, that cause climates to change.
   b) Determine whether each factor results in an increase or a decrease in temperature.
   c) Explain how each factor affects global warming.

2 Explain how changes in technology and data collection could distort assumptions made about long-term climate change.

THE HOLE IN THE OZONE LAYER

In polar regions the ozone layer is being depleted at a greater rate than anywhere else on earth. The darkness of winter lasts for six months in these regions. During this time ozone is neither created nor destroyed by the sun's

ultraviolet radiation. Under normal circumstances the ozone layer is dormant during the polar winter. However, the CFCs in the atmosphere are continuously at work, depleting the ozone layer. When the sun returns, ozone production begins again. But by now, the existing ozone layer is so thin that more UV rays are able to penetrate the atmosphere.

MODERATING OZONE DEPLETION

Just as there are natural systems that moderate the effects of increased greenhouse gases on global warming, there are also natural systems that moderate the destruction of ozone. If the chlorine atom that causes all the damage could be stabilized, the problem would be solved. There are, in fact, atmospheric chemicals that do just that. The **chlorine sinks** created by chemical reactions render chlorine monoxide stable and prevent it from reacting with ozone molecules. Con-

The hole in the ozone layer was first noticed over the Antarctic when technology became available to measure the ozone layer. In fact, it is not a hole at all, but a thinning of the layer of ozone protection. Scientists predict that ozone depletion will not occur to the same extent over the Arctic and lower latitudes because polar stratospheric clouds are not present there.

sider the reaction that occurs between chlorine monoxide (ClO) and nitrogen dioxide (NO_2), a common greenhouse gas. Chlorine nitrate ($ClONO_2$) is the new chemical: $ClO + NO_2 \rightarrow ClONO_2$. Similar reactions occur between methane and chlorine oxide. As a result of these reactions, the chlorine is locked up in stable compounds and cannot affect the ozone layer.

Scientists once believed that these sinks would contain chlorine atoms and that the damage to the ozone layer would be minimal. This theory turned out to be inaccurate. In 1986, researchers at the

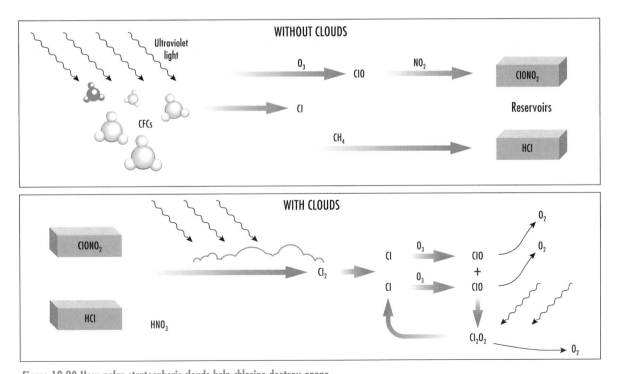

Figure 19.20 How polar stratospheric clouds help chlorine destroy ozone

Source: Illustration by Ian Warpole from "Polar Stratospheric Clouds and Ozone Depletion," by Owen B. Toon and Richard P. Turco, *Scientific American*, June 1991.

NOAA (National Oceanic and Aeronautic Academy) at Harvard University observed a negative correlation between ozone level and the presence of polar clouds in the stratosphere. They discovered that when the clouds were present, ozone depletion was more rapid. They made an assumption that a chemical reaction on ice particles in the clouds was freeing chlorine monoxide. Later, researchers discovered that what actually happens is very different. The extreme coldness of the air causes nitrous acid to precipitate from the clouds. This releases the chlorine

> Polar stratospheric clouds form as a result of extremely low temperatures and certain greenhouse gases. If temperatures climbed or the amount of greenhouse gases dropped—as the outcome of either less volcanic activity or fewer greenhouse gases— the impact of polar stratospheric clouds on the ozone layer would be diminished.

monoxide from the chemical compound $ClONO_2$, which is created when the chlorine combines with nitrous oxide. In turn, this reaction reduces the amount of nitrogen available to stabilize ClO. When volcanic activity intensifies, as it did with the eruption of Mt. Pinatubo, polar stratospheric clouds increase because of the additional nitrous acid and sulphuric acid in the atmosphere. In Antarctica, ozone levels dropped significantly after the Mt. Pinatubo eruption spewed millions of tonnes of acid into the stratosphere.

CONSOLIDATING AND EXTENDING IDEAS

1 Global warming is a direct result of industrial processes. Power plants using fossil fuels (coal, oil, natural gas) for electricity produce more CO_2 gases than any other source. Investigate alternative forms of power sources: wind, water, solar, and nuclear power. Do any or all of these offer savings in terms of air quality?

2 a) Explain why ozone depletion is a greater problem at the higher latitudes than it is closer to the equator.
   b) Why is ozone depletion a greater problem in Antarctica than in the Arctic?
   c) Why would ozone depletion be highest in Canada in late winter or early spring?

3 At the 1992 Earth Summit in Rio de Janeiro, 154 nations pledged to cut emissions of greenhouse gases back to 1990 levels by the year 2000. By 2005, this pledge would accelerate the decrease in emissions to below 1990 volumes. Since 1992, Canada's Clean Air Alliance, a joint effort of government, industry, and environmentalists, has developed a national strategy to reduce emissions pumped into the air. However, the real gains to be made are in the developing world, in populous countries such as India and China where energy efficiency is low and consumption is growing most rapidly.

Do some research to determine ways in which developed countries might help developing countries to reduce emissions. Consider types of foreign assistance (see Chapter 9, page 123) that might be particularly useful in addressing this issue (for instance, technical aid to reduce emissions and boost production in petroleum technology).

CHAPTER 20

The Waters of the World

Water is the most abundant natural resource in the world. Seventy-two per cent of the earth's surface is covered in water; 97 per cent of this is sea water, which contains dissolved salts. The remaining 3 per cent is freshwater, and 99 per cent of that is frozen in polar ice-caps and glaciers. In the past, the use of this seemingly endless supply of water was unlimited. No one questioned the exploitation of this life-sustaining system—for agriculture, industry, or domestic use. In recent decades, however, people have become increasingly aware that the water supply is neither limitless nor immune to pollutants.

THE HYDROLOGIC CYCLE

The total amount of water in the world is constant and cannot be increased or diminished. Earth's hydrologic cycle has existed for millennia. It is one of the planet's global cycles, and

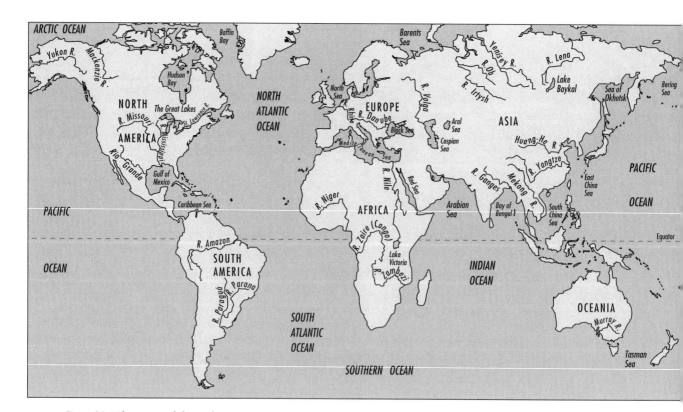

Figure 20.1 The waters of the earth

these cycles reflect equilibrium—nature's balance.

Water from the surface of the earth escapes into the air through the process of evaporation. When the water vapour in the atmosphere reaches a certain concentration, it condenses into a liquid and returns to the earth in the form of precipitation. Falling snow in the arctic regions builds glaciers that eventually either melt or plunge into the ocean as icebergs. Each day, close to 1 trillion tonnes of water fall in one form or another on the planet. Most of this precipitation falls directly into the oceans, but about 40 000 km^3 annually fall on land. This water circulates towards the seas, flowing over land as brooks, streams, and rivers, and collecting in lakes and ponds or artificial reservoirs and irrigation channels along the way.

River water is constantly moving, forming currents that take the water to the oceans. Water also filters through the earth's surface until it reaches an impermeable layer, creating subterranean streams and reservoirs. Groundwater is freshwater stored through the natural processes of the hydrologic cycle in aquifers beneath the earth's surface. It represents thirty times more water than exists in all the world's lakes and rivers combined. Groundwater levels fall with use and are replenished by precipitation.

One result of global warming is the melting of surface glaciers, which in turn causes a rise in sea level and flooding of lowland port cities. Another is the change in patterns of precipitation and evaporation. Rainfall is redistributed globally so that previously fertile regions become arid wastelands.

GLACIER RETREAT THREATENS WATER SUPPLY

When Banff naturalist Mike McIvor first eyed the Athabasca Glacier in 1961, the ice was a stone's throw from Highway 93.

Today, the glacier is several hundred metres smaller, and it's just one of many whose rapid retreat has scientists worried about the future of Prairie rivers. . . .

New dating methods indicate that large ice fields in the Rockies have shrunk by around 25 per cent in the last 150 years, experts say, and global warming may be to blame. And the trend has accelerated in the last half-century.

"There's no question there, all you have to do is look at the aerial photographs," says Larry Dyke of the Geological Survey of Canada, a federal agency. "We've got air pho-

tographs back to the late '40s for all of the cordillera (mountain ridges) in BC and the Yukon. There's been quite a marked retreat." . . .

Many important rivers depend partly on the release of water from glaciers.

"If the glaciers do melt away, as they're threatening to do, then the continued flow of those streams, especially in the dry season, is threatened," said Jim Bruce, a member of the Intergovernmental Panel on Climate Change. "It will affect water supply, not just for people, but water supply to preserve the ecosystems in the deltas and in the marsh areas on which the wildlife depends."

Bill Kaufmann. From the *Calgary Sun*, 5 August 1996.

ECONOMIC DEVELOPMENT AND ENVIRONMENTAL RESPONSIBILITY

There are many possible outcomes of human activities that alter parts of the hydrologic cycle, none of which are beneficial or advantageous. The dumping of pollutants nearly destroyed the Great Lakes in North America, and continues to kill off lakes, ponds, and inland seas around the world.

Human activities that have placed stress on water resources include the building of dams to regulate river flow during seasons of intense rain and the construction of reservoirs to provide agricultural lands with irrigation water during arid periods. Hydraulic power stations are often built into dams to supply an important source of energy. In the past, societies obtained water for domestic use and crop cultivation by diverting and damming rivers and tapping into groundwater. But damming and diversion of rivers have lead to the destruction of river and canyon habitats. The tapping of aquifers to irrigate agricultural crops has drained this underground source of water to such an extent that rivers and wetlands that feed many of the aquifers have become exhausted and can no longer sustain wildlife.

River water used in agriculture can become polluted by fertilizers and pesticides used to boost production. When the water returns to the river, it may carry toxins such as manufactured organic chemicals and heavy metals downstream. In the lower stretches of the world's rivers, these pollutants may accumulate on relatively flat land and destroy aquatic plant and animal species of the delta regions. Humankind's settlement pattern of

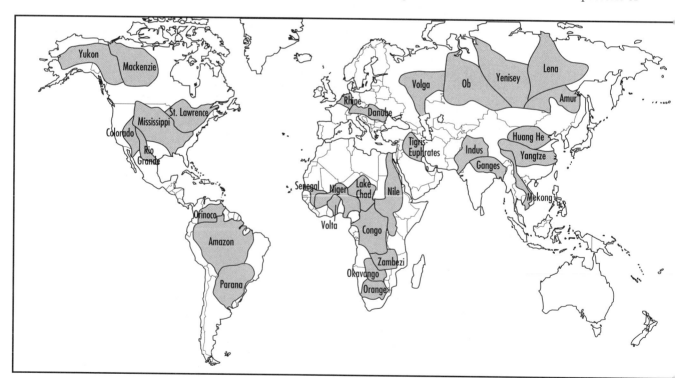

Figure 20. 2 The world's major river basins

Source: *World Resources, 1992–93* by the World Resources Institute. Copyright ©1993 by the World Resources Institute. Used by permission of Oxford University Press, Inc.

building port cities at the mouths of estuaries has led to permanent change in patterns of water flow and the ultimate blight of coastal habitats.

While the pollution of rivers may result partly from accidental or natural causes, it is largely the result of contaminants in city sewage water (human waste, detergents, solvents), chemicals used in agriculture (fertilizers, insecticides), and industrial by-products (effluents from oil/coal burning power plants, nuclear reactors, paper mills, petroleum refineries, chemical factories).

In many coastal regions, toxic chemicals from industry and untreated or partially

Russia's Baltic and Black Seas are among the most polluted on the planet. The Baltic has been poisoned particularly with toxic chemicals—uranium, thorium, mustard gas, and nuclear wastes—dumped by the military. The Black Sea, traditionally a Russian holiday resort, is catastrophically contaminated with municipal, agricultural, industrial, and military wastes. Approximately 90 per cent of its water is **anoxic**.

treated domestic sewage from cities pour into rivers and streams and flow directly into the ocean. Along their path to the ocean and once there, these substances menace life in the water and along the shores. Each year, millions of tonnes of toxic wastes are dumped in coastal waters worldwide. Particularly hazardous, PCBs and human excrement are potent mutagens and carcinogens. After reaching the oceans, some non-biodegradable pollutants return with the tides to collect on beaches, while others fall to the bottom of the ocean and mix with previous decades of sludge hazardous to marine life.

CASE STUDY

The Mystery of Lake Nakuru

Kenya's Lake Nakuru used to turn pink every year, not from toxic chemicals or rogue algae, but from a dazzling carpet of flamingos—up to 2 million of them—settling into the lake's shallow waters to feed on its rich supply of small aquatic plants, mollusks, and crustaceans. The dramatic spectacle, and the abundance of rhinos, leopards, and other animals around the lake, prompted the Kenyan government in 1968 to make Lake Nakuru a national park. Tourists followed, as many as 200 000 a year, boosting the economy of the

nearby town of Nakuru and its pride in the lake's flamingo show.

Yet today, Lake Nakuru is mostly brown, the color of the cracked mud left behind by its slowly receding waters. Desultory groups of flamingos break the monotony, but the mass migration has receded too. . . . Despite the concerted efforts of an imposing international coalition, no one knows for sure why the lake is drying up, why the flamingos are leaving—or what to do about it. Nakuru's mystery stands as a small but telling example of the difficulties in understanding the complex relationship between relentless urban sprawl and the environment.

Explanatory theories—from global warming to flamingo fickleness—abound. But the prime suspect in the avian drama is Nakuru's booming growth. With good rail and road links to Nairobi, 100 miles [160 km] to the southeast, Nakuru has attracted twenty-five major industries, from textile manufacturing to soft-drink bottling. The town's population has grown . . . from about 30 000 in the 1950s to around 360 000 [by mid-1996].

Yet urban growth has brought troubling environmental side effects. The town's already inadequate water supply is pumped mainly from underground boreholes, which could be taxing groundwater supplies to the lake. The public sewer network covers barely a third of households, and solid waste is dumped where it can leach into local groundwater. Despite safeguards, some of Nakuru's industries emit toxic materials.

The sins of the town are quickly visited upon the nearby lake and the wildlife that depend on it. Sampling in

Figure 20.3 Lake Nakuru—past (*top*) and present (*bottom*)

1994 revealed that the northeastern portion of the lake's waters, near the town, had levels of dissolved oxygen too low for healthy fish or plant life. Heavy metal, pesticides, and oil in the lake sediment have increased dramatically since 1975.

Kenya's Wildlife Services, the United Nations, and the World Wildlife Fund are conducting studies to understand better Lake Nakuru's delicate ecology and the impact of the town. Japanese assistance has been used to increase Nakuru's sewage-treatment capacity by more than 70 per cent. And the town has a program to build waste-collection sites. Nakuru's town council has even taken the controversial decision to ban industries, such as chemical processing, that pose an easily recognizable environmental threat. . . .

Yet it is difficult to make the right trade-offs between economic growth and the environment until the cause of the flamingo exodus is discovered. Dr. Warui Karanja, a Nairobi University lecturer in ecology, charges that the new sewage-treatment works, far from being part of the solution, are in fact the root of the problem. Noting that Lake Nakuru—which means "a place of dust devils" in the Masai language—used to dry up regularly before the growing town started draining its sewage into it earlier this century, [he] argues that the lack of effluent from the capacious new plant is lowering nutrient levels as well as causing the lake to go dry again.

Tim Zimmerman. From *U.S. News & World Report,* © 5 August 1996. Reprinted by permission.

CONSOLIDATING AND EXTENDING IDEAS

1 a) Many aspects of life on earth are governed by cycles. Investigate either the carbon cycle or the nitrogen cycle to determine their impacts on life on our planet.
   b) Create diagrams to compare the hydrologic cycle with the cycle of your choice.

2 Read the case study "The Mystery of Lake Nakuru," and write answers to the following:
   a) Summarize the possible causes and effects associated with "the mystery."
   b) Suggest other ways in which the town council of Nakuru could address the problems of environmental pollution.
   c) Propose ideas aimed at creating "environmentally friendly" economic growth for Nakuru.
   d) Label the location of Lake Nakuru on a map of Kenya, and then on a map of Africa. Referring to recent atlases, label other protected areas (reserves, parks, monuments, sanctuaries) and species on your two maps.

3 How important are the waters of the earth to you? Use lyric formats (haiku, limerick, ode, ballad, song, rap, epic, elegy) to reflect on this precious resource.

IDENTIFYING THE ISSUES

Some of the most topical environmental issues today include

- land-based pollution from runoff, logging, agriculture, dam construction and irrigation, and urban, commercial, and industrial activities;
- increasing coastal pollution;
- accelerated destruction of coastal marine habitats;
- declining marine mammal and fish populations as a result of overfishing and pollution;
- the "bleaching" of coral reefs;
- oil spills;
- acid precipitation;
- adequate freshwater supply.

ACID PRECIPITATION

In the middle latitudes acid precipitation affects many wilderness areas downwind from urban centres. When mineral ores containing high concentrations of sulphur are smelted or when coal with high sulphur content is used to generate electricity, sulphur is released into the atmosphere. Nitrous oxides are also emitted into the atmosphere when hydrocarbons are burned in automobile engines. These two chemicals combine with water in the atmosphere to form dilute sulphuric and nitrous acids. Areas downwind from steel mills and urban centres are then subjected to this acid precipitation.

The physical geography of a region either intensifies or reduces the effects of acid precip-

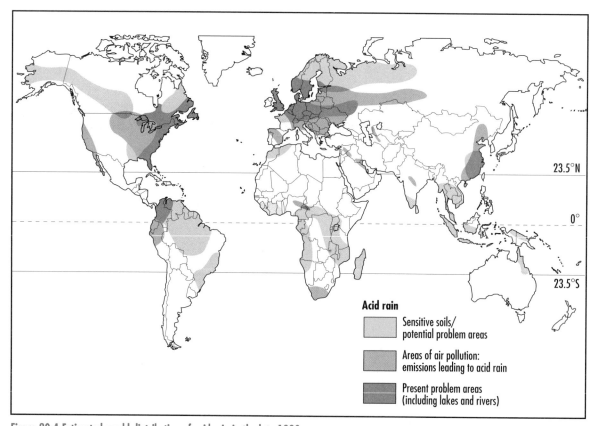

Figure 20.4 Estimated world distribution of acid rain in the late 1980s

Source: Adapted from *Atlas of the Environment,* ©Banson Marketing Ltd., 1990, 1992. Reprinted by permission of Helicon Publishing Ltd.

itation. Soil with high concentrations of calcium or other alkalines neutralizes the acid. For this reason, acid precipitation does not present a problem in southern Ontario or the Prairies, where sedimentary rocks such as limestone and dolomite form the parent materials from which soils are derived. If, on the other hand, the soil is formed from igneous rocks, which are either neutral or slightly acidic, there is no neutralizing effect, and in fact the problem of acid precipitation can be intensified. Regions that experience acid precipitation include the Canadian Shield, the Appalachians and Adirondacks in the United States, Germany's Black Forest, the Alps, many regions in the former Soviet Union, the highlands of Scandinavia, and most of Japan.

In temperate regions, most of the ecological damage that results from acid precipitation occurs in early spring. The snow that has accumulated over the winter quickly melts in a matter of days or weeks. The sudden rush of stored acidity into streams and lakes kills fish, amphibians, and reptiles while eggs are hatching and immature individuals are struggling to survive. As the food chain breaks down, aquatic birds abandon the area and the entire aquatic ecosystem dies. Acidity also enables lake water to dissolve heavy metals such as mercury and cadmium that occur naturally or as a result of industrial pollution. These heavy metals cause genetic mutations and nervous system disorders all the way up the food chain.

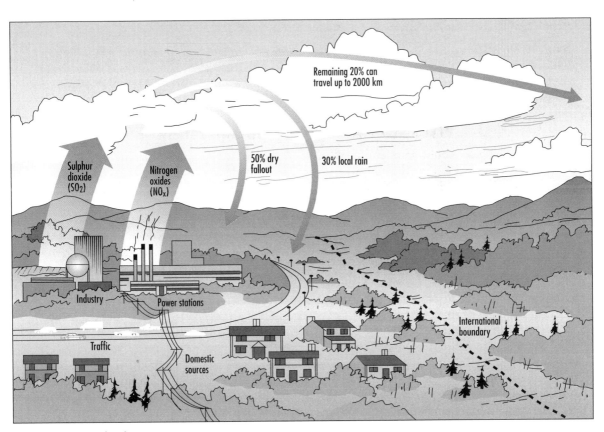

Figure 20.5 Causes of acid rain

Source: *Understanding Our Environment*, 2d ed., by S. Dunlop and M. Jackson. Copyright © Oxford University Press Canada. Reprinted by permission of Oxford University Press Canada.

In addition to the loss of animal life, trees lose their ability to assimilate nutrients from the soil and they, too, slowly die. In Switzerland avalanches have become more common because of acid precipitation. Avalanches are a form of mass wasting that results when snow builds up on steep mountain sides. In the past, trees held the snow in place and the risk of avalanches was low in the valleys, where temperatures were warm enough for trees to grow. Today, vegetation is dying on mountain slopes because of the fumes from diesel trucks travelling through mountain passes. The reduction of vegetation on the lower slopes of the Alps is allowing snow to slide more readily into mountain passes. Artificial snow barriers are being built at great expense but the results are not as effective as the original vegetative cover. Of course, the magnificent natural beauty of the region is also being destroyed because of the increased vehicular traffic. As a solution, Swiss authorities have restricted truck traffic and increased the use of piggy-back electric trains. Truck trailers are shipped on flatbed cars through the mountains and are unloaded, using trucks, at their destinations.

The United Nations Environment Programme (UNEP) estimates that damage from acid deposition could be as high as 1 to 2 per cent of a country's GNP. One answer to the problem of acid rain is to pour buffering agents such as limestone into acid lakes. The increased

---

Deformed frogs have been turning up in Minnesota, Wisconsin, and in Quebec's St. Lawrence River Valley. In the summer of 1996, a research team from the University of Minnesota's Morris campus and a pollution agency received an emergency grant from the Minnesota legislature to study the frog problem.

DEFORMED FROGS WORRYING SCIENTISTS

...The team—which could scarcely keep up with the reports pouring in from all over the state—found frogs with missing legs, extra legs, misshapen legs, paralyzed legs that stuck out from the body at odd places, legs that were webbed together with extra skin, legs that were fused to the body.

They also found frogs with missing eyes. One memorable specimen was a one-eyed frog that turned out to have the second eye growing inside its throat.

Significantly, the mink frog, the species with the highest incidence of deformity at around 50 per cent of the total, is the species that spends the most time in the water. American toads and wood frogs, the least aquatic species, had deformity rates of less than 5 per cent.

Frogs with compromised limbs cannot feed themselves or escape from predators. . . It is rare to find an adult frog with a substantial limb abnormality.

Early evidence to explain the frogs' deformities points to something in the water where frogs breed and develop, in which they spend every stage of life. Their skin is highly permeable; what gets in the water can get into the frogs. . . .

Also on the table as possible causes are viral or bacterial disease; the presence of various heavy metals in the water known to cause [congenital disabilities]; farm pesticides; acidification of the water; and increasing ultraviolet radiation as the earth's ozone layer is depleted.

William Souder. From *The Washington Post*, 12 October 1996.

alkalinity raises **pH levels**. The cost, however, is prohibitive in view of the number of lakes and the amount of limestone needed. The best solution is to lower the emissions that cause acid precipitation. The burning of fuels containing low levels of sulphur reduces the amount of sulphur dioxide that enters the atmosphere. Scrubbers are devices that remove sulphur content from exhaust fumes in chimneys. In New Brunswick and Prince Edward Island, **fluidized bed combustion** has been used to remove up to 90 per cent of the sulphur and nitrogen in coal used in thermal power plants. The coal is burned with a finely ground mixture of limestone. The limestone reacts with the impurities to remove them in solid form, so that they cannot escape into the air. In Nova Scotia, coal-water fuel technology uses water to wash the sulphur and nitrogen from the coal.

Politics plays a significant role in addressing this serious problem. The environment

"He wasn't always bald. It's acid rain."

Herman® is reprinted with permission of LaughingStock Licensing Inc. All rights reserved.

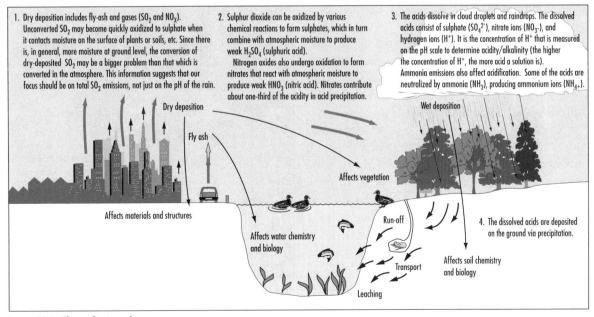

1. Dry deposition includes fly-ash and gases (SO_2 and NO_x). Unconverted SO_2 may become quickly oxidized to sulphate when it contacts moisture on the surface of plants or soils, etc. Since there is, in general, more moisture at ground level, the conversion of dry-deposited SO_2 may be a bigger problem than that which is converted in the atmosphere. This information suggests that our focus should be on total SO_2 emissions, not just on the pH of the rain.

2. Sulphur dioxide can be oxidized by various chemical reactions to form sulphates, which in turn combine with atmospheric moisture to produce weak H_2SO_4 (sulphuric acid).
Nitrogen oxides also undergo oxidation to form nitrates that react with atmospheric moisture to produce weak HNO_3 (nitric acid). Nitrates contribute about one-third of the acidity in acid precipitation.

3. The acids dissolve in cloud droplets and raindrops. The dissolved acids consist of sulphate (SO_4^{2-}), nitrate ions (NO_3^-), and hydrogen ions (H^+). It is the concentration of H^+ that is measured on the pH scale to determine acidity/alkalinity (the higher the concentration of H^+, the more acid a solution is). Ammonia emissions also affect acidification. Some of the acids are neutralized by ammonia (NH_3), producing ammonium ions (NH_4^+).

Dry deposition

Fly ash

Wet deposition

Affects vegetation

Affects materials and structures

Affects water chemistry and biology

Run-off

4. The dissolved acids are deposited on the ground via precipitation.

Transport

Affects soil chemistry and biology

Leaching

Figure 20.6: The acid rain cycle
Source: Federation of Ontario Naturalists.

knows no boundaries—pollution that originates in one country can easily float into an adjoining country. This is especially true in southern Ontario where pollution originating in the United States is blown into Ontario by the westerlies. Unfortunately, the United States and Canada do not have the same attitude towards the control of sulphur and nitrogen emissions.

Acid precipitation leads to growth slowdown in deciduous trees and can destroy entire forests. It causes lake acidification, ultimately decimating plant, fish, mammal, and bird populations. Acid rain eats paint, corrodes steel, and accelerates erosion of national monuments. It also affects lung capacity in human beings and increases the incidence of disease. Until governments cooperate to address this problem—by introducing and enforcing regulatory programs and legislation, and by offering incentives to industry to clean up its practices—toxic acid rains will continue to fall.

OIL SPILLS

The world's oceans are among its greatest treasures. They provide protein-rich food to millions in the form of fish, plants, sea mammals, and other sea foods that they produce. But they are also a source of wonder and beauty. If the oceans are polluted, habitats will be lost for untold species of animals and plants.

Oil spills are one of the greatest threats to the world's oceans. They occur when giant oil tankers lose their cargo. Supertankers may be more than 300 m long, as much as 30 m from deck to the bottom of the hold, and can hold more than 200 000 t of crude oil. It may take a supertanker 400 m to stop, therefore, when it is faced with an unforeseen obstacle such as an iceberg, there is potential for disaster. Crude oil floats on the surface of the water and is carried from one place to another by ocean currents, tides, waves, and wind. Oil spills have a devastating effect on marine life, especially sea birds.

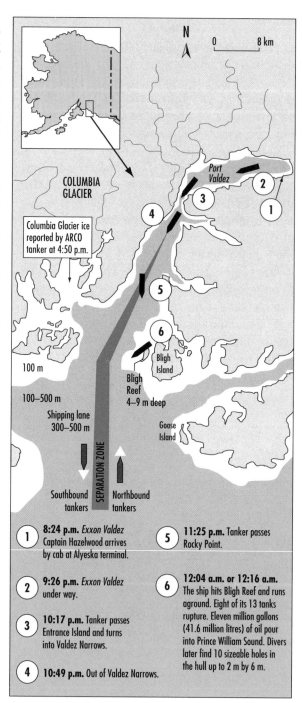

Figure 20.7 The Alaskan oil spill of 1989

Source: Alan B. Nichols, "Alaskan Oil Spill Shocks the Nation," *Journal of Water Pollution Control Federation* 61 (July 1989). © 1989 Water Environment Federation. Reprinted with permission.

As they try to clean off the oil from their feathers, they ingest the tar and die. Fish can usually avoid the pollution because they live below the surface of the water. But fish whose habitat is close to the surface collect tar in their gills, and suffocate. Sea mammals such as sea lions, sea otters, whales, and porpoises have to surface to get air. Their nostrils and blowholes become clogged and they can suffocate. Natural habitats are destroyed as the tar washes up on rocks that serve as rookeries and nurseries for sea birds and marine mammals.

Tankers routinely pollute the sea with oil residue. Once their holds have been emptied of cargo, a thick, oily tar lines the hold. The walls are hosed down and the oily water is dumped into the ocean. Although this practice is outlawed by UNEP, many tankers continue to pollute the oceans in this way. Usually, they do so in the open seas where they are unlikely to be caught. Unfortunately, this is also where the greatest number of fish live.

Today, new technology can help detect oil spills and reduce the damage they inflict. Satellites now monitor ocean shipping. Oil slicks can often be traced to leaky tankers as they travel the

The *Exxon Valdez* accident was the worst environmental disaster ever to hit the Pacific coast of Alaska. Human error was blamed when the massive vessel transporting crude oil from northern Alaska ran aground on 24 March 1989. Thousands of local people worked day and night to mop up the oil. It cost Exxon over US $1 billion to clean up the area. But wildlife and habitats will take years to recover fully.

seas. The ships can be identified and shipping companies fined for improper maintenance. Satellites also play a role in navigation. They are able to detect ships, icebergs, and other hazards so that an approaching ship can be notified of the danger.

Engineers have now developed a pollution-free method of reclaiming oil residues from the holds of ships. Instead of water, jets of oil are used to wash oil tar off the walls. The tars are then off-loaded with the main cargo.

To reduce the risk of accidental oil spills, companies are building stronger, sturdier tankers. Reinforced double holds help to increase the strength of the ships' hulls. The use of smaller tankers

Figure 20.8 Oil spills are a threat to sea birds, fish, and sea mammals.

ensures that if a spill does occur, less oil will escape into the sea. Holds are now divided into contained sections so that if one section ruptures, the leak can be restricted to that section.

If a slick does occur, one solution is to allow nature to clean up the mess. Naturally occurring bacteria eat the oil, waves disperse it, and in a relatively short period of time (forty to fifty years) most ecosystems return to their normal states. But many people consider this amount of time too long. Scientists are working on strains of bacteria that will consume the oil more quickly.

THE WATER SUPPLY

In the decades to come, shortages of water, the uses of lakes, rivers, and oceans, and pollution of water supplies will likely lead to new issues surrounding territoriality and ownership. The competition for this ubiquitous natural resource will have national and international consequences. When we looked at resource issues in Chapter 10, water rights between Israel and the Palestinians were emphasized as one of the paramount issues in the peace process. Increasing urbanization is presenting a challenge to local governments—Mexico City, São Paulo, Lima, and

A Canadian invention called a "slick-licker" consists of a conveyer belt that skims the oil off the surface of the water and collects the oil for reclamation. Of course, it is only practical for relatively small oil spills. The 3M Corporation has also developed a device called the "Giant Sponge" to clean up oil spills.

According to the United Nation's medium-term projection of demographic growth, the number of people affected by water scarcity will soar to 3.3 billion in 2025 and to 4.4 billion by 2050. Countries such as Ethiopia, Somalia, Kenya, Rwanda, Burundi, Malawi, and South Africa will confront extreme situations of water scarcity.

Buenos Aires are four Latin American megacities where the provision of a freshwater supply is a daily struggle. Urban needs compete with rural needs; as more water is required in swelling cities to quench the demands of households and industry, more water is also needed in the countryside to irrigate food crops grown to supply the cities.

The most serious resource problem that populations will face twenty-five years into the future will be the shortage of freshwater. The problem is one of uneven distribution. Over half of the world's area of freshwater, and 15 to 20 per cent of its volume, is in Canada. The most water-rich nation in terms of runoff from rainfall is Iceland. The world's most water-poor nation is Egypt; China and India also face severe water shortages.

Competition between different interests for dwindling, unspoiled natural resources is vigorous today in all developed areas of the world. In the future, competition between neighbouring countries may lead to water-grabbing at national borders, perhaps even to war. Water-rich nations may be willing to exert their economic power, as did petroleum-rich countries during the 1970s and 1980s. Water will become a political issue of the next millennium.

CONSOLIDATING AND EXTENDING IDEAS

1 In the past twenty years the number of people living in cities has tripled. This has placed increasing strain on urban water resources. Research the water issues of megacities in developing nations, for example, São Paulo, Buenos Aires, Lima, and Mexico City.

2 While we tend to be better informed about oil spills that occur in the oceans, oil spills and diesel-based wastes on land present enormous challenges in terms of clean-up. Hydrocarbon wastes dumped or spilled near thousands of oil wells in Alberta pose a danger of seeping into and contaminating the groundwater system.

   In a couple of paragraphs, suggest the impacts of oil spills that seep into the groundwater, and possible ways of reducing the risks associated with oil drilling.

3 Just as this century has been devastated by wars over oil, there is increasing concern that the next century may experience wars over water. Unlike other natural resources, the total amount of water in the world is constant and can be neither increased nor decreased. Water can, however, be degraded. Over a three-week period, read your local/regional newspapers and magazines and cut out articles about water, water pollution, sewage treatment, and the costs of water supply. Review your clippings, then write a brief report that summarizes your findings.

4 As a class, speculate about the national and international consequences that arise from
   • acid rain deposition;
   • oil spills in oceans;
   • freshwater shortages.

COUNTRY PROFILE: CHILE

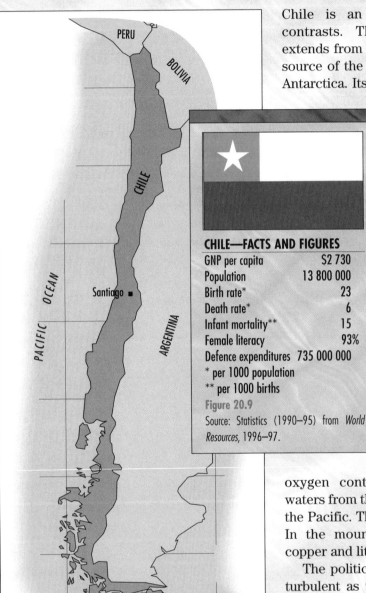

CHILE—FACTS AND FIGURES

GNP per capita	$2 730
Population	13 800 000
Birth rate*	23
Death rate*	6
Infant mortality**	15
Female literacy	93%
Defence expenditures	735 000 000

* per 1000 population
** per 1000 births

Figure 20.9

Source: Statistics (1990–95) from *World Resources*, 1996–97.

Chile is an incredible land of many contrasts. This long, narrow country extends from the subtropical land near the source of the Amazon all the way south to Antarctica. Its physical diversity is extraordinary. In the north is the world's driest desert, the Atacama, where rain falls so infrequently that virtually nothing survives. In the south the land is wet and mountainous, not unlike the northern coast of British Columbia. The middle region is a temperate land, bordered on the east by the towering Andes—the second highest mountain chain in the world. This region has a Mediterranean climate and is renowned as one of the world's best wine-producing areas. Off-shore, cold ocean currents with a high oxygen content mix with nutrient-rich waters from the many rivers that empty into the Pacific. The result is a rich fishing area. In the mountains abundant deposits of copper and lithium are mined.

The political scene in Chile has been as turbulent as the volcanic mountains in its interior. In 1970 Salvador Allende, leader of the Communist party, was elected president. A military coup, led by General Augusto Pinochet and with some indirect support

Figure 20.10 Santiago, the capital of Chile, lies at the northern end of the fertile Central Valley, within view of the Andes.

from the US government, overthrew Allende in 1973. He was killed during the coup. Pinochet imposed a harsh, repressive military rule until he was forced to call elections in 1989. The Chilean people voted to end his dictatorship, and a democratically elected president took office in March 1990. But Pinochet continues to be in charge of the army. The government is powerless to act on its own because it is virtually controlled by the constitution Pinochet left in place. A precarious peace exists between the military and the civilian government. Despite this situation, the country is prospering economically.

With the fastest-growing economy in Latin America, Chile was invited to join the North American Free Trade Agreement with Canada, the United States, and Mexico in 1996. The country has a strong economic link with Canada through the export of fresh fruits and vegetables to northern cities such as Edmonton and Hamilton during the winter. As the world's largest producer of copper and lithium, Chile is a welcome trade partner in the industrial north, where these raw materials are needed. Chile has a high literacy rate and a good health-care system, and its political stability is improving. On the verge of becoming a developed country and an economic powerhouse, Chile's continuing prosperity depends on the government's ability to remain in control. A powerful army is waiting in the wings, ready to intercede if the government falters.

CHAPTER 21

SOLVING ENVIRONMENTAL ISSUES: EARTH, AIR, AND WATER

Individuals, cultures, and governments through-out the world have different attitudes towards issues of environmental biodiversity, disposal and storage of toxic wastes, deforestation, global warming and ozone depletion, acid rain, and water supply. By carefully examining these opposing viewpoints, we will be able to understand the limitations of our own opinions, to consider the strengths of other viewpoints, and to reformulate our opinions.

When it comes down to dollars and cents, economists and legislators, hard-pressed to reduce government spending and maximize profits, do not want to spend money on solving

Figure 21.1 The results of land, air, and water pollution: *left*, garbage along a highway in Venezuela; *top right*, smog over Paris; *bottom right*, contaminated water in the St. Lawrence River

issues if they are not really significant over the short term. Owners and employees in business and industry are similarly concerned with profits over each fiscal year. Scientists, environmentalists, and a growing number of private citizens, on the other hand, are apprehensive about the long-term trends that we are starting to experience today. They want to be proactive, rather than reactive, by addressing these issues and working towards solutions before environmental damage becomes irreversible.

SOLVING THE ISSUES: THE EARTH

Human activities are destroying much of the living world. We face mounting dilemmas as we try to dispose of toxic waste—waste composed of chemicals created by science and industry. Alien to the environment, these chemicals cannot be broken down by nature. Each year, our cities and rural areas consume increasing amounts of energy. This depletes nonrenewable resources and exacerbates pollution. Agricultural practices that are foreign to nature reduce wild varieties in species to a single crop, and use non-biodegradable chemicals to enhance growth and production. Habitat destruction, the devastation of the rainforests, and shrinking biodiversity demand that humankind bring an end to the heedless exploitation of the ecosphere.

For every technological intervention there is an ecological reaction. We need to think of solutions in broad and integrated ways. Manufactured chemicals must be more tightly controlled and their use must be registered. Where they exist, waste-management laws need to be strictly enforced. Nations without such laws must develop legislation to prevent dumping. Through technology and research, chemical producers and users must find less toxic alternatives.

Recycling can save extractive costs in terms

Figure 21.2 Alternative renewable energy sources: *top,* solar energy collectors in Saudi Arabia; *bottom,* a wind-powered electric generating facility in Hawaii

of energy. It takes twenty times as much energy to make aluminum cans from scratch as it does from recycled cans. Recycling contributes to cleaner air and water, and with less solid waste for disposal, pollution in landfills and oceans can be reduced. "Precycling"—making purchasing choices to support responsible products and packaging—links today's consumerism with tomorrow's environment by making recycling easier and by reducing the garbage destined for landfills.

The development and use of alternative renewable energy sources, for example, wind, water, and solar power, offer us twofold savings: reduced depletion of nonrenewable resources, such as oil, gas, and coal, and improved air quality for all living things. Developed and developing nations need to consume energy prudently and to regulate its uses. Through laws and building regulations we can force industry to use energy more effectively. In North America, legislation has been passed compelling the automobile, refrigerator, and incandescent light bulb industries to create products that use energy more efficiently.

Many recent and current agricultural practices have resulted in short-term increased production, but have led to long-term soil degradation, polluted groundwater, and, ultimately, to decreased crop production because of less arable land. Agriculture must become a sustainable industry. Crop rotation and the use of animal and vegetable fertilizers, instead of chemical fertilizers, herbicides, and pesticides, can perpetuate the use of cultivable lands indefinitely by improving soil nutrition.

Some strategies, such as forest clear-cutting, waste natural resources. People in positions of responsibility must be educated to develop policies that protect the tropical, montane, and temperate forests of the world. Forest-management, reforestation, and timber-harvesting programs that encourage sustainable

Figure 21.3 Half of all the species on earth are found in tropical rainforests.

use of this resource have to be created, shared, and enforced worldwide. Government and non-government intervention are often necessary to prevent habitat destruction, and endangerment and extinction of species.

The tropical rainforests are an important source of biodiversity. They are believed to be home to at least half of all living species. It is perhaps unrealistic to expect that economic growth and development will stop because this natural habitat is endangered. It is also unrealistic to expect that political intervention is the most effective means for bringing about economic diversity. The article "Can Marketing Save the Rainforests?" on the following page discusses the attempts of American companies to save this ecosystem.

CAN MARKETING SAVE THE RAINFORESTS?

It is a far cry from when many environmentalists thought that the best way to save wild habitat was to declare it off-limits to the destructive habits of humans. Now, ecologists hope to convince those who live in and near rainforests that the trees and plants in their backyards can provide a continuing source of income.

In 1989, Charles Peters, of the New York Botanical Gardens' Institute of Economic Botany, completed a study that finally provided rainforest conservationalists with the data they needed to prove that a rainforest is worth more intact than logged or burned for cash crops or cattle pasture, the usual fate of the 50 million acres [20 235 000 ha] of tropical forest lost worldwide each year.

Peters and two colleagues found that the fruits and rubber that could be gathered from 1 ha of Peruvian rainforest could net a harvester $422 a year, *every* year, which adds up to a far more secure future than the $1000 that hectare would yield, just once, if it were cut for timber.

Now, in Guatemala, Peru, the Philippines, Ecuador, Zambia, Indonesia—in nearly every country where there are still large stands of rainforest remaining—experts are combing the forests to see what might be harvested and sold.

The number of products with origins in the rainforest already on the market seems remarkable—until you consider that half of all the species on earth are found in tropical forests. Oranges, lemons, bananas, pineapples, chocolate, coffee, avocados, resins used in paints and varnishes, latex, bamboo, rattan, many dusky-hued hardwoods used to make furniture, and more than forty prescription drugs have their origins in wild tropical rainforests.

The members of the Xapuri Agroextractive Cooperative are rubber tappers in the state of Acre in Brazil. To supplement the income they earn from latex tapped from rubber trees in the Amazon, they now gather the fruit of the Brazil nut trees from the forest floor. Inside each softball-sized fruit are up to twenty nuts, which are brought to a nearby processing plant. There the nuts are dried, shelled, and sorted for shipment to foreign markets.

In 1989, Cultural Survival, an indigenous-rights group based in Boston, bought a shipment of Brazil nuts from the rubber tappers and sold them at 5 per cent above the market price to a new company called Community Products, which used them in a nut brittle it dubbed "Rainforest Crunch." Cultural Survival returned 100 per cent of the profit from the sale to the Xapuri Cooperative. Suddenly, a movement was born. Cultural Survival now sells fifteen commodities from ten to fourteen different countries to sixty-six companies. The program has thus far generated about $3 million. All profits are returned to the product harvesters.

Another environmental group, Conservation International (CI), also works in the production and marketing of rainforest products. CI first identifies what it calls a "hot spot" ecological system, a natural area rich in biodiversity that faces development threats. Working with the people nearby, CI designs programs that involve community development, conservation science, long-term management, and policy work.

CI's first product extracted under this multifaceted program was tagua nut, the hard seed of a rainforest palm tree. Tagua nut, sometimes called "vegetable ivory" for its creamy white color and easily carved texture, is sold as buttons, jewelry, and carvings. CI linked up with a community development group outside the Cotacachi-Cayapas Ecological Reserve in Ecuador to establish a locally run business that purchases tagua and resells it to factories on the coast, which in turn shape the tagua into disks for buttons. CI then found three US button manufacturers who would buy the semi-processed tagua disks, as well as companies who would buy the finished buttons. In the first year of CI's "Tagua Initiative," 7 million buttons worth $500 000

were sold. Conservation International is developing similar programs with a variety of products in ecological hot spots in Columbia, Guatemala, Peru, and the Philippines.

The majority of rainforest product companies and organizations are led by people with sincerely good intentions, who are trying to merge heart and business in a positive way. It's a step toward maintaining the forests for the future.

Diane Jukofsky. From *E: The Environmental Magazine*, July/August, 1993.

CONSOLIDATING AND EXTENDING IDEAS

1 Many people link their personal behaviour with their principles regarding protection and conservation of the environment. Add your suggestions for environmentally friendly practices to the list below.
   • Avoid products with excess packaging.
   • Reuse shopping bags (canvas or string).
   • Buy products made from recycled materials.
   • Choose products that are packaged in recyclable containers.
   • Do not dispose of hazardous household wastes in the municipal trash; take them to collection centres.
   • Plant trees and gardens.
   • Use organic fertilizers and pesticides.

2 Read the article "Can Marketing Save the Rainforest?" on page 339.
   a) What types of products originate in tropical rainforests?
   b) Can logging of the rainforest be prevented if indigenous peoples and local farmers are able to make a steady income by harvesting nuts, oils, and other renewable resources? Explain your viewpoint.

SOLVING THE ISSUES: GLOBAL WARMING

It is impossible to restore carbon dioxide, methane, and other greenhouse gases to pre-Industrial Revolution levels of the eighteenth and nineteenth centuries. The only option for solving the problems of global warming is to reduce the production of greenhouse gases now and in the future. Still, a considerable amount of damage has already been inflicted on the environment. More responsible attitudes will not only diminish the warming trend but will also help to alleviate other environmental issues.

ATTITUDES IN THE DEVELOPED WORLD

Diversified economies contribute to global warming through the sheer volume of their consumerism. Every product we buy contributes to global warming; energy is needed to produce the goods, transport them to market, and heat and light the retail outlets that sell them. Thus, each pair of new jeans, each new CD, each new high-tech toy means that more and more greenhouse gases are being emitted into the atmosphere. If we as consumers were more conscious of the effects of our actions, global warming could be curbed.

Conservation is the most effective means of limiting our consumption of energy-consuming products. By reducing the amount of greenhouse gases we produce either directly or indirectly, we can help lessen the impact of our lifestyles on global warming. Walking or cycling to school or work, instead of driving, lowers gasoline consumption and, at the same time, is conducive to a healthy lifestyle. If walking or cycling is not possible, we can opt for public transit or car pooling. A well-insulated and well-ventilated home reduces heat loss in winter and heat retention in summer. We can also conserve energy in the way we use household appliances. For example, wash larger loads of clothing and use more cold water; hang clothes to dry rather than running the dryer; and run the dishwasher only with a full load.

Television is another way in which we contribute to global warming. Many of us like to unwind each evening in front of the TV. For numerous Canadians this involves watching several hours of TV each day. We even leave the television on when we're not watching it! In many homes, there are several television sets to ensure that family members can each watch their favourite programs. So how does this contribute to global warming? Like so many of our modern conveniences, televisions consume energy. The more televisions we have and the more hours we leave them on, the more we are adding to the problem.

Business and industry must also play a part in moderating the harm their activities inflict on the environment. Simply turning off the lights in office buildings at night lowers the amount of electricity consumption. Since some of our electricity comes from the burning of oil or natural gas, this practice would reduce emissions. Reusable packaging would not only lower production costs, it also would decrease the amount of energy needed to produce the packaging. Companies that transport goods must ensure that vehicles are properly tuned up and that drivers do not exceed speed limits, as these two factors result in increased gasoline consumption. Buildings should be designed to take advantage of passive heating and cooling systems. Office buildings with blinds or insulated glass are cooler in summer and warmer in winter. These are only a few ways in which companies can reduce energy consumption and expenses.

The emergence of "green" companies that manufacture environmentally friendly products has been a trend in recent years. These companies often manufacture goods from recycled materials and eliminate excessive—and therefore wasteful—packaging. When packaging is minimized, so too are carbon dioxide emissions because less energy is used to manufacture the packaging. Many consumer products such as cleansers and soaps are now available in refillable packages. In addition to reducing the amount of harmful emissions, this practice also keeps costs down, for both the consumer and the manufacturer.

HEALTHY HOUSE

Total operating costs of the 1700 square foot [518 m²] house will be less than $800 a year.

... Home builder Rolf Paloheimo and his family are looking forward to living in [architect] Martin Leifhebber's latest revolutionary design—a house in the heart of Toronto that will have no connections to the city's water, electricity, or sewage lines. ...

The Paloheimos ... will rely on rainwater and snow-melt for their water, sunshine for much of their heat and electricity, and a miniature sewage plant in the basement to convert solid waste to compost. Water from morning showers will be collected, recycled through a natural cleaning process, and used later to wash dinner dishes.

And that may become normal living for many if our cities are to become more efficient with higher densities, healthier living, and environmentally friendly, says Leifhebber, one of the country's leading environmental architects. ...

Solar panels on the top and the sides of the house will send 10 kW·h of electricity a day in summer to be stored in batteries. The DC power in the batteries is converted to AC 115 V power to run electrical hardware in the home. ... If there is no sun for several days, a methanol-fuelled auxiliary generator in the basement takes over. ...

The four-level home with three bedrooms and 1700 square feet [518 m²] of living area will get about 66 per cent of its winter heat from passive solar energy entering the home via large triple-glazed windows on the south side ... There is also radiant heat from solar-heated water circulating through pipes in the concrete floors and ceilings. ...

Summer cooling will be done with a large vent at the bottom of the house to drink in cool night air that will flow through to a large vent at the top of the house. ... And the water in the 20 000 ℓ cistern will be circulated through piping below the frost line outside the house to cool it, thus creating a cool reservoir in the home through the summer. ...

Pat Brennan. From the *Toronto Star*, 2 December 1995. Reprinted with permission— the Toronto Star Syndicate.

WHAT MAKES IT WORK (see Figure 21.4):

1 Rooftop solar panels generate electric energy that can be stored for later use.

2 Airtight wall construction cuts heat loss, eliminates drafts, and minimizes entry of moisture and pollution.

3 Thermal windows are situated to provide natural light and passive solar heat, cutting power and heating costs.

4 Building envelope has high levels of insulation to improve energy efficiency.

5 Low-volume toilets, low-flow shower heads, and aerator faucets conserve water.

6 Material used to furnish and decorate home emit few chemicals and vapours.

7 Rainwater is collected, filtered, purified, and stored for drinking and washing. It is recycled for use in appliances and toilet.

8 Waste water from appliances and toilet is filtered and treated before being discharged.

9 First-floor design is ideal for a home office. Working at home reduces the need for transportation, reducing pollution.

10 Design is appropriate for building on infill lots, promoting efficient use of land and reducing urban sprawl.

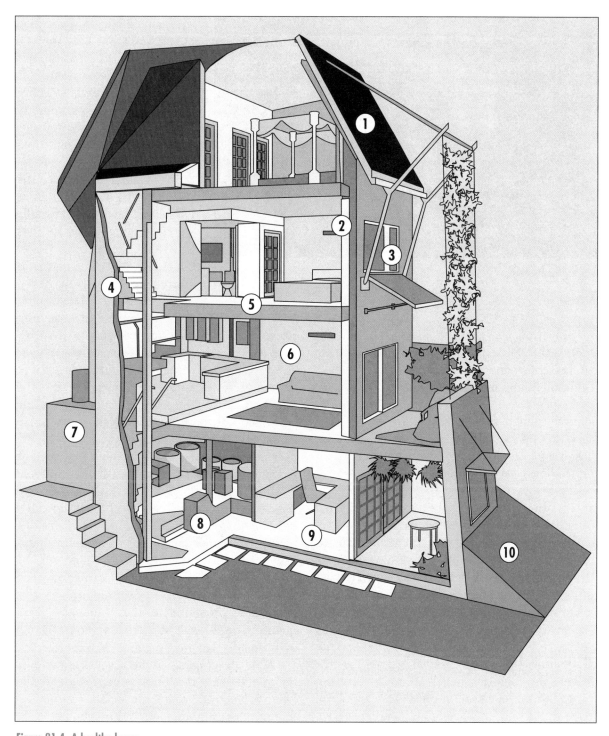

Figure 21.4 A healthy house

Figure 21.5 A "green" product
The box is made from 100 per cent recycled fibre; the super-concentrated powder results in a 30 per cent reduction in packaging.

Governments need to play their part to ensure that all segments of society curb their energy consumption. In regions as diverse as Ontario, Minnesota, and Denmark, clean air laws, incentives to encourage home insulation, and recycling have been successful energy-conservation policies. However, decisions are often based on votes at the next election rather than on policies that would achieve long-term results. Governments try to provide what the voters want. If people are preoccupied with other issues, such as debt reduction, sovereignty, inflation, or unemployment, it is more likely that legislation on other issues, such as the environment, will be placed far down the list of priorities. It is the responsibility of voters to make elected officials aware of their concern for the environment.

ATTITUDES IN THE DEVELOPING WORLD

Developing countries contribute to the problem of global warming in a different way. Until recently, people in developing countries have had a minimal impact on global warming. They do not use the enormous amounts of fossil fuels

that populations in diversified economies do. Few people own automobiles, and the use of fuel to heat homes is uncommon. They also have fewer material possessions, all of which require energy to be produced. The only contribution most people in subsistence economies make to the production of greenhouse gasses is the fuel (usually wood or dung) used to cook the evening meal.

However, the situation is starting to change. As developing countries industrialize, the consumption of fossil fuels rises. Rural peasants are moving to the city. They buy bicycles and other consumer goods, all of which need fossil fuels either directly or indirectly. As urbanization, education, and the standard of living increase, people are adopting consumer attitudes similar to those in North America. Often, consumers in developing nations look to "modern" conveniences such as refrigerators, but in some cases they are sold old-product technology, which creates the same problems in the developing world as it did in the developed world thirty years ago. Media advertising encourages people to consume hamburgers, drink cola, and wear jeans. Through the media, people observe a North American lifestyle of excessive consumption, yet this lifestyle is promoted as the "good life." Frequently, traditional attitudes of conservation and frugality are put aside.

Many developing nations are also experiencing population growth that is unprecedented in history. More and more consumers continue to pump increasing amounts of greenhouse gases into the atmosphere. People migrate from overcrowded areas to marginal agricultural regions, where they burn rainforest to obtain plots of land for subsistence agriculture. The carbon stored in these trees is released into the atmosphere. Despite all these factors, the developing world's contribution to greenhouse gases pales in comparison to that of the developed world.

In countries such as India and China, economic growth is phenomenal. As these economies are transformed from primary industry to secondary industry, there is a growing need to develop industrial infrastructures such as roads, railways, factories, smelters, and power plants. But these facilities demand enormous energy consumption. The huge levels of carbon dioxide produced by these newly emerging economic powers are likely to increase the greenhouse effect and global warming. Yet these countries are merely following the same path as long-established economic powers such as Britain and the United States, which owe their early industrial growth to the burning of coal to produce energy.

Studies by the World Resources Institute for 1993 show that half of all commercial energy was consumed by developed countries. Energy consumption in these countries has grown by 30 per cent over the past twenty years. By contrast, energy consumption dropped by 17 per cent from 1989 to 1992 in the former Soviet Union and in Eastern European countries adjusting after the fall of Communism. In 1993, a rise in energy consumption indicates some improvement as their economic systems redevelop. The largest area of energy consumption growth is in the developing world. Asia accounts for 60 per cent of the energy used by developing countries in response to the burgeoning economies of Korea, Thailand, Malaysia, Singapore, and especially China. But increased development in South America and Mexico also shows signs of substantial growth in energy consumption. The International Energy Agency (IEA) and the US Department of Energy (DOE) both forecast enormous growth in the Far East well into the next century. Both agencies project a doubling of energy demand in Asia by the year 2010. In Latin America a rise of 50 to 77 per cent is predicted. By the year 2010, energy consumption in the so-called developed world will more or less equal that of developing nations. The fact that fossil fuels will continue to account for about three-quarters of the energy demand is a disastrous development as far as global warming is concerned. Is it appropriate that developing nations be told to limit their industrial growth to reduce greenhouse gas emissions?

The solution seems to be the development of alternative energy sources. The burning of biomass in the form of wood, dung, charcoal, and straw accounts for about 10 per cent of the world's energy resources. While these products are renewable, they still generate just as much carbon dioxide per unit of energy as do fossil fuels. So their consumption is not a solution to global warming. Large hydroelectric developments provide about 6 per cent of the world's energy requirements; they contribute no carbon dioxide and are clean energy sources. But as we have seen (pages 284 to 286), there are other environmental factors that make these projects harmful to local ecosystems. There is great potential for future hydroelectric development in many regions thirsty for electricity. China, for example, has the potential to generate an estimated 2 000 000 MW of electricity from hydro developments. Currently, China has developed only 60 000 MW of hydroelectricity. The development of alternative energy sources such as wind power, geothermal energy, and solar power has great potential, but at the present time energy produced from these sources is very small—about 2 per cent of total energy supplies. Growth will likely be limited unless governments and consumers support the development of these sources. The World Energy Council (WEC) forecasts that non-traditional energy sources will account for only 4 per cent of total energy sources by 2020 if current policies continue. "However, WEC projects that under an 'ecologically driven' scenario, which assumes vigorous energy conservation measures and deliberate economic and regulatory incentives to accelerate the penetration of renewable energy resources, these new technologies could contribute as much as 12 per cent of global energy by 2020." (*World Resources, 1996–97*)

In developing countries, governments are often caught up with the need to reduce foreign debts and increase exports. We have seen how

China has an abundant supply of coal to fuel its new industries. But coal-powered energy is one of the greatest contributors to greenhouse gases. While China has yet not developed its hydroelectric power sites, which would provide a clean-burning fuel, it has virtually no other fuels. Thus, coal-burning power plants are being built across the country to continue to power China's industrial expansion.

the former Soviet Union practised unsustainable development over the past seventy years in order to expand industrial output. There is real danger that in some developing regions unsustainable practices will continue with one objective in mind—economic development.

Governments have to act responsibly. Legislation that encourages people to reduce their energy consumption needs to be passed. Today, international organizations such as the United Nations and the World Bank are studying development projects much more carefully to ensure that a greater extent of funding goes to sustainable projects.

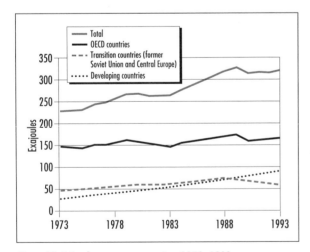

Figure 21.6 Total energy consumption, 1973–1993
Source: *World Resources 1996-1997* by the World Resources Institute. Copyright ©1996 by the World Resources Institute. Used by permission of Oxford University Press Canada.

Figure 21.7 Coal-generated energy is boosting China's industrialization.

STATLAB

1 Study the technology indicators on pages 450 to 457 in Appendix C.
   a) List four or five categories that relate to global warming. Be prepared to explain your choices.
   b) Create a scattergraph that compares each of the indicators to GDP. What correlation exists?
   c) Describe how economic development may be incompatible with reducing greenhouse gas emissions.

2 a) Choose at random twenty developing countries and twenty developed countries.
   b) Prepare an organizer or a chart showing statistics for the categories chosen in 1 (a).
   c) Calculate median and mean values for developing and developed countries.
   d) Compare mean and median values for developed and developing countries.
   e) Which countries are more responsible for global warming—developed economies or developing ones? Explain.

3 a) Which ten countries are the biggest contributors to global warming on a per capita basis? Explain how you arrived at this conclusion and support your answer with statistical evidence.
   b) Where does Canada stand in this ranking?

4 Prepare chloropleth (graded shading) maps to show the levels at which countries contribute to global warming.

SOLVING THE ISSUES: OZONE DEPLETION

The depletion of the ozone layer from CFC emissions will continue well into the next century, no matter what actions we take today. Once CFCs enter the atmosphere, it can take as long as ten years for the gases to reach the stratosphere and the ozone layer. In other words, today's damage may be the result of gases that were emitted ten years ago! Furthermore, it is uncertain exactly how long these indestructible chemicals stay active in the atmosphere. Estimates range from between fifty and one hundred years. This means that the situation will only get worse before it improves.

ELIMINATING REFRIGERANTS

The problem of ozone depletion is exacerbated by the number of old refrigerators, freezers, and air conditioners still in use in our society. As these appliances age, they begin to leak and the destructive CFCs they contain enter the atmosphere. While it is now possible to have CFCs removed from old refrigerant units, many people are reluctant to do so because of the cost and inconvenience.

Business and industry are also looking for new ways to solve the problem of ozone depletion. Coolants that are less harmful to the ozone layer have been developed and new technologies are continuously being researched. Today's coolants still contain chlorine but are more stable in the atmosphere. Hydrochlorofluorocarbons (HCFCs) have about 3 per cent of the ozone-depleting potential of old CFCs. These safer HCFCs are expected to be phased out as even safer alternatives are invented in the near future.

While it is impossible to eliminate the damage CFCs are already inflicting, what can be done to diminish this

> Developing alternatives to CFCs not only makes good environmental sense, but good business sense as well. The use of CFCs has been banned by many countries, so industries that come up with safe alternatives stand to make a substantial profit.

problem in the future? Repairing the ozone layer is almost impossible (see the article "In Search of a Magic Bullet" below). But if we reduce the amount of CFCs we are emitting today, the ozone layer will repair itself. As the amount of CFCs in the atmosphere drops, UV rays will be able to produce ozone to earlier levels. This is the natural process of **photolysis.**

IN SEARCH OF A MAGIC BULLET

Why can't technology rescue the world from the mess that technology has created? Isn't there a quick fix? Scientists know there isn't, but that doesn't stop them from musing about fanciful schemes for mechanically or chemically refurbishing the ozone layer in short order. By discussing and critiquing these ideas, researchers hope to educate the public about the dangers of climate engineering as well as learn for themselves the feasibility of various solutions.

"One of the common solutions is, 'Why don't we just ship LA's ozone up?'" says chemist Sherwood Roland. "Well, 30 per cent of the ozone is in the stratosphere, and it drifts down from there to the lower atmosphere rather than the other way around. The energy that would be needed to move the ozone up is about two and a half times all of our current global power use. If you could take every power plant in the world . . . the energy would be insufficient . . .

Considering that there are almost 320 million tonnes of ozone in the stratosphere, it would take 350 000 trips by specially outfitted 747 freighters to replace even a tenth of the protective gas. Alternatively, climate engineers could shoot multi-ton bullets made up of frozen ozone into the upper reaches of the atmosphere. But the technology for designing and building the tens of thousands of big guns that would be required does not yet exist . . .

At Princeton University, physicist Thomas Stix has suggested using lasers to blast the CFCs out of the air before they can reach the stratosphere and attack the ozone. His idea is to tune the lasers to a series of wavelengths so that only the offensive molecules would be destroyed. Admittedly, the energy requirement would be exorbitant, but Stix believes a twenty-fold improvement in the overall efficiency of this approach could make it feasible. Even so, tens of thousands of lasers would have to be designed, tested, and built before the first CFC molecule could be zapped. If this is the best idea for reviving the ozone layer, an ounce of prevention is worth more that many tons of cure.

From *Time*, 17 February 1992: 46. ©1992 Time Inc. Reprinted by permission.

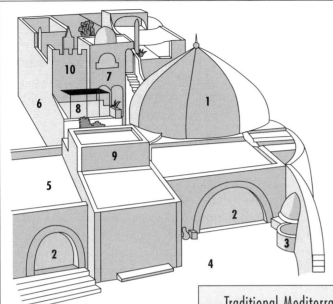

1 White stucco
2 Arches without doors
3 Fountains
4 Open courtyards
5 Flat roofs with low walls around them
6 Thick walls
7 Open grills instead of solid doors
8 Canopy over rooftop eating area
9 High ceilings
10 Few windows, small in size

Figure 21.8 Mediterranean architecture
Source: R. Chasmer, *Patterns in the Physical World* (Toronto: McGraw-Hill Ryerson, 1990), 92.

Traditional Mediterranean architecture provides a natural form of air conditioning in hot, dry climates. Thick, white stucco walls absorb the sun's rays during the day and keep the rooms cool, while at night the stored solar heat is re-radiated into the home. Ceilings and windows are high so that rising hot air can be pushed out the windows by ceiling fans. Arches with grills instead of doors, and windows with louvres instead of glass enable air to circulate freely. What other features contribute to cooling these buildings?

ALTERNATIVE ARCHITECTURE

The developed countries bear the greatest responsibility for the thinning of the ozone layer because North American and European consumers are the main users of CFCs. People in the developed world use air conditioners, consumer goods produced with CFCs, and computer products cleaned with CFC sprays. These technologies are unavailable to most people in developing countries.

We can reduce our use of CFCs in several ways. With appropriate building design and construction, air-conditioning systems can be replaced by alternative pas-sive-cooling systems. Tinted windows reduce the amount of light entering buildings, thereby keeping temperatures lower. Houses that are fully insulated are not only warmer in winter but also cooler in summer. Windows can be opened at night and closed during the day when temperatures are at a maximum. Many houses are equipped with forced-air furnaces. In the summer it is possible to run the furnace fan without turning on the furnace. This circulates cool air from the basement throughout the house.

Revolutionary new architectural designs make use of passive cooling. For example, cold water 10° to 15°C below air temperatures can be drawn from deep within the earth and pumped from a well into a house. A heat exchanger in the house can use the cold water to reduce the air temperature inside the home. The warm water that is generated from this process is then returned to the aquifer. In winter, this water is warmer than the air and can be used to heat the house. This system is sustainable since the water is recirculated. However, it is much more expensive to install than conventional air conditioning and is reliant upon an adequate groundwater supply. But there are no fuel costs, and electrical costs for the pump and heat exchanger are low compared to other systems.

Another innovation is underground housing. This is most commonly found in remote mining regions of Australia where temperatures are extremely high. Old mines have been converted into comfortable homes that are well insulated by metres of earth. These homes remain cool and comfortable despite outside temperatures that regularly exceed 30°C.

GOVERNMENT INITIATIVES

The Canadian government has shown leadership in developing policies to understand the problem and reduce CFC use. Environment Canada monitors the ozone layer using weather balloons, ozone spectrophotometers, and satellites. The World Ozone Data Centre collects data from an international network of monitoring stations. In 1980, the Canadian government banned spray cans that used CFCs.

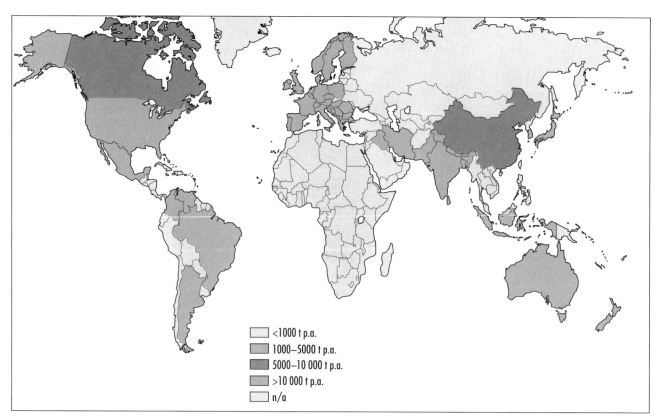

Figure 21.9 Consumption of CFCs annually (thousands of tonnes)
Source: UNEP, 1993.

This one action reduced CFC production by 45 per cent. In 1987, Canada joined 130 nations in signing the Montreal Protocol, an agreement that set timelines for the elimination of CFCs. The agreement called for a 50 per cent reduction of CFC use by the year 2000. These time lines were changed to speed up the process of phasing out the most dangerous chemicals by 1 January 1996, but developing countries were given an extra ten years to phase out production of CFCs. There are loopholes in the international agreements, however, that allow for the export of CFCs to developing countries, which can still use the chemicals until 2006. There is concern that this postponement may cause more damage than originally projected. While they accounted for only 20 per cent of global production, CFC exports surged by 1700 per cent from 1986 to 1993 in developing countries. By contrast, CFC production dropped by 74 per cent in the developed world during the same period.

CFCS
HOLE-STOPPERS

At the meeting of the American Chemical Society in 1930, Thomas Midgeley, a chemist on the research staff of General Motors, demonstrated the safety of his newly discovered refrigerant by inhaling a lungful and using it to blow out a candle. Related compounds of the gas turned out to have other uses: as a foam-blower, a cleaning solvent, and an aerosol propellant. Phasing out of the production of chlorofluorocarbons (CFCs) therefore deprives the world of an exceptionally useful product. . . .

Since 1986 world production [of CFCs] has declined by an impressive 40 per cent. From now on, though, the costs of cutting back will rise. Users must choose between the recycling of existing stock of CFCs, or switching to substitute chemicals. The more existing equipment has to be scrapped before the end of its life or retro-fitted to use substitutes, the higher the cost to users.

In Europe there has been little recycling so far. Mike Harris, regulatory manager for ICI's fluorocarbon business, says that his company had hoped to recover 5 per cent of the CFCs and halons it sells, but has so far retrieved less than 1 per cent. In America progress has also been sluggish, although the 1990 Clean Air Act [in the US] requires the recovery of CFCs from cars from the beginning of [1992] and from stationary sources from July [1992]. The Alliance for Responsible CFC Policy, which represents America's producers, wants the government to allow large users to store CFCs so that they could be used again in air-conditioning systems . . . rather than replacing them with equipment that can use substitutes. . . .

The main substitutes for CFCs are being developed by some of the nineteen companies that now make the stuff. Tony Vogelsberg, environmental manager at Du Pont, the world's largest manufacturer of both CFCs and CFC substitutes, reckons that because the substitutes are far more complicated to make, they may always be two or three times as expensive as the CFCs they replace. Most manufacturers therefore expect demand for substitutes always to be less than for CFCs. . . .

Where substitutes are available, some are themselves ozone-depleting, although less seriously than CFCs. William Reilly, [former] head of America's Environmental Protection Agency, recently frightened users by hinting that the EPA may try to ban the production of these substitutes (called HCFCs) by 2005. Such threats have discouraged some companies from investing in their manufacture. Du Pont deliberately decided two years ago [1990] not to make HCFC-141B, one of the main substitutes. . . .

Electrolux, a Swedish company, has tried to develop a CFC-free refrigerator. It can use HCFC-134A . . . as a coolant, but has had difficulties with the insulating foam, which typically contains more CFCs than the cooling system. . . .

Replacing CFCs has gone much faster in the electronics industry, which has used CFC-113 as a cleaner. In some countries the industry has already stopped using it, switching to water-based cleaners or to technologies which make cleaning unnecessary. A recent report by the OECD on the electronics industry in Asia's industrializing countries found evidence that, in this price-competitive industry, financial sticks and carrots could speed up change. The American CFC tax, which applies to products containing or made with CFCs as well as to the chemicals themselves, has spurred Asian electronics firms to find CFC replacements.

Quotas, the bluntest of all government instruments, have also helped. The OECD found that the three countries where CFC consumption had fallen or hardly risen—Singapore, Hong Kong, and Taiwan—had all restricted imports to 1986 levels. Singapore auctions half its import quota to CFC users, thus creaming off revenue that might otherwise go to producers. The government uses the money to subsidize recycling and replacement.

From *The Economist*, 7 March 1992: 76. ©1992 The Economist Newspaper Group, Inc. Reprinted with permission. Further reproduction prohibited.

CONSOLIDATING AND EXTENDING IDEAS

1 Read the article "Healthy House" on page 342.
   a) Explain the ways in which this house could reduce the effects of global warming.
   b) What would you like/dislike about living in such a house?
   c) In which ways do people have to change their attitudes towards a sustainable lifestyle?

2 Even if the production of CFCs is eliminated, why will depletion of the ozone layer continue to be a problem in the next century?

3 Study the graphs in Figures 21.10 (a) to (c).
   a) Describe the trend for each graph.
   b) What correlation do you observe between CFC concentrations and ozone levels?
   c) How do you account for this correlation?
   d) Why are CFC levels increasing even though production is dropping?
   e) What positive trends, if any, can you determine?

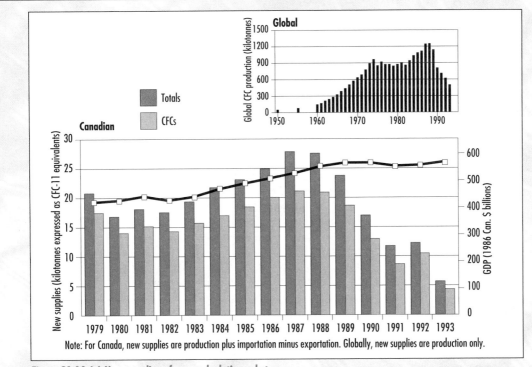

Figure 21.10 (a) New supplies of ozone-depleting substances

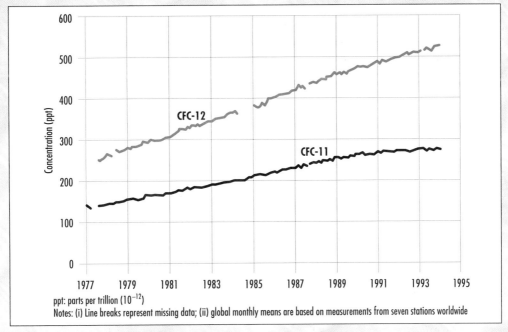

Figure 21.10 (b) Global atmospheric concentrations of ozone-depleting substances

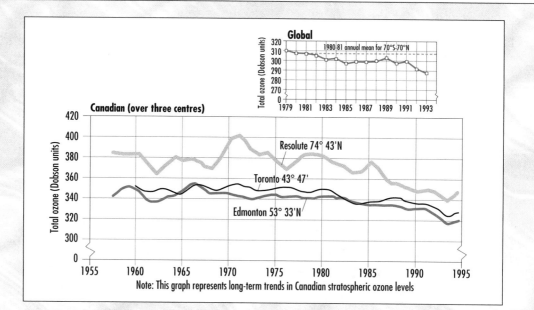

Figure 21.10 (c) Stratospheric ozone levels
Source: *Stratospheric Ozone Depletion*, National Environmental Indicator Series, SOE Bulletin No. 94–96, Environment Canada, 1994.
Reproduced with permission of the Minister of Public Works and Government Services Canada, 1997.

4 Read the article "CFCs: Hole-stoppers" on page 351.
  a) Summarize the information in this article in point form.
  b) Circle the points that indicate a resolution of the problem.
  c) Underline the points that indicate setbacks.
  d) Add up each category and determine if the article is positive or negative about the issue.

SOLVING THE ISSUES: WATER

WATER POLLUTION

Environmentalists insist that we should pay more attention to what are called the "net effects" of any technology. Our ability to predict the far-reaching impacts of any technology on environment and society must improve. In our role as custodians of the earth's resources, we need to consider the pluses and minuses of any technological change or proposed economic development. We must assess risks and weigh these against benefits over the short and long term. Because technology shapes our future, people in developed countries must invent new technologies with positive net effects and export these to the rest of the world.

Improved technology offers some solutions to problems of water pollution. Water purification and treatment plants remove contaminants from sewage water so that it can be reused by industry or safely returned to rivers and seas. Sand filters and screens are used to gather solid wastes out of sewage. Chlorination ensures that pathogenic microbes are removed from water. Water can be disinfected through carbon filtration, ozone treatment, and ultraviolet radiation. In Sweden, lakes and streams are being treated with lime in order to neutralize the acidity created by acid precipitation.

The problem of polluted rivers and lakes is being addressed through the modernization and extension of sewage systems in cities, and through reformed legislation for controlling industrial pollution. International efforts to improve safety standards, control accidental spills, and monitor oil tanker traffic by satellite have decreased the number of accidental oil spills at sea.

WATER SUPPLY

Many areas in the world have shortages of usable water. The natural aridity of some regions (for example, central Africa, the Middle East) and increased demand by growing populations contribute to water shortages. To answer problems of distribution, long-distance transfers of water have been proposed. Technological "fixes" such as desalination of sea water and improved treatments for waste water are being developed to solve water shortages.

In the article on the following page, Umberto Columbo, director-general of the Agency for New Technologies, Energy, and the Environment in Rome, and president of the European Science Foundation, argues for the conservation of water by individuals, governments, and industries.

Figure 21.11 Sewage pollution in the seaside resort of Blackpool, England

HUMANKIND'S WATER NEEDS REQUIRE SEARCH FOR COST-EFFECTIVE RESOURCE MANAGEMENT

Promoting Conservation

Water is an essential "common good" that until fairly recently was thought to be super-abundant, and therefore used wastefully, especially in the wealthy industrialized nations. But in reality water is a relatively scarce resource: extremely scarce in some parts of the world. And as the world's population grows and its standard of living gradually rises, the demand for water, and therefore its cost, is bound to increase.

Of course, large-scale projects will always be needed to transport freshwater from places where it abounds to places where it does not. In addition, cost-effective ways must be found to exploit new water resources: by desalinating sea water, for instance, and purifying polluted water. But more to the point are small-scale actions that, with relatively small investment, can promote conservation of this precious resource.

Sound water management starts with land-use and watershed management. Drainage must be designed to collect runoff, especially from torrential rains that lead to erosion and landslides. The creation of adequate vegetation cover, irrigation systems, and small interlinking ponds makes it possible to store water for dry seasons and prevent erosion.

Another aspect is the modernization of water treatment, recycling wherever possible, and always aiming to prevent waste. Cascade use must be properly managed: in geographical terms, from higher to lower localities; in pollution terms, to lesser contaminating uses; in terms of priorities, taking adequate account of each country's typical needs, from domestic uses to farming, power generation, and industry.

Saving Water in Agriculture

Today agriculture accounts for around two-thirds of all the water consumed worldwide. With the population growing steadily, it would be unthinkable to try to limit world food production or farm productivity. Irrigation, together with fertilizers and pesticides, is the principal means of increasing farm productivity, and in the past few decades it has been instrumental in fighting hunger in the Third World, especially in Asia.

Water availability is a particularly dramatic problem in many parts of Africa, where the combination of an arid climate, drought, soil depletion, and deforestation has aggravated the malnutrition of an exploding population. Here, as elsewhere, the problem must be thought through in new terms that take account of the complex interaction among the factors of food, water, land conservation, and preservation of genetic diversity.

Irrigation systems can be redesigned to give plants exactly the amount of water they need, without flooding the ground. The ultimate technology today consists of computerized drip irrigation, using underground humidity detectors to reduce water inputs to the minimum required for optimal production.

Plant geneticists are developing less-water-demanding crop strains; for instance, rice cultivars that need not be submerged in water. Plants that tolerate brackish water are also being developed, though in this case one should contemplate crop rotation, which is essential to prevent soil depletion.

Other important aspects are the replacement of chemical pesticides with biological pest-control techniques, and the application of fertilizer directly to plant roots, which reduces water pollution and the oxygen-starvation of lakes.

Saving Water in Industry

Industrial water requirement can also be rationalized. Enormous amounts are consumed, and heavily polluted, by paper mills, tanneries, and many processes in the chemical, textile, and hydrometallurgical industries. But it is becoming increasingly feasible to design solutions that require much less water. Two fundamental aspects that should be provided for in all industrial activities are the purification of waste water to make it suitable for other uses, and the creation of closed cycles, whereby waste water is recycled in the industrial process itself.

Numerous steps can be taken to rationalize our household and sanitary uses of water, for instance, by installing alternate-action toilet tanks, or water taps that turn on only when a person's hands are underneath them. The main issue is to educate the public and adopt water-conservation policies.

Important opportunities are opening up for wide-ranging mammoth actions that require no earthshaking decisions or mammoth investment, but that comprise an organized and integrated package of often-intangible initiatives that involve participation by individual citizens. This is a promising path that should guide future strategies for managing water resources.

Umberto Columbo. From *The Christian Science Monitor*, 27 May 1992.

WORKING TOGETHER TO SOLVE THE ISSUES

We must learn to use resources responsibly; to realize that our activities will have global repercussions now and in the future. We need to focus on uses that are sustainable—uses that do not deplete nonrenewable resources and irreparably harm our global environment. People throughout the planet must recognize that maintenance of healthy ecosystems is our ethical responsibility, and our actions must equal our commitment to preserving the environment.

CONSOLIDATING AND EXTENDING IDEAS

1 Desalinization is becoming a popular option in areas with water shortages. Countries that can afford the relatively high costs of the process, for example, Kuwait and Saudi Arabia, use their desalinization plants to supplement their limited freshwater resources.

   Using electronic media retrieval tools and the Internet (if available), find out more about desalinization. Try to discover how the process works and how feasible a solution it is today. Predict its importance as a solution for freshwater supply in the future.

2 Refer to Figure 21.12.
   a) Identify countries that are "water-rich."

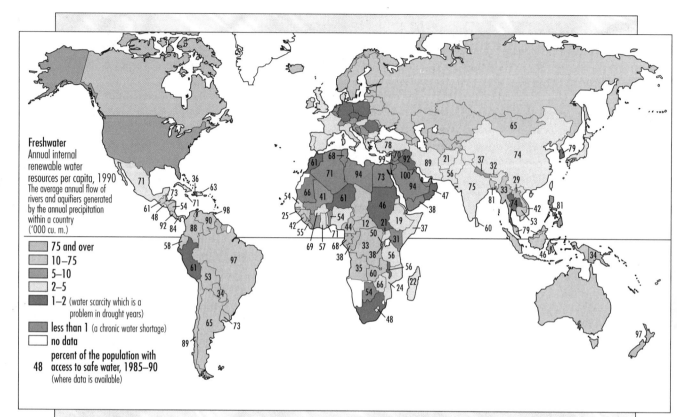

Figure 21.12 Freshwater

Source: *Canadian Oxford School Atlas*, 6th ed. © Oxford University Press Canada. Reprinted by permission of Oxford University Press Canada.

b) What indicator did you use to determine "water richness"?

c) Identify countries that are "water-poor."

d) What indicator did you use to determine "water poverty"?

e) Make feasible suggestions for redistributing freshwater resources around the globe.

3 Read the article "Humankind's Water Needs Require Search for Cost-Effective Resource Management" on page 356.

a) Why does the author suggest that problems with the world's water supply are going to worsen?

b) What factors contribute to problems of water availability in Africa?

c) Chart agricultural and industrial solutions to problems of water supply.

d) Speculate about how the author's proposed solutions might affect earth's hydrologic cycle.

PART 6

GEOPOLITICAL ISSUES: GLOBAL CONFLICT AND COOPERATION

CHAPTER 22

CONCEPTS IN GEOPOLITICS

Politics influences many aspects of life. It affects national and international issues, and sometimes even personal issues. In its formal sense, the study of politics involves government, policy, and legislation. In its most basic form, politics is about power and, more specifically, the way in which power is exercised. Power is the ability of a person or group to control or influence another. Usually, it is associated with governments. The people in power enact legislation that governs people in a particular region. Some authorities define **geopolitics** as the study of how geography relates to the foreign policy of states and to international conflict.

Yet politics is more than just government. Wealthy people and organizations often have a

Figure 22.1 Government leaders from Congo, Mali, Benin, and France at the Francophone Summit in December 1995

great deal of power. The OPEC (Oil Producing and Exporting Nations) cartel wielded enormous international power when it forced the developed world to pay more for oil in the 1970s. The selection of Atlanta as the site of the 1996 summer Olympics, some observers argue, was linked to the power of the huge multinational corporations that stood to benefit from holding the games in that American city.

Sometimes, large groups that have been mistreated by governments become powerful, even though they lack traditional political or economic power. They win public sympathy because they are able to demonstrate effectively against those in power. Mahatma Gandhi led millions of Indians against the enormously powerful British government in the 1940s and gained independence for India in 1947. Martin Luther King led African Americans in their struggle for civil rights in the 1960s. In 1988 Aung San Suu Kyi became a leader of Burma's pro-democracy struggle. Although her party won the 1990 elections, the military regime refused to surrender power. Suu Kyi was placed under house arrest in 1989 until public pressure and international outrage forced the government to release her in 1995. Awarded the Nobel Peace Prize in 1991, today she leads the people of Burma (officially called Myanmar since 1989) against a repressive government.

The way in which your school is governed can be compared to Canada's government in the first half of the last century. At the head of your school is the principal. Like the queen or the governor-general, the principal is the key decision maker. The principal does not make these decisions alone, however. Teachers act as advisors similar to the Cabinet or Executive Assembly in Canada in the 1830s. Together, the principal and teachers develop policies to run the school within the bounds of provincial legislation and with advice from school boards, parent groups, and the business community. The students are the citizens of the school. Like the citizens of a country, they must follow the rules and policies established by the school government. They elect a student council similar to the Legislative Assembly of the 1830s. The members of this Assembly expressed the views of the people but they had virtually no power. The government could choose whether or not to act on their suggestions. Usually, the only decisions the student council makes involve social and sporting activities. Its members do not try to change the way the school is run. Of course, Canada's political system changed after the Rebellions of 1837. Today, Canadian voters elect politicians who represent the people. If the politicians do not carry out the wishes of the electorate, they will not be re-elected. Do you think student councils should be given the power to make policies in high schools?

To answer this question we must examine the terms **individualism** and **collectivism.** Individualism refers to the rights of the individual over the group. Collectivism refers to the rights of all members of society. In the case of education, if an individualistic view is taken, students will have unlimited opportunities to grow and learn. This approach definitely benefits stu-

> In the mid-1990s, many provincial governments in Canada moved to reduce the funding of education and health services with a view to balancing provincial budgets that had been in the red for decades. A debate waged for months over the obligation of government to support essential services in a time of restraint.

dents, but is it fair to all elements of society? If unlimited funding is given to education, the government has either to reduce the money allocated to other government-supported services or to incur debt. Neither option is viable since government should support the interests of all members of society impartially. On the other hand, individuals have undeniable rights and freedoms guaranteed under Canada's Charter of Rights and Freedoms.

Collectivists believe that the interests of the individual are subordinate to those of society as a whole. To enjoy the benefits of living together, including law, order, and protection of property, individuals must relinquish a certain amount of freedom to further the collective welfare. For instance, while individual industrialists may benefit financially if they do not have to install pollution controls in their factories, it is in a nation's best interests to regulate emissions and wastes to protect the environment.

Figure 22.2 Alberta nurses at a health-care protest in Edmonton in February 1997

CONSOLIDATING AND EXTENDING IDEAS

1 School policies are made for the benefit of both the students and the school as an institution. Consider these typical school rules and determine whether they benefit students, the school, or both. Explain your answers.

a) Smoking is not permitted on school property.

b) Students require permits to park their cars in the school parking lot.

c) Compulsory exams are held for all courses.

d) Weapons, drugs, and alcohol are not permitted on school property.

e) Students must be on time for class.

f) Disruptive students are sent to the office.

g) There is a standard dress code for all students.

2 Society also has laws that benefit both citizens and the state. Consider these typical laws and policies and determine whether they benefit the people, the state, or both. Explain your answers.

a) imposing a sales tax on all items

b) incarcerating people convicted of murder

c) establishing laws against drinking and driving

d) establishing a legal voting age

e) requiring citizens to perform jury duty

f) imposing a prison term on people who sell government secrets to other countries

g) requiring people to pass driving tests before they can obtain a driver's licence

POLITICAL SYSTEMS

A set of beliefs and values that guides a society's decision making, orders its activities, and provides it with ideals and goals is an **ideology**. An ideology colours our perception of the world, reflects beliefs about human nature and society, and influences attitudes about how society should be organized and governed. Ideology presents an understanding of why a group of people choose a specific type of government and economy.

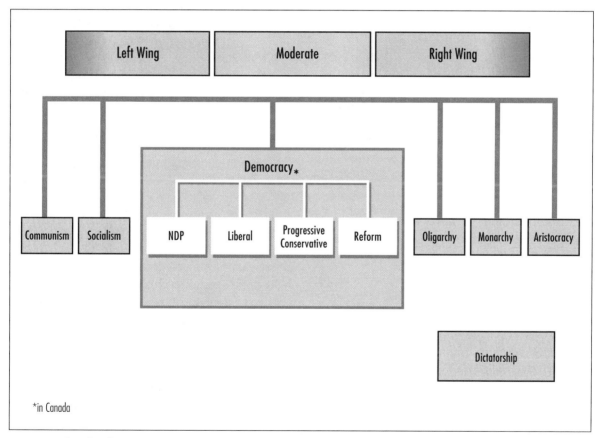

Figure 22.3 The political continuum

The nations of the world are governed by a variety of political systems. Ideology offers the key to classifying the various types of political and economic systems. The simplest method of classification is to arrange the different systems along a continuum. This achieves two objectives: it indicates where the various systems are in relation to each other and suggests how elements of one system may overlap into another.

Figure 22.3 is a political continuum. Systems based on freedom of the individual are called democracies; systems based on government control are most often dictatorships. In terms of political ideology, this continuum is based on change. The left encourages change while the right resists it. The end points represent extremes of ideology. The extreme left supports swift, often violent, change. The terms *revolutionary* or *radical* describe this position. The extreme right resists change, with violence if necessary. The terms *reactionary* or *counterrevolutionary* explain this position. The extreme right supports change only when it means reverting to the way things were "in the good old days."

If a society believes in the granting of greater freedom (political, economic, religious rights) to its people, it is often called *liberal*, or **left wing**. If a society seeks to maintain existing institutions and is slow to accept change, it is called *conservative*, or **right wing**. The centre represents a moderate position that may range from moderate left to moderate right. Moderates may support change on some issues while resisting it on others. Dictatorships can be either to the extreme left, as with Communism, or to the extreme right, as with fascism.

> Canada is a democracy. The federal and provincial governments are voted into power through general elections held at least every five years. If the government fails to earn the support of the people, it can be voted out of power in the next election.

DEMOCRACY

The **democratic** form of government seeks to find a middle ground between the two extremes—the rights of the people are protected but laws to safeguard the state are also enacted. Through the electoral process, voters decide the degree to which individualism or collectivism is to be exercised. If it is the people's will that the needs of the state should take precedence over the needs of individuals, a political party with a right-wing ideology will obtain a majority of seats in government. On the other hand, if the voters favour the expansion of policies that increase individual freedom and responsibility, left-wing parties will take over.

Democracy originated in ancient Greece in the fifth century BCE. All adult male citizens met in the town square once a week to vote on the town's policies. Women were not allowed to vote. This would not be considered a democracy now because of the exclusion of women. The democracy of ancient Greece was **direct democracy**; every citizen voted on every piece of legislation in the city-state. Today, direct democracy is practised only in some cantons of eastern Switzerland. This form of government is not feasible in most countries because of the size of voting populations.

RIGHT-WING POLITICAL SYSTEMS

The earliest political systems were for the benefit of the state over the people. In feudal societies, one person owned a specific region, including the land, the resources, and even the people. The lord was the state, and everybody worked for the benefit of the state. In many parts of what is now Britain, serfs had to give up one-seventh of all the crops they produced, work in the lord's fields, and serve as soldiers in time of war. They had no say in the system and were not even allowed to move away. This system is called an **aristocracy**. Although it is not commonly found in the world today, it was the way of life in China and Europe in the Middle Ages.

In time, one lord emerged as the most powerful, and unified a large region into a **monarchy.** The king or queen owned all the land under his or her dominion, including the resources and the people. It was generally believed that the monarch had divine power bestowed by God. Nobody dared question the monarch's authority. Lords of smaller regions owed allegiance to this ruler. They had to pay taxes and provide military as well as political support in times of conflict as much of the Middle Ages saw one European monarch fighting another. While monarchies still exist in the world, most operate under a **parliamentary system** in which the monarch is a figurehead who has little or no real power.

Sometimes a government is conducted and controlled by only a few people. An **oligarchy** is a government formed by a few influential individuals, who become the ruling **élite**. These governments often come to power through military take-over; the army suppresses the rights and freedoms of the citizens in order to gain and maintain control. Such military regimes can become more concerned with retaining power for their own benefit than with addressing national issues and concerns. They often justify their actions by claiming that they are acting in the best interests of the state, but their actions often do not support their rhetoric. There are many examples of oligarchies. In the 1980s, the Marcos government in the Philippines siphoned off millions of dollars for the benefit of the president, his family, and his supporters. Uganda,

Democratic Republic of Congo, Myanmar, Haiti, and a host of other countries have been ruled by oligarchies.

A **dictatorship**, rule by one person of exceptional power, is a type of right-wing government that dates back to ancient Rome. A dictator was an individual who was given absolute power by the Republic in times of crisis. When the emergency was averted, the dictator returned to life as a private citizen. In modern times, dictators such as the Soviet Union's Joseph Stalin, Germany's Adolf Hitler, Iraq's Saddam Hussein, and Chile's Augusto Pinochet have seized power and refused to relinquish it.

During the 1920s and 1930s, the period between the two World Wars, depressed economic conditions and political instability in Europe led to the rise of public support for ideologies at either extreme of the political spectrum. To the extreme left, Communism, as it developed in the USSR, first under Lenin then under Stalin, offered economic recovery and a return to law and order following the chaos of the Russian Revolution. Benito Mussolini stepped into the chronic political instability of Italy and became the undisputed fascist dictator of that country from 1926 to 1943.

Fascism is a political ideology at the extreme right of the political continuum; its key elements include glorification of the state and aggressive nationalism. Fascism in Italy drew support from all classes of society (especially the lower class and the economic élite*) and focused all activities—social, economic, and political—on nationalistic goals. Within the fascist state individuals were first and foremost members of the nation, to which they owed their property, allegiance, and lives. Ultimately subordinate and obedient to the authority of their supreme leader, Italians devoted themselves to Mussolini's creation of a strong Italy

---

*All references to socio-economic status are presented in a historical context; these terms are not intended to label societal groups.

and an Italian Empire (Yugoslavia, Ethiopia, and Albania were targets for Italian expansionism). Fascists condemned democracy, spurned socialism and Communism, and supported their single-party system guided by their charismatic leader. Intimidation and violence were used to silence critics, and all aspects of people's lives were regulated and controlled by the state.

When dictators and fascists seek total control over all facets of society, they are called **totalitarian**. Fascist Italy, Nazi Germany, Stalin's Soviet Russia, and present-day Iraq, Libya, and Myanmar are examples of such regimes.

LEFT-WING POLITICAL SYSTEMS

Communism originated as an extreme left-wing ideology. Its founder, Karl Marx, conceived of a dictatorship of the **proletariat** in which all workers shared power and had a say in their own governance. It was envisaged as a social system in which property was vested in the community and each member worked for the common benefit. Marxist ideals were first put into practice following the Russian Revolution of 1917.

Sweden has been one of the world's most progressive socialist states. Swedes have enjoyed cradle-to-grave social programs that include everything from hospital care to education programs to retirement benefits. While the economy was expanding in the 1970s and 1980s, such massive social spending was possible and desirable. With the declining global economy in the 1990s, it has become increasingly difficult to maintain these high social standards. Like many Western societies, the Swedish government has drastically reduced government spending, which has resulted in cuts to many social programs.

Democratic socialism grew out of the belief that all people had the right to a good life. During the 1920s and 1930s, political movements in several Western democracies led to the establishment of the welfare state. The intent was that the community as a whole would own and control production and distribution of the nation's wealth. Subsidies and tax benefits provided everything from baby bonuses to universal medical care and pension plans. Emanating from the liberal movement in the previous century and supported by an increasingly well-educated working class, socialism democratically sought to improve living conditions for working class people. The advantages for the working classes were financial security, a stimulated economy, and peace of mind. The downside, of course, was the enormous expense of so many government social programs. Socialist and labour parties came to power in many Western democracies including Britain (the Labour Party), France (under President De Gaulle), and the United States (the Democratic Party).

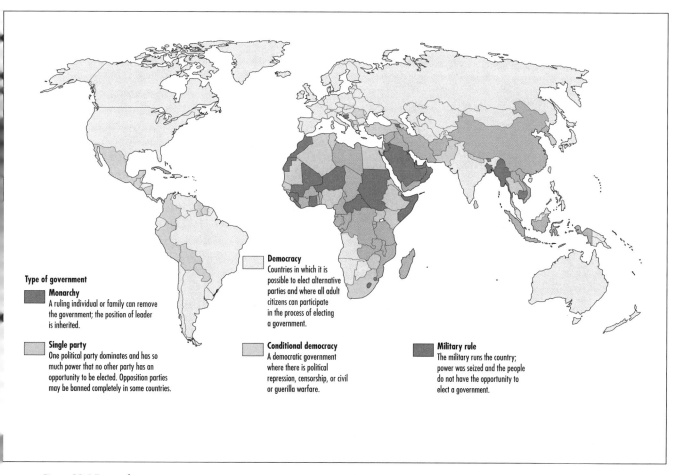

Type of government

Monarchy
A ruling individual or family can remove the government; the position of leader is inherited.

Single party
One political party dominates and has so much power that no other party has an opportunity to be elected. Opposition parties may be banned completely in some countries.

Democracy
Countries in which it is possible to elect alternative parties and where all adult citizens can participate in the process of electing a government.

Conditional democracy
A democratic government where there is political repression, censorship, or civil or guerilla warfare.

Military rule
The military runs the country; power was seized and the people do not have the opportunity to elect a government.

Figure 22.4 Types of government

Source: G. Matthews and R. Morrow, *Canada and the World: An Atlas Resource,* 2d ed. (Toronto: Prentice Hall, 1995), 97–98.

THE PARTY IS OVER

Anybody tempted to rhapsodize about democratic freedom in the new Russia should take a moment to ponder the fate of Valeria Novodvorskaya. . . . She has managed to infuriate the authorities in every regime in Moscow for three decades. Arrested by the KGB for anti-Communist views in the 1960s and locked up in a psychiatric hospital, and arrested again in the 1980s, she was one of the best-known dissidents of the Soviet era.

Today she is once again a police target. Prosecutors have charged her with "humiliating the national honour and dignity" by writing unpatriotic articles in various intellectual journals. She has been threatened with an eighteen-month prison sentence and is prohibited from travelling outside the Russian capital until the case is over.

Figure 22.5 On the campaign trail—the Russian presidential election of 1996

None of this has made much impact on Western journalists or Western embassies in Moscow. Dissidents have fallen out of fashion. Their lingering presence, an embarrassing reminder of the authoritarian tendencies of post-Soviet governments, is generally ignored.

In the Soviet era, dissidents such as Valeria Novodvorskaya were lionized by the Western media. But after the collapse of the Soviet Union where capitalism and democracy were installed as the official ideology, the West lost interest in the political prisoners who still crowd the jails of many ex-Soviet republics.

It was . . . on December 25, 1991, that Mikhail Gorbachev acknowledged the death of the Soviet Union. A few hours later, the Soviet flag was lowered at the Kremlin and the world proclaimed a new world of freedom.

The uncomfortable reality, however, is that atrocities have not disappeared. All across the former Soviet Union, from Minsk to Tashkent, from Almaty to Baku, authoritar-

ian leaders and powerful security agencies are exercising the same nasty old habits of harassing and arresting their critics.

The situation in Russia is not nearly as bad as the climate of fear in other ex-Soviet republics such as Azerbaijan, Turkmenistan, Belarus, and Uzbekistan. But even in Russia, freedom is slow to arrive. The saga of Ms. Novodvorskaya is proof of this.

As a teen-age student in 1967 in the Leonid Brezhnev era, she was arrested by the KGB for writing an anti-Communist poem and tossing it from an opera-house balcony. They sent her to a psychiatric hospital, where she was diagnosed a schizophrenic. It was standard punishment for Soviet dissidents, but it failed to silence her.

When Mr. Gorbachev was the Soviet leader, at a time when political parties were still illegal, she helped create the Democratic Union—the first openly anti-Communist political party in Soviet history. A few weeks after the founding of the party in 1988, she was arrested again.

Two years later, the police made another attempt to silence her. This time she was charged with "insulting the president" by describing Mr. Gorbachev as a "red fascist." She was finally released from custody after the failed coup by Communist hardliners in 1991.

Now the KGB has returned to complete its unfinished business. In its new guise as the Federal Security Service, the agency has persuaded a Moscow prosecutor to file criminal charges against Ms. Novodvorskaya for her outspoken criticism of Russia's policies in Chechnya and the Baltic states. . . .

Nor is this an isolated example. The anti-nuclear environmentalist Alexander Nikitin was jailed for ten months on treason charges this year. His environmental colleagues were interrogated by the police. The former KGB captain and dissident sympathizer Viktor Orekhov, sent to prison on trumped-up weapons charges last year, was not released for eight months when he fell ill with hepatitis.

These cases have never gained a fraction of the attention given to jailed dissidents in the Soviet era. The West is determined to portray the Soviet Union's collapse as a good-news, happy-ending story. The worst abuses in the fifteen former Soviet republics—even the bloody war in Chechnya, with its 100 000 deaths—have rarely provoked any sustained criticism from Western leaders.

For decades, the West called for freedom in the Soviet Union. Today it is clear that economic freedom, not political freedom, is what the West really wanted. As long as the former Soviet republics remain capitalist and open to foreign investment, they are allowed to be as repressive and authoritarian as they want.

Notorious police states such as Turkmenistan and Azerbaijan have been legitimized and courted by Western political and business leaders, who visit them to seek profits from their lucrative oil and gas industries. There is barely any mention of their authoritarian systems and jailed dissidents.

This year [1996] has been another bad year for democracy. In the past three months, two more ex-Soviet republics have fallen into repressive hands. Belarus, which had an open election in 1994, has become an authoritarian system in which President Aleksandr Lukashenko controls the parliaments and courts. Armenia, once a leader in the democratic movement with an ex-dissident as president, sent tanks into its streets to disperse protesters after a vote that was widely condemned as a sham.

Almost half of the former Soviet republics have become police states. Most of their strongmen are relatively popular and might actually win a democratic election, but they prefer to avoid the risks and inconveniences of elections and free debate. They clamp down on the opposition, impose tight controls on the media, ban opposition rallies, and imprison dissidents.

The trend toward authoritarian rule has accelerated as leaders borrow tactics from one another. The former Soviet republics of Central Asia, for example, introduced the idea of calling a referendum to extend and entrench a leader's rule. The tactic was borrowed by Mr. Lukashenko, who exploited it to install himself as the supreme ruler of Belarus.

In Russia, the biggest and most important of the former Soviet republics, the principle of democracy is slowly taking root. This winter [1996–97], for the first time, regional elections are being held in most parts of the country. Two parliamentary elections and one presidential election have taken place in the past three years. The actual process of voting and vote-counting was generally considered free and fair with a few key exceptions in several districts where there was evidence of widespread vote-rigging.

But the toughest test of democracy is whether the government can hand over power peacefully if it is defeated in an election. On this, Russian democracy is still in doubt.

Many observers question whether Boris Yeltsin would have voluntarily stepped down if he had lost the presidential election this summer [1996]. Just a few months before the vote, a powerful clique of Kremlin aides—led

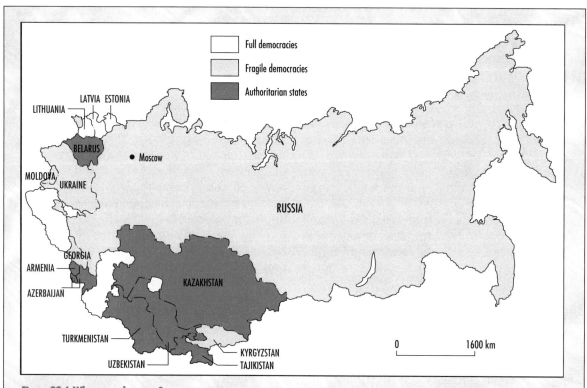

Figure 22.6 Where are they now?

by the ex-KGB bodyguard Alexander Korzhakov—had been trying to cancel the election. The Kremlin's US political consultants later revealed that the election would have been cancelled a month before voting if it appeared that Mr. Yeltsin was going to lose. . . .

Stephen Cohen, a Princeton University professor and author on Russian issues, gives a bleak assessment of democracy in Russia. "Boris Yeltsin's re-election campaigns was one of the most corrupt in recent European history," he wrote recently. "The parliament has no real powers and the appellate court little independence from the presidency . . . Russia's market and television industries are not truly competitive or free but substantially controlled by the same financial oligarchy whose representatives sit in the Kremlin as chieftains of the Yeltsin regime.". . .

Mr. Yeltsin . . . supports dictatorships in neighbouring countries such as Belarus, Armenia, Kazakhstan, and Turkmenistan. . . . His security services helped to arrest dissidents from these countries. Their Kremlin courts maintain Russia's sphere of influence and military bases in strategic regions of the former Soviet Union. . . .

While democracy struggles to grow, Moscow has helped to perpetuate authoritarian systems in Europe and Asia. This is how Russia has ended up: not as a beacon of freedom, but as a supporter of tyranny. It's a long way from the dreams of five years ago.

Geoffrey York. From *The Globe and Mail*, 21 December 1996. Reprinted with permission from The Globe and Mail.

CONSOLIDATING AND EXTENDING IDEAS

1 a) Using CD-ROM systems, research each of the following forms of government. Try key word searches.
  • aristocracy
  • monarchy
  • oligarchy
  • democracy
  • dictatorship
  • fascist
  • socialist
  • communist

b) Review collectivism versus individualism.

c) Create an organizer or a retrieval chart (see Figure 22.7) that details the framework, the advantages, and the disadvantages of each system.

FORM OF GOVERNMENT	INDIVIDUALIST/ COLLECTIVIST	ADVANTAGES	DISADVANTAGES
Aristocracy			
Monarchy			
Oligarchy			
Democracy			
Socialism			

Figure 22.7 Forms of government

2 Study the map in Figure 22.4
  a) What patterns can you establish from the map for each continent?
  b) What patterns can you establish from the map for each hemisphere?
  c) For either Africa or Asia, make a list of countries that all have the same type of government. For example, in Asia monarchies exist in Saudi Arabia, Jordan, Kuwait, United Arab Emirates, Oman, Bhutan, and Cambodia.
  d) Study the GNP or birth rate or literacy rate for each country (see Appendices A and B), and calculate the average for the category you have chosen.
  e) Compare your results with those of other students who have chosen the same category but a different form of government.
  f) What generalizations can you make about different forms of governments?
  g) What overgeneralizations could this type of analysis encourage?

3 Study the article "The Party Is Over" on page 367.
  a) What attitudes are presented in this article?
  b) What facts support the opinions expressed?
  c) What is the difference between economic freedom and political freedom?
  d) According to the article which form of freedom does the West want? Why?
  e) What predictions does the article make?

IDENTIFYING GEOPOLITICAL ISSUES

Like other global issues, politics is often influenced by cultural, environmental, economic, and resource factors. Culture sometimes determines a society's political philosophy. Deeply held religious convictions may result in a **theocracy**—a political system in which legislation is enacted by religious leaders. The Aztec Empire and present-day Iran are both examples of theocracies. Ancient cultural traditions may dictate other forms of government, such as a monarchy. Some countries have developed governments that centre on the rights of the individual. In France and the United States, revolutions against right-wing monarchies occurred because these societies believed that the rights of the individual took precedence over all else.

Environmental concerns are often at odds with politics. War is a political issue, but it ultimately causes great harm to the environment. During the Iraq-Kuwait conflict in 1991, the retreating Iraqi army destroyed oil wells by setting them on fire. The damage that resulted to the fragile Persian Gulf ecosystem was devastating. High temperatures from the fires and a sickening black rain of oil killed everything in the region. Birds, coral reefs, and fish were all destroyed in the aftermath of the war.

Thomas Homer-Dixon, coordinator of the Peace and Conflict Studies Program at the University of Toronto, predicts that environmental degradation will increasingly contribute to mass violence in the future. He believes that future conflicts will result for three reasons, all relating to how people obtain the resources they need to live. A **supply-induced scarcity** arises when the environment has been so misused that it can no longer provide people's basic requirements. A **demand-induced scarcity** develops when the rate of population increase

Figure 22.8 A rebel training camp in Chiapas, Mexico

exceeds the growth of essential resources. The third environmentally induced conflict relates to **structural scarcity**. There may be enough resources for the population but for political reasons the problem lies in their distribution. According to Homer-Dixon, environmental scarcity will not be enough to trigger the violence; other related factors will most likely be the dominant cause for the conflict. Nevertheless, the stress placed on people who are competing for their day-to-day necessities will exacerbate political instability.

Within developing countries violence will probably take the form of rebel insurgencies. Examples abound in the late 1990s. In the state of Chiapas in southern Mexico, marginalized Native peoples are in conflict with a government hundreds of kilometres away in Mexico City. They are fighting for the right to have equal access to the essentials of life. Homer-Dixon has

studied the effects of environmental degradation on urbanization and city violence in Karachi, Pakistan, and Johannesburg, South Africa. The South African capital has the highest murder rate of any city in the world. The reason is indirectly related to environmental degradation in Native homelands such as Natal. Under the repressive white supremacist government, the black majority was forced to settle in large numbers in these homelands during apartheid. Overuse of the land stripped away the natural vegetation and eroded the topsoil. To compound the problem, local chiefs sold the lumber rights to multinational companies. Today, regions such as Natal are completely barren, and the people can no longer obtain the essentials of life. This supply-induced scarcity was intensified by a high rate of population growth. Now that the segregationist laws of the former regime have been repealed, people are free to leave the home-

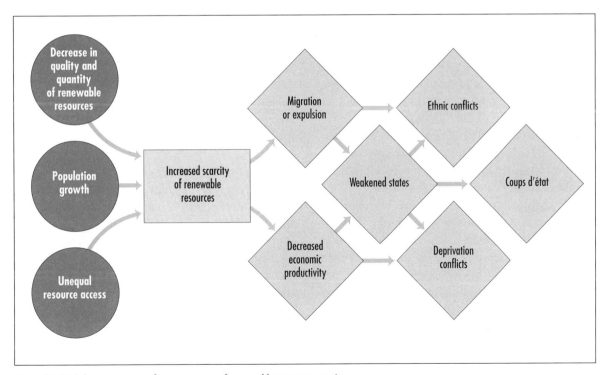

Figure 22.9 (a) Some sources and consequences of renewable resource scarcity

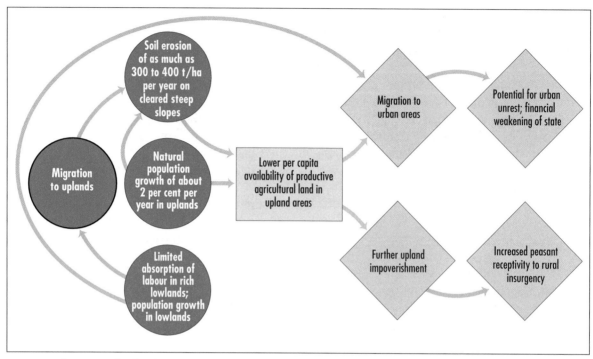

Figure 22.9 (b) An example: The Philippines

lands. It is estimated that 750 000 people migrate to the cities of South Africa each year. Once there, they are often at odds with the local residents because of conditions in their squatter settlements. The services are inadequate, every scrap of land is cultivated, local streams are polluted with human excrement, and building materials are stolen from construction sites.

Economic and resource development are also affected by politics. A great deal of government policy is directed towards the development of the economy and natural resources. When the economy is strong, people enjoy prosperity and governments stay in power. If economic uncertainty results from government mismanagement, democratic governments are usually voted out of office. When totalitarian governments mismanage the economy, the reaction may be more extreme. The people are held in check for as long as possible, but eventually they gain the upper hand and the government is defeated, often through a bloody revolution. This has been the case in many Latin American and African nations.

Resource development has become an important part of many government strategies, especially in developing countries. With staggering foreign debts, high government expenditures (often for the military), and a depressed economy, many countries sell off their natural resources to earn export income. The wholesale destruction of rainforests, unsustainable agricultural practices, and the mining of environmentally sensitive areas for economic gain over the short term have left some countries devastated to the extent that they may never recover their natural environments and resource bases.

The Ontario government decided to open the Temagami wilderness area to logging in the 1990s. The reason for this decision was twofold: to increase employment in an economically depressed area and to gain additional tax revenues from the lumbering company. Environmental groups opposed the decision to clear-cut one of the last remaining old-growth forests in Ontario. While this is a local issue, similar conflicts between politics and the environment have occurred in areas of deforestation around the world, for example, Thailand, Malaysia, Mozambique, and Brazil.

In 1995, Canada and Spain were engaged in a controversy over fishing rights on the Grand Banks. The Canadian government charged that Spanish fishing fleets were overfishing the area. In an effort to enforce fish quotas, Canadian authorities seized Spanish trawlers in international waters beyond the 370 km fishing zone. Although Canada had no authority to carry out this act of aggression, the action was supported by many other nations that saw the benefits of allowing fish stocks to recover.

would have been the start of a third and perhaps final world war. By the 1970s, both superpowers had enormous nuclear arsenals capable of destroying every living thing on the planet many times over.

The Soviet Union was part of the Allied invasion of Germany at the end of the war. All the territory conquered by Germany during the Second World War was recaptured, and by the beginning of 1945 the Soviets had entered Poland, Hungary, and Romania as well as east Germany. Communist governments were set up in these and other eastern European countries. Some of the Allied generals wanted to march past Berlin, into Soviet-held territory, and push the Soviets back into their own country, but a war-weary world felt otherwise. Stalin, the Soviet leader, ended up with six Soviet-dominated satellites between his country and the Western powers. Winston Churchill, prime minister of Great Britain, described the artificial border between Communist Europe and Western Europe as the **Iron Curtain**. For the next forty-five years, this line between Communist-held territory and the rest of Europe was guarded and virtually impregnable. Ideologically and politically, the East was separated from the West.

The North Atlantic Treaty Organization (NATO) was formed in 1949 as a military alliance to prevent the Soviet Union from extending its power into Western Europe. It included the United States, Canada, Great

SELECTED POLITICAL ISSUES

THE FALL OF THE SOVIET UNION AND THE RISE OF THE UNITED STATES

During the Cold War that followed the Second World War, the United States and the Soviet Union dominated the world political scene. The suspicion and distrust that characterized relations between the two nations resulted in the build-up of two great war arsenals. The two superpowers posed such a military threat that if war had actually broken out between them, it

Britain, Spain, Portugal, Italy, France, Germany, Greece, Turkey, Norway, the Benelux countries, and Iceland. On the other side of the Iron Curtain were the Warsaw Pact countries—the Soviet satellites/buffer states. With the collapse of the USSR, the demise of the Warsaw Pact in 1991, and the end of the Cold War, Warsaw Pact countries became free to ally with whomever they wished. Some of these countries, emerging democracies in central and eastern Europe, and newly independent states derived from the former Soviet Union, have become members of NATO's new Partnership For Peace Program.

NATO has undergone fundamental changes in the 1990s. Its structures and policies now include a new Strategic Concept to meet security challenges in Europe, a Cooperation Council (NACC), extension of membership, the Partnership for Peace Program, and a Nuclear Planning Group. It remains a defensive political and military alliance of countries that links the security of North America to that of Europe.

Today, the United States goes unchallenged as the most powerful nation on earth. Not only does it have large armed forces and military bases all over the world, it also has superior technology. This includes nuclear weaponry, so-called **smart bombs** that can lock on to a target and destroy it, surveillance satellites observing every part of the planet, and even stealth bombers that cannot be detected by radar. Even though the US has encouraged global peace and is not necessarily abusing its power, is this nation's hold on world power healthy for international peace and prosperity? When one country is so much more dominant than any other, there is a possibility that it will abuse its power.

> Initiated in 1994, NATO's Partnership for Peace Program acts to expand political and military cooperation throughout Europe. Active members take part in search-and-rescue missions and in peacekeeping and humanitarian operations. Included in the twenty-five central and eastern European nations that have joined the program are Azerbaijan, the Czech Republic, Moldavia, Russia, Armenia, Bulgaria, Estonia, Hungary, Latvia, Lithuania, Poland, Romania, Slovakia, Slovenia, Ukraine, and Uzbekistan.

NATIONALISM

Nationalism is the sense of belonging to a group that is linked together through shared history, language, race, and values. It is a powerful force in many parts of the world. The concepts of nation and state do not necessarily go hand in hand. A state is an independent political unit. It has clearly defined and internationally accepted boundaries that can be identified on a map. A nation is more of a cultural concept. It consists of a large number of people, bound together by a common heritage. A nation is not invariably a state, and we cannot always identify it on a map.

Sometimes nations want political recognition as a state. They seek to govern themselves according to their own traditions and values. Such nationalist movements can be found all over the globe. Many European groups want to secede from existing countries and become independent. Catholics in Northern Ireland wish to join Catholics in Eire. Independence movements in the former Yugoslavia resulted in one of the bloodiest civil wars in European history as Serbs, Croats, and Bosnians fought for independent territories.

Figure 22.10 Residents of Quebec ex-pressed their views prior to the referendum.

Many people in the province of Quebec don't think of themselves as Canadians, but as Québécois. This sense of nationalism predates the battle on the Plains of Abraham where the British finally defeated the French over the control of North America. Québécois want to be considered a "distinct society." They are concerned about losing their culture and language. As their population continues to drop, many Québécois believe they will become less significant in the Canadian Confederation. While some Québécois promote separation, not everyone in Quebec agrees. When the province held a referendum in October 1995, 51 per cent of the people voted to reject separation.

When the Soviet Union was dissolved in 1991, the many nations that existed within the Soviet political entity formed their own independent states. Ukraine, Latvia, Lithuania, and Estonia regained independence after years of political domination by the Soviet Union. New countries such as Armenia and Azerbaijan were formed, giving recognition to nationalist cultures that had survived Russian domination for centuries. A civil war in Chechnya resulted when people living in the Caucasus Mountains wanted independence like other cultural groups in Armenia and Azerbaijan farther south. If Chechnya succeeds in forming an independent state, this will set a precedent for the approximately eighty other cultural groups that populate the Caucasus Mountains. Each group may seek to become independent even though some groups number less than 10 000 people.

In south Asia, the fight for independence has persisted for years. Tamils want independence in Sri Lanka and southern India. Sikhs would like to create their own nation in eastern India, and Kashmiris constantly struggle with the Indian authorities. China prevents nationalist groups such as Tibetans and Uygur Muslims of the isolated western regions from forming their own nations.

Nowhere is the nationalist struggle more evident than in Africa where there are over 5000 tribal groups. Within

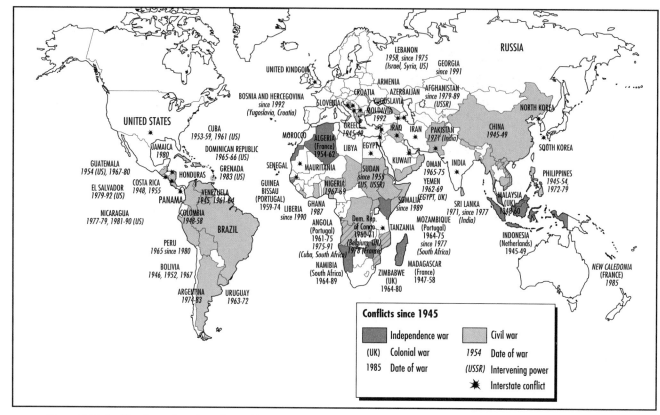

Figure 22.11 Conflicts/wars throughout the world

Source: *World Government,* rev. ed., by Peter Taylor. Copyright ©Andromeda Oxford Limited 1990. Used by permission of Oxford University Press, Inc.

some African countries animosity persists between various groups, and the potential for nationalist clashes is apparent. Conflicts between Tutsi and Hutu in Burundi; between Eritreans and Ethiopians; and between Hausa, FulBe, and Yoruba in Nigeria are just a very few examples. Bitter struggles for nationhood between cultural groups are tearing many of the world's countries apart.

SELF-DETERMINATION FOR ABORIGINAL PEOPLES

The arrival of European settlers during the period of colonialism forever ended the way of life of aboriginal peoples. In North and South America, Africa, Australia and central Asia, Native peoples were displaced by European imperialist expansion. The benefits of colonialism were primarily for the colonizers. Their objective was to find new resources for the growing industrial base at home and to provide new markets for the goods they produced there. This led to a permanent change in the way of life of Native peoples. Many were forced to abandon their traditional lifestyles to work for the Europeans. The land on which they had lived was taken over by the settlers. They were forced onto marginal lands where they were unable to practise their traditional sustainable lifestyles. In time, Native populations dwindled, their cultures declined, and their languages were threatened. Today, many Native groups in Canada and other countries are demanding political recognition.

The United Nations Commission on Human Rights addresses the concerns of Native groups. The numbers of aboriginal peoples are difficult to document in developing countries where many tribes are isolated, but estimates of under 5 million people have been made (*World Government* [New York: Oxford University Press, 1994], 15). Two and a half million live in the United States, 800 000 in Canada, 400 000 in New Zealand, 200 000 in Australia, and about 200 000 in the Amazon bush. Estimates of aboriginal peoples in the marginal lands of Asia are unavailable. In the Americas and in Australia and New Zealand, Native leaders are gaining political power. This is not the case in Asia where small enclaves are continually being driven into ever smaller wilderness areas or are forced to assimilate. Africa is a different case as most Native groups have reclaimed self-determination from European imperialists.

HIGHLIGHTS OF THE NUNAVUT AGREEMENT*

Figure 22.12 Celebrating the Nunavut Accord in Igloolik

Below are the key terms in the Land Claims Agreement and the Nunavut Political Accord:

- The much-debated extinguishment clause: in exchange for the rights and benefits in the agreement, Inuit surrender all aboriginal claim to lands and waters anywhere in Canada.
- Any other existing or future rights that Inuit may have as aboriginal people are not affected.

- The federal government will pay $1.15 billion to Inuit over fourteen years. The Nunavut Trust will be set up by Inuit to receive this capital (tax-free) and to be responsible for protecting, managing, and investing it.
- Inuit will own 353 610 km^2 of land, approximately 18 per cent of the entire Nunavut Territory.
- On 36 257 km^2 of their deeded land, Inuit will own the subsurface rights to oil, gas, and minerals.
- In the event of oil, gas, or mineral development on Crown land within Nunavut, the federal government must pay Inuit a share of the royalties: 50 per cent of the first $2 million and 5 per cent thereafter.
- Inuit retain the right to hunt, trap, and fish throughout Nunavut. No licence is needed to hunt for "basic needs."
- The land claim agreement requires the governments of Canada and NWT to negotiate a political accord establishing Nunavut as a new territory.
- The Powers of the Nunavut Legislative Assembly will be the same as those held by the government of the NWT.

- A Nunavut Implementation Commission will be formed immediately to design the new government, plan the first election, select the new capital(s), and administer training.
- The Nunavut Wildlife Management Board will have responsibility for wildlife management throughout Nunavut. All wildlife quotas and restrictions currently in effect remain until the NWMB changes them.
- The Nunavut Impact Review Board will screen projects throughout Nunavut to determine whether there is a need for review of their impact on the ecosystem or on socio-economic conditions.
- The Nunavut Planning Commission will review land-use plans, ensure that developments conform to land-use plans, and identify clean-up requirements.
- The Nunavut Water Board, working closely with the NIRB and NPC, will have specific responsibility to license the use of water (other than for navigation) and to approve the disposal of waste into water.
- In areas where land-fast ice has traditionally been used by Inuit, provisions dealing with wildlife, resource sharing, conservation, and land-use planning all apply.
- Future developers must negotiate impact and benefit agreements with Inuit before major projects (including parks and hydro-electric generation) may proceed. These agreements may include requirements for the training and hiring of Inuit, and economic and social benefits for Inuit.
- Nunavut will become a reality on 1 April 1999. On that date, the new government will go into operation.

From an article by David F. Pelly first published in *Canadian Geographic*, March/April 1993: 29. Reprinted by permission of the author.
*See also page 286.

INTERNATIONALISM

Internationalism is the policy of countries working together for the good of all people regardless of race or nationality. The United Nations is perhaps the best example of internationalism in action. This international political organization has expanded its mandate from ensuring world peace to improving the quality of life throughout the world.

Many nationalist wars around the world have been keeping the UN peacekeeping forces busy. Expenditures for peacekeeping operations amounted to US $3.36 billion in 1995. While this seems like a great deal of money and is a substantial increase from the mid-1980s, it is really a small price to pay. It amounts to 0.5 of 1 per cent of global military spending and is less than New York City spends on its police, fire, and correctional services. In the late 1980s, the number of peacekeepers employed by the UN was less than 15 000. By 1993, the force had expanded to 80 000 troops from eighty different countries.

They were deployed primarily in Bosnia, Croatia, and eastern Slovenia. By the end of 1996, the number of troops had fallen to 26 000.

In 1991, UN forces led by the United States removed the invading Iraqi army from Kuwait after the country pleaded for help. The following year, a UN multinational force rescued Somalia from famine and clan violence. The UN's Security Council authorized deployment of the force to bring law and order back to the country for the benefit of its citizens. The peacekeeping operation ended in 1995 without a settlement. Factional fighting in Somalia continues.

Peacekeeping is based on the premise that ensuring continued peace during a cease-fire between warring factions or nations provides a greater opportunity for settling differences through negotiation. In these situations, United Nations forces are neutral. Finding a solution to the conflict depends on the willingness of the two sides in the conflict to reach a settlement.

MEMBERS

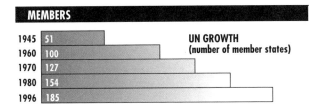

UN GROWTH
(number of member states)

Year	Members
1945	51
1960	100
1970	127
1980	154
1996	185

UN membership embraces virtually the whole world but continues to expand because of decolonization in the 1960s and, more recently, the arrival of smaller nation-states, especially after the break-up of the USSR.

- Regardless of size or population, each of the member states has one vote at the UN General Assembly, from the most populous (China 1.3 billion) to the smallest (Palau 17 000).

Official representations of the UN's structure show the principal organs as satellites revolving around the General Assembly. But a more realistic diagram would be as shown on the right.

THE RHETORIC...

POWER STRUCTURE

THE REALITY...

- The decisions of the Security Council are binding on member states; those of the General Assembly are not. The five main victors in the Second World War have permanent membership on the Security Council and an individual veto over its decisions. These five nations in practice also appoint the Secretary-General, though officially the appointment is made by the Assembly on the Council's "recommendation."

PEACEKEEPING

As 1996 ended, 26 000 military personnel and civilians were serving in 16 peacekeeping operations at a total annual cost of about $1.6 billion. Only a year before, in 1995, 60 000 personnel were serving in 17 UN peacekeeping missions—including 3 in the former Yugoslavia—at an annual cost of $3.5 billion.

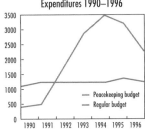

Top 5 contributors of troops to current UN peacekeeping missions (as of 1 September 1996)	
PAKISTAN	1719
BANGLADESH	1184
RUSSIAN FEDERATION	1177
INDIA	1203
BRAZIL	1203

- Since 1945, 110 nations have contributed personnel at various times; 71 are currently providing peacekeepers.
- The small island nation of Fiji has taken part in virtually every UN peacekeeping operation, as has Canada.

COSTS

The UN is routinely criticized as a big bureaucracy wasting vast amounts of public money. Yet its budget and staff numbers are small considering what the world expects this organization to deliver.

UN Peacekeeping and Regular Budget
Expenditures 1990–1996

[Line graph showing Peacekeeping budget and Regular budget, y-axis 0 to 3500, x-axis 1990 to 1996]

- The total operating expenses for the entire UN system—including the World Bank, the IMF, and all UN funds, programmes, and specialized agencies—come to $18.2 billion a year. This is less than the annual revenue of a major corporation such as Dow Chemical, which took in more than $20 billion in 1994.
- The budget for the UN's core function is $1.3 billion a year—about 4 per cent of New York City's annual budget and nearly $1 billion less than the yearly cost of Tokyo's fire department.
- The UN system, which includes the Secretariat and 28 other organizations such as UNICEF, employs 53 333 people. Three times as many people work for McDonald's, while Disney World and Disney Land employ 50 000 people.
- The UN system has $4.6 billion a year to assist countries in their economic and social development. This is the equivalent of $0.80 per human being. In 1994 the world's governments spent about $778 billion in military expenditures—the equivalent of $134 per human being.
- Now in its 52nd year, the UN's ability to function is hampered by financial problems. Unless member states act quickly to pay their debts to the organization in full—a total of over $3 billion as of February 1997—the UN will remain in a precarious financial situation.

Figure 22.13 United Nations: the Facts

Source: Adapted from *New Internationalist*, December 1994. Copyright © New Internationalist. Reprinted by permission of Guardian News Service Ltd. Updated statistics from UN Home Page on the Internet.

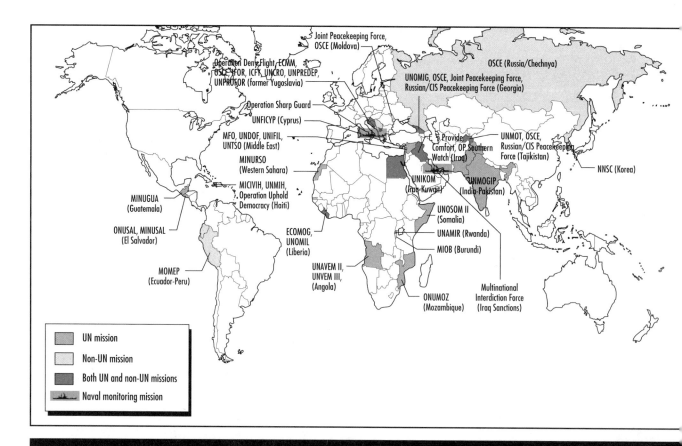

UN missions

MINUGUA (UN Verification Mission in Guatemala)	UNIKOM (UN Iraq-Kuwait Observer Mission)
MINURSO (UN Mission for the Referendum in Western Sahara)	UNMIH (UN Mission in Haiti)
MINUSAL (UN Mission in El Salvador)	UNMOGIP (UN Military Observer Group in India and Pakistan)
ONUMOZ (UN Operation in Mozambique)	UNMOT (UN Mission of Observers in Tajikistan)
ONUSAL (UN Observer Mission in El Salvador)	UNOMIG (UN Observer Mission in Georgia)
UNAMIR (UN Assistance Mission in Rwanda)	UNOMIL (UN Observer Mission in Liberia)
UNAVEM II (UN Angola Verification Mission II)	UNOSOM II (UN Operation in Somalia II)
UNAVEM III (UN Angola Verification Mission III)	UNPREDEP (UN Preventive Deployment Force in Macedonia)
UNCRO (UN Confidence Restoration Operation in Croatia)	UNPROFOR (UN Protection Force: Bosnia, Croatia, Macedonia)
UNDOF (UN Disengagement Observer Force: Israel, Syria)	UNPROFOR (UN Protection Force: Bosnia)
UNFICYP (UN Peacekeeping Force in Cyprus)	UNTSO (UN Truce Supervision Organization: Egypt, Lebanon, Syria)
UNIFIL (UN Interim Force in Lebanon)	

Figure 22.14 Peacekeeping/observer/enforcement missions in 1995

Source: *Armed Conflicts Report 1996*, annual publication of Project Ploughshares, Conrad Grebel College, Waterloo, ON N2L 3G6; tel.: 519-888-6541; fax: 519-885-0806; e-mail: plough@watservl.uwaterloo.ca

In **peacemaking**, the goal of the UN forces is to achieve peace by imposing economic sanctions or by fighting with one side against the other. The UN initiates military action to force a peaceful solution to the conflict before it escalates. The UN's intervention in the Persian Gulf War involved taking sides in the conflict and imposing a solution. In the decades to come, there is no doubt that this organization's peacekeeping and peacemaking international forces will continue to help resolve conflicts.

The United Nations has as many problems today as it had when it was founded over fifty years ago. One of the organization's most serious problems, above and beyond its involvement in international conflicts, involves funding. No member country is permitted to pay less than 0.01 per cent of the regular budget. In 1996 this amounted to just over US $103 000—the fee for eighty-seven countries at the bottom of the scale. The United States undertook to pay 49 per cent of the annual budget when the UN was founded. At present, its contribution is 25 per cent of the annual budget. But the United States is notorious for not paying its annual dues and owes the organization $1.6 billion for past and 1997 payments (over 50 per cent of the total debt of all nations in arrears). Other countries significantly in arrears include Russia, Japan,

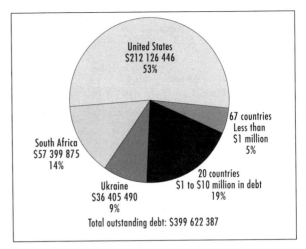

Figure 22.15 Distribution of outstanding debt to the UN (as of 31 August 1995 for prior years)

Ukraine, France, South Africa, and Italy. As of February 1997, member states owed the UN over $3 billion—for peacekeeping, the regular UN budget, and international tribunals.

Another issue relates to the cost of the United Nations' bureaucracy. In 1996 operating costs were about $1.3 billion, while an additional $1.6 billion was spent on peacekeeping. Although there is no doubt that this is a great deal of money, the budget is relatively small when one considers the importance of the UN.

FLASHLIGHTS OVER MOGADISHU

Damned if it acts, damned if it doesn't— after the event we can all be wise about UN intervention. But Mohamed Sahnoun was the man sent in 1991 to assess what kind of intervention the UN should make in Somalia and who saw his advice ignored as US troops staged their invasion for the world's cameras.

My first appeal to Somalis was ironically from Nairobi, Kenya, because I knew that anyone with a small transistor radio tends to turn on at six o'clock to listen to the Somali service of the BBC [British Broadcasting Corporation]. "I'm here on behalf of the United Nations and the international community," I said. "I'll try to get as much assistance as I can to fight the famine but you have to find within yourself forces and people who can help rebuild

Somalia, can stabilize the country, and create peace."

The reaction was extraordinary. People came from all over to see me, people who were starving or came out of hiding especially to respond to my message. These were teachers and police officers, professors and community leaders. Many had tears in their eyes saying, "Mr. Sahnoun, we want to help, we understand your message." I was in contact with women's leaders who wanted to create an association. "All we need," they said, "is some way to show we are useful. If you bring in the supplies and ask us to manage them then we can become a real alternative to the warlords [who controlled the distribution of food and other essential items]."

Figure 22.16 Thousands of people died in clashes between UN forces and Somalis.

I reported this to the UN because it seemed to me to be of utmost significance: it showed that the warlords were not the solution. I told the Secretary-General's office that while clearly we had to talk to the warlords, we had a real alternative in these community leaders. There was a civil society out there just waiting to be empowered if only emergency relief had come in on a large scale and enabled them to start organizing themselves. There was the potential for a bottom-up approach that would provide a real challenge to the warlords.

But this message was not understood in New York by [then Secretary-General] Boutros Boutros-Ghali and his assistants. What they always want is big fixes, spectacular solutions. This time it was no different—they wanted the warlords to meet in Nairobi and have them shake hands for the flashlights. And that is the road the UN chose to go

down. By avoiding the grassroots approach it gave power and prestige to the warlords, built up the sense that they were the only people who could resolve the problem. As a result, the potential community leaders became dispirited, realizing they had no alternative but to ally themselves with one warlord or another instead of seeking within themselves for solutions to Somalia's problems. . . .

When I arrived in Somalia in March 1992 it was a shock. We could not even land at the airport in Mogadishu [the capital]—we had to land on a small strip in the bush to the north. And driving in the Landrover we passed many thousands of displaced people who had fled the capital and were now living in the worst conditions imaginable.

The capital itself was deserted—the only people to be seen were carrying guns, and even they were starving. The only UN agency at work there was UNICEF, and they were limited to the capital. There were just a few voluntary agencies—notably the International Committee of the Red Cross and Save the Children—doing a fantastic job with virtually no means.

It was then that I contacted community

leaders and recommended to New York [UN headquarters] that we needed a massive humanitarian intervention . . . Instead of this massive operation what did we get? A trickle of food. . . . When in June 1991 the neighbouring country of Djibouti organized a conference of the different factions and asked the UN for help, the response was blunt: "We are not going to involve ourselves in Somalia; the issue is too complex."

Procrastination continued even after I'd been dispatched on my mission. By August 1992, I realized that the only way to make things happen was to involve the media. The *New York Times* and *The Guardian* were approached and they wrote the first articles. After that CNN came and then *60 Minutes* (a flagship US documentary) and the snowball was rolling. Once media coverage became overwhelming and people could see on their TV screens what a tragedy it was, donor countries finally began airlifting supplies.

Then all of a sudden, after months of delay and distance, the urgent priority of the UN in New York was to send forces. I knew how sensitive the situation in Somalia was and urged them not to send troops until the conditions for that had been negotiated. The warlords had been against any UN military force from the beginning and even community leaders had been suspicious. But I persuaded the community leaders that 500 troops were needed to stop looting and banditry at the port and the airport—really a police force rather than an army.

Then in New York, when the 500 were not even operational, they started making statements about sending 3000 troops to Somalia. This made the Somalis very nervous. . . . They started asking, . . . "Are you plotting to put Somalia under UN trusteeship?"

In the end the UN had 30 000 troops in Somalia. They were completely rejected by the population. It turned into a total mess and at least 6000 people died in clashes between UN forces and Somalis. Things became so dire that we even talk about a Somalia Syndrome now.

This must never happen again. The UN tends to think the more blue helmets you have, the more likely you are to solve the problem. This is absurd. Naturally, blue helmets are needed sometimes but we need to create conditions for their deployment. Military intervention should be well timed, the ground should be prepared, the troops should be efficient and know exactly why they are there. You can't bring in UN troops for humanitarian purposes and then change the mandate as they did in Somalia when they started pursuing the warlord Aideed. If you do you have a total mess. . . .

People talk about UN intervention as if the only intervention possible is military. But we urgently need to intervene at a much earlier point one—that removes the need for blue helmets later. People talk about *stopping* the genocide in Rwanda but we should be looking at how we could have avoided it altogether. It would have been possible if the UN had used all its powers to enhance the negotiations that took place under the auspices of the OAU in Tanzania; if the UN had seen to it that there was a proper embargo on arms to Rwanda. . . .

The UN is still not prepared today to have a preventative policy, only to react. The result is that the cost of peacekeeping has jumped tenfold over the last two years [1993–94], while at the same time development assistance to the Third World has decreased from $60 million to $50 million. So at the same time as people talk about building up the UN, they allow it to withdraw from development assistance and by doing so might be creating the conditions for future Rwandas. . . .

Mohamed Sahnoun. From *New Internationalist,* December 1994. Copyright © New Internationalist. Reprinted by permission of Guardian News Service Limited.

FREE TRADE

Increasing global interdependence is well evidenced by flourishing world trade markets (see Figure 22.17). As discussed earlier (Chapter 14), trends in global trade since 1950 have led towards the formation of free-trade zones, common markets, customs unions, and economic unions. Proponents of free trade argue that unregulated competition should be allowed so that countries can provide the best products to consumers at the lowest possible cost. They insist that artificial trade barriers and restraints merely protect industries that would not survive in a truly competitive marketplace. Global competition would encourage constant improvement in the quality of goods and in production technology.

In the past, trade arrangements have tended to regional interdependence. Countries have joined various trade associations/organizations to reap mutual benefits in terms of higher prices for commodities and natural resources and better political relations with neighbouring nations. In Africa, regional trade arrangements include the Southern African Customs Union (1910), the Central African Customs and Economic Union (1964), the East African Economic Community (1967), the Economic Community of West African States (1975), the Preferential Trade

When Canada entered into NAFTA, there was much debate over the consequences to the Canadian economy. For over 100 years Canadian industries had been protected by tariffs. Those who opposed free trade argued that Canadian companies could not compete with their American counterparts without the benefit of protectionist policies. Those who favoured the deal believed that free trade offered Canadian companies greater access to American markets, which would lead to their growth if they were willing to be competitive. The reality was that if Canada continued a policy of economic isolationism, Canadian products would not be competitive with those from other countries and export markets would dry up.

Area for Eastern and Southern Africa (1981), and the Indian Ocean Commission (1982). In Asia and the Pacific, countries belong to the Association of Southeast Asian Nations (1967), the Economic Cooperation Organization (1985), the South Asian Association for Regional Cooperation (1987), and the imminent Asian Pacific Economic Cooperation Agreement. European nations have organized into the Benelux Union (1948), the European Free Trade Association (1960), and the European Union (1994). Trade arrangements in the Middle East include the Arab Common Market (1964), the Cooperation Council for the Arab States of the Gulf (1981), and the Economic Cooperation Organization (1985). In the Western Hemisphere, countries have aligned in the Central American Common Market (1960), the Andean Pact (1969), the Caribbean Community (1973), the Latin American Integration Association (1980), and the North American Free Trade Agreement (1994).

Of the 192 nations that exist in the world today, 189 of these are associated in at least one regional trade arrangement. This means that four more nations are involved in economic relationships than are members in the United Nations.

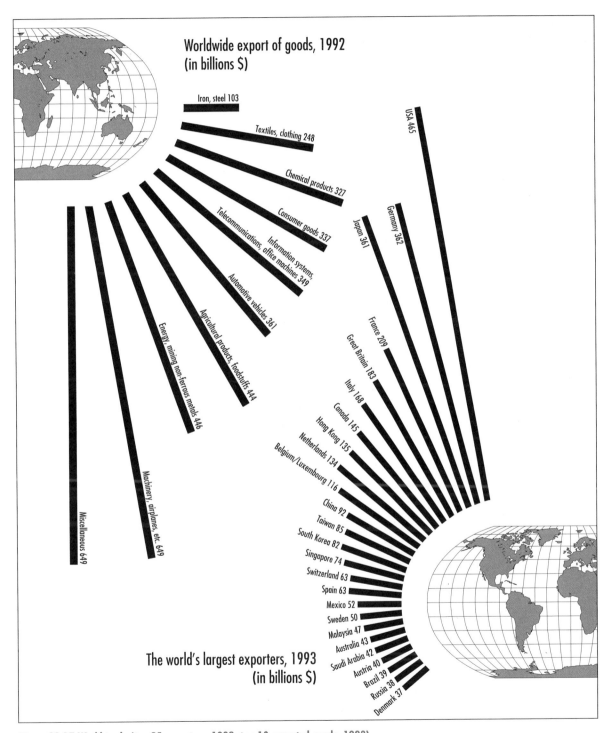

Worldwide export of goods, 1992 (in billions $)

- Iron, steel 103
- Textiles, clothing 248
- Chemical products 327
- Consumer goods 337
- Information systems, office machines 349
- Telecommunications
- Automotive vehicles 361
- Agricultural products, foodstuffs 444
- Energy, mining non-ferrous metals 446
- Machinery, airplanes, etc 649
- Miscellaneous 649

The world's largest exporters, 1993 (in billions $)

- USA 465
- Germany 362
- Japan 361
- France 209
- Great Britain 183
- Italy 168
- Canada 145
- Hong Kong 135
- Netherlands 134
- Belgium/Luxembourg 116
- China 92
- Taiwan 85
- South Korea 82
- Singapore 74
- Switzerland 63
- Spain 63
- Mexico 52
- Sweden 50
- Malaysia 47
- Australia 43
- Saudi Arabia 42
- Austria 40
- Brazil 39
- Russia 38
- Denmark 37

Figure 22.17 World trade (top 25 exporters, 1992; top 10 exported goods, 1993)

CONSOLIDATING AND EXTENDING IDEAS

1 Outline how political issues relate to each of the following:
   - culture
   - environment
   - economics
   - resources

2 Examine Figure 22.17 to predict continental areas that will continue to experience economic success over the next ten years, and over the next twenty-five years. Consider future demands for certain products, depleting world resources, and the impacts of "free-trade marketing." Present your predictions in a class discussion.

3 a) Brainstorm in a group a list of political issues that have not been discussed in this chapter.
   b) Select one of these issues and collect articles from a variety of periodicals and newspapers. Consider using electronic information retrieval systems available in most large reference libraries. Conduct an inquiry using the approach described on page 28 in Chapter 4.

4 The United Nations recently published the *Fiftieth Anniversary Secretariat* (1996), a document that identifies goals and achievements of the UN from 1945 through 1995. Refer to Figure 22.18
   a) In an organizer or a chart list each goal/objective under the appropriate heading: Cultural Issues, Resource Issues, Economic Issues, Environmental Issues, Geopolitical Issues.
   b) Highlight in one colour the "Top Ten"—the goals/achievements that you believe have been most successful.
   c) Identify the five goals/objectives of the UN that will be most pervasive into 2025. Highlight them in another colour.
   d) In several paragraphs, explain how the issues in (b) and (c) interrelate and how you would prioritize future actions of the UN for the next fifty years.

5 Read the article "Flashlights over Mogadishu" on page 383.
   a) It could be said that Mohamed Sahnoun was in a position to know the best approach to solving the problems in Somalia. Give evidence from the article to support the view that he was an authority on the issue.
   b) What approach did Sahnoun recommend?
   c) How was he eventually able to get action?
   d) How did the UN respond?
   e) Why was he critical of the approach taken?
   f) What recommendations does Sahnoun make for improving the UN's prospects of solving international conflicts?

- maintaining peace and security
- promoting development
- preventing nuclear proliferation
- handing down judicial settlements of major international disputes
- aiding Palestinian refugees
- alleviating chronic hunger and rural poverty in developing countries
- eradicating smallpox
- fighting parasitic diseases
- promoting investment in developing countries
- providing food to victims of emergencies
- limiting deforestation and promoting sustainable forestry development
- promoting workers' rights
- promoting stability and order in the world's oceans
- promoting the free flow of information
- establishing children as a "zone of peace"
- improving women's literacy
- facilitating academic and cultural exchanges

- making peace
- promoting human rights
- promoting self-determination and independence
- ending apartheid in South Africa
- focusing on African development
- pressing for universal immunization
- orienting economic policy towards social need
- clearing land-mines
- curbing global warming
- cleaning up pollution
- reducing fertility rates
- improving global trade relations
- introducing improved agricultural techniques and reducing costs
- empowering the voiceless
- generating worldwide commitment in support of the needs of children
- promoting democracy
- protecting the environment

- strengthening international law
- providing humanitarian aid to victims of conflict
- promoting women's rights
- providing safe drinking water
- reducing child mortality rates
- reducing the effects of natural disasters
- protecting the ozone layer
- preventing over-fishing
- protecting consumers' health
- fighting drug abuse
- promoting economic reform
- improving air/sea travel
- protecting intellectual property
- improving global communications
- improving education in developing countries
- safeguarding and preserving historic cultural and architectural sites

Figure 22.18 Goals and achievements—the UN at fifty, (1945–1995)

CHAPTER 23

GLOBAL CONFLICT: DEFINING THE ISSUE

International conflict takes many forms. In the international arena institutions have been established to deal amicably with conflicts. Most large countries have embassies in the capital cities of other countries. These offices are governed by the laws of the country they represent rather than by the countries in which they are located. The host country promises to treat the embassy as if it were part of the country it represents. If a dispute arises between two countries, the embassy serves as a safe haven for citizens of its country. The embassy provides sanctuary, arranges for lost passports, and offers advice to travellers. When problems develop between two countries, diplomats in the embassy meet with representatives of the host country's

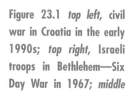

Figure 23.1 *top left,* civil war in Croatia in the early 1990s; *top right,* Israeli troops in Bethlehem—Six Day War in 1967; *middle left,* Allied soldiers during the Normandy Invasion—Second World War; *middle right,* civil war in Rwanda in October 1990; *bottom left,* capture of Fort George—War of 1812; *bottom right,* a parishioner walking through the remains of her church in Sarajevo, 1992—Bosnian War

government to explain government policy, act as a liaison between both governments, and work out diplomatic solutions. If a situation is irresolvable, embassy staff are instructed by their government to close the embassy and return home. In 1975, the American embassy in South Vietnam was swiftly evacuated by helicopter as North Vietnamese troops were storming the gates. In most instances, conflicts can be handled diplomatically without violence.

Military alliances between countries may indirectly prevent conflict because a group of countries is stronger than one isolated nation. If Canada were targeted for invasion by another country, the invader would have to deal with our allies—the United States, Great Britain, and a score of other countries. This type of political strength may reduce the chance of armed conflict between two nations. When Iraq invaded Kuwait, Iraqi leader Saddam Hussein calculated that other Arab countries with which he had alliances would not come to Kuwait's defence. In this case, he miscalculated.

In the past most conflicts were between countries. Embassies, membership in the UN, and international alliances helped mediate disputes and reduce the chance of armed conflict. The nature of conflicts in the 1990s has been different. It has tended towards hostilities between nationalist groups within countries. Without mechanisms for mediation, factions, cultural groups, and religious sects within nations are resorting to armed conflict in order to resolve their differences. The conflict in Somalia between rival warlords, the war in the Balkans between nationalist forces in former Yugoslavia, the Chechnya Revolt in southern Russia, and the civil war in Burundi have all resulted from nationalist uprisings.

HISTORICAL PERSPECTIVES

War is the ultimate global conflict. It breaks out when opposing sides are no longer willing to participate in reasoned discussion and constructive debate in an effort to solve their differences. Wars are fought over four basic issues: politics, culture and religion, economics, and competition for resources. Of course, wars cannot be neatly categorized. Every war is fought for a variety of reasons. Sometimes one issue is more contentious than others; other times a variety of issues comes equally into play.

POLITICS

Throughout the ages, wars have been waged over differences in political ideologies. Some of these conflicts have been **revolutionary wars** in which the people wanted to free themselves from the tyranny of their rulers. In the American Revolution of 1776 and the French Revolution of 1789, citizens overthrew their respective monarchies in favour of republican democracies. **Civil wars** occur when two or more factions within a country disagree with the way a country is being governed. Differences in political ideology result in one part of the country wanting to separate from the rest of the country. The American Civil War is a prime example. The agricultural South did not agree with the policies of the industrialized North. The South wanted to secede from the American union because of conflicting opinions about slavery and economic issues. The Civil War was the bloodiest conflict the world had seen up to 1865.

Other wars may begin as civil wars but escalate to major confrontations between several powers. In 1954 Vietnam gained its independence from France after a bitter nationalist war. Vietnam was divided along the seventeenth parallel into North and South Vietnam. The South remained a democratic enclave in a mostly Communist region. The Vietnamese War began when Ho Chi Minh, the Communist leader of the North, sought to reunify the country militarily.

China and the Soviet Union supported his regime and provided military and financial aid to help the North invade the South and unify the country under Communism. The United States sided with South Vietnam. What started as an internal conflict in Vietnam became the most expensive war the Americans have ever fought.

Many developing countries experience **political coups**. A political leader gains control of the military and takes over the government by force. This happened in Burma when General U Saw Maung staged a coup in 1988. The democratic movement had gained momentum under the leadership of Aung San Suu Kyi, daughter of Burma's first leader. On 8 August, troops began a four-day massacre in which at least 10 000 demonstrators were killed. Just when democracy seemed to be making a comeback, the military regained control of the country. Similar military coups have occurred in Argentina, Chile, Uganda, and Liberia.

In addition to internal conflicts, wars develop because of disagreements between neighbouring countries. Both World Wars and innumerable conflicts before and after these wars resulted because countries were not able to solve disagreements. The following sections explore the various reasons for war.

> Canada is one of the few nations in the world that has not experienced a revolution or civil war in its history. Differences of opinion between the people of Canada and the ruling élite from Britain culminated in several armed rebellions in 1847. The rebellions were quickly put down by the militia. The British Colonial Office realized the people had serious concerns, and dispatched Lord Elgin and Lord Durham to Canada to resolve the matter. Not wanting to repeat the disastrous loss of the American colonies in 1776, Lord Durham initiated reforms that satisfied the citizens and avoided a revolution. Canada gained independence from Britain in 1867 through peaceful negotiations.

CULTURE AND RELIGION

There are many diverse cultures in countries around the world. Nationalist wars erupt when cultural groups within a state want to become independent. The people are proud of their culture and want to be recognized as a nation.

What is the difference between a **state** and a **nation**? A state, or **country**, is an area on the earth's surface that has definite, widely acknowledged boundaries. States have **sovereignty** over their people. This means they have the authority to govern their citizens without interference from other political entities. Sovereign states are also entitled to sign treaties, exchange ambassadors with other countries, and join the United Nations. Countries express their sovereignty through a variety of ways: flags, national anthems, currencies (in most cases), postage stamps, and passports for citizens. They also establish capital cities where government representatives meet to deal with issues of state and where diplomats from other countries set up embassies for international cooperation.

The term *nation* is a more difficult concept to define. Although it is frequently interchanged with the term *country* (as it is in this textbook), *nation* signifies more than the existence of boundaries or political institutions. A cultural group shares a common language, heritage, culture, and set of values. In order to be considered a nation, the group must have a political presence. This means that over a considerable length of time it has been the dominant group occupying a specific territory in sufficient numbers. Unlike a state, a nation does not have most of the trappings of sovereignty, nor does it have official international recognition.

In the nineteenth century the doctrine of nationalism evolved. It makes the assumption that everybody belongs to a nation and each nation has the right to its own sovereign state where it can express its unique culture. The problem is that this is an ideal that is difficult to meet in a world with fewer than 200 countries, but with an estimated 1300 potential nations, some of which number only a few hundred people. It is perplexing to imagine a world made up of hundreds of mini-states. Conflict arises when nationalism develops within a country. A group seeks sovereignty over the territory it occupies; in other words, the group wants to form its own country. Sometimes groups rebel if their country's government does not allow them some degree of local autonomy or recognition. Often concessions are made in order to appease the nation. Such is the case in Scotland, a nation in Britain that has grudgingly accepted English domination for hundreds of years (so far). Nations may still want to separate, even if they are given concessions, because they want to be recognized internationally, to preserve their culture, and to have complete control over how they are governed.

Occasionally, nations and states are more or less synonymous. For example, Japan is a country with few diverse cultures other than the Ainu, the aboriginal people on the north island of Hokkaido. Most people in Japan are of Japanese nationality, so nationalist conflict is nonexistent. Usually, there are a variety of nations within a country. In Canada, there are First Nations and French Canadian peoples, all of whom wish to express their nationalism through political autonomy.

Religious differences can also lead to conflict. When a religious group has values and attitudes that another group finds repugnant, war may ensue. As with nationalism, if a country is dominated by members of one religious denomination, less divisiveness develops. Misunderstandings do not arise as readily as when people have different paradigms based on religious beliefs. At present, nearly a billion people—one-fifth of the world's population—are Catholic. Adherence

Figure 23.2 Basque nationalists in Guernica, Spain, demonstrate in front of a copy of Picasso's famous painting *Guernica*.

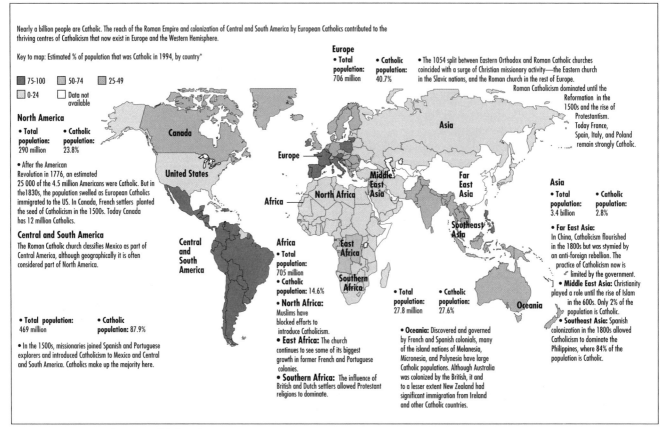

Nearly a billion people are Catholic. The reach of the Roman Empire and colonization of Central and South America by European Catholics contributed to the thriving centres of Catholicism that now exist in Europe and the Western Hemisphere.

Key to map: Estimated % of population that was Catholic in 1994, by country*

- 75-100
- 50-74
- 25-49
- 0-24
- Data not available

North America
- **Total population:** 290 million
- **Catholic population:** 23.8%

• After the American Revolution in 1776, an estimated 25 000 of the 4.5 million Americans were Catholic. But in the 1830s, the population swelled as European Catholics immigrated to the US. In Canada, French settlers planted the seed of Catholicism in the 1500s. Today Canada has 12 million Catholics.

Central and South America
The Roman Catholic church classifies Mexico as part of Central America, although geographically it is often considered part of North America.

- **Total population:** 469 million
- **Catholic population:** 87.9%

• In the 1500s, missionaries joined Spanish and Portuguese explorers and introduced Catholicism to Mexico and Central and South America. Catholics make up the majority here.

Europe
- **Total population:** 706 million
- **Catholic population:** 40.7%

• The 1054 split between Eastern Orthodox and Roman Catholic churches coincided with a surge of Christian missionary activity—the Eastern church in the Slavic nations, and the Roman church in the rest of Europe.

Roman Catholicism dominated until the Reformation in the 1500s and the rise of Protestantism. Today France, Spain, Italy, and Poland remain strongly Catholic.

Africa
- **Total population:** 705 million
- **Catholic population:** 14.6%

• **North Africa:** Muslims have blocked efforts to introduce Catholicism.
• **East Africa:** The church continues to see some of its biggest growth in former French and Portuguese colonies.
• **Southern Africa:** The influence of British and Dutch settlers allowed Protestant religions to dominate.

- **Total population:** 27.8 million
- **Catholic population:** 27.6%

• **Oceania:** Discovered and governed by French and Spanish colonials, many of the island nations of Melanesia, Micronesia, and Polynesia have large Catholic populations. Although Australia was colonized by the British, it and to a lesser extent New Zealand had significant immigration from Ireland and other Catholic countries.

Asia
- **Total population:** 3.4 billion
- **Catholic population:** 2.8%

• **Far East Asia:** In China, Catholicism flourished in the 1800s but was stymied by an anti-foreign rebellion. The practice of Catholicism now is limited by the government.
• **Middle East Asia:** Christianity played a role until the rise of Islam in the 600s. Only 2% of the population is Catholic.
• **Southeast Asia:** Spanish colonization in the 1800s allowed Catholicism to dominate the Philippines, where 84% of the population is Catholic.

Figure 23.3 Catholicism's global reach
Source: *Calgary Herald*, 25 January 1997. Reprinted with permission of Knight-Ridder/Tribune Information Services.

to one faith is a uniting factor among peoples of very diverse cultures.

Religious wars within countries and between sovereign states are frequently the most bloody since combatants are driven with a religious fervour, often believing that their combative efforts will be rewarded in the hereafter. During the Reformation in Europe, religious intolerance led to many clashes. Persecution of Protestants led to many sects such as the Huguenots and Mennonites emigrating to the more tolerant New World. After the re-establishment of the Church of England as the dominant religion in England during the Restoration Period, many Protestants from Britain immi-

grated to New England in the early seventeenth century. Similar religious migrations occurred in Canada. Doukabours, a religious sect from Russia, settled in Alberta in the 1800s. Quakers from Germany settled in Pennsylvania in the eighteenth century. After the Second World War Jews migrated to their new homeland of Israel, established in 1947.

One recent conflict in which culture and religion played an important part was the war in the Balkans. At the heart of the war in the former Yugoslavia were deep-rooted cultural and religious conflicts. Several distinct cultural groups live in this eastern European region known as the Balkans. Located between the Muslim

Cultural groups in former Yugoslavia

Serbians	38%
Croatians	20%
Muslims	9%
Albanians	8%
Slovenes	8%
Macedonians	6%
Montenegrins	3%
Hungarians	2%
Other	6%

Romania

1945	Communist government established.
1974	Nicolae Ceausescu became president.
1987	Workers demonstrated against austerity program.
1989	Uprising against Ceausescu's regime.
	Ion Iliescu and National Salvation Front came to power.

Yugoslavia

1980	Tito died; collective presidency assumed power.
1981	Demonstrations by Albanian nationalists living in Kosovo began.
1987	Slobodan Milosevic seized leadership of Serbia.
1988	Milosevic used Kosovo situation to stimulate Serbian nationalism.
1989	Milosevic used national army to suppress strike in Kosovo.
1989	Slovenia condemned use of troops in Kosovo.
1991	Slovenia and Croatia declared independence.
	Milosevic refused to recognize the collective presidency, thus declaring Serbia independent.
	Civil war started.
1992	Bosnia-Hercegovina and Macedonia declared independence.
1995	Serbia, Croatia, and Bosnia-Hercegovina endorse a US-brokered peace plan for Bosnia.

Albania

1946	New Communist republic under Enver Hoxha.
1967	Delcared first atheist state.
1985	Hoxha died.
1990	Single-party system abandoned.
1991	First free elections held.

Bulgaria

1946	Communist state established.
1954	Todor Zhivkov became leader.
1989	Turkish minority began leaving country. Zhivkov ousted.
	Formation of opposition parties.
1992	Zheylyu Zhelev became first freely elected president.

Figure 23.4 The Balkans

Source: *A Map History of the Modern World* 2d ed., by I. Hundey, M. Magarrey, and B. Catchpole. (Toronto: Irwin, 1995). Reprinted by permission of Stoddart Publishing Co. Limited.

nations of the Middle East and the Christian nations of central Europe, the Balkans have been repeatedly conquered by invading peoples. In such a mountainous region, relatively isolated villages began to develop their own cultures and customs that were distinct from other groups in the region. After invasions by Slavic peoples from the north and Greeks from the south prior to the tenth century, the region became part of the Christian Byzantine Empire in the eleventh century. The Muslim Turks of the Ottoman Empire conquered and occupied the region until the empire collapsed at the end of the First World War. Some of the peoples, such as the Albanians and the Bosnians, converted from Christianity to Islam during Ottoman rule, but most indigenous peoples retained their religion, language, and culture. Under Turkish rule, they were allowed to be fairly autonomous, as the Turks were more interested in the tax revenues they could obtain than in assimilating the people. As a result, the various religious and cultural groups lived in isolation.

Political instability became synonymous with the region because of the cultural, linguistic, and religious diversity of the region. In 1914 the Balkans became the powder keg that led to the First World War. After the war, the Serb-Croat-Slovene Kingdom was created, uniting Serbs, Croats, Slovenes, Bosnians, and Macedonians. The territory was torn by constant political unrest because no one group would accept a monarch from another group. Conquered by Germany in the Second World War, the different factions united under Josip Broz Tito to fight against Nazi rule. After the war, Tito governed the new nation of Yugoslavia under an autocratic Communist rule and managed to keep cultural divisions under control through his charisma and military might. When Tito died in 1980, dissension was averted by allowing the six republics and two autonomous territories to govern the country alternately. This system succeeded in maintaining peaceful co-existence until Communism collapsed in the late 1980s and ancient hostilities flared again. In the early 1990s, Serbs, Croats, and Bosnians were engaged in a bloody civil war in which all sides accused the others of **ethnic cleansing**. With atrocities being committed by leaders on all sides, the United Nations sent in a peacekeeping mission. In 1995, the UN forces were replaced with NATO troops to ensure that peace is maintained. In 1997, an uneasy peace hangs over the war-torn Balkans.

The Crusades were a series of wars against non-Christians and dissidents, sanctioned by the popes between 1096 and 1291. European rulers wanted to recover Palestine from the Muslims. They were motivated by religious zeal, the political desire to obtain more land, and the economic ambitions of the major Italian cities to further trade opportunities with the Middle East.

ECONOMICS AND RESOURCES

Economics and resources are often motivating factors for international conflict. When a country wishes to increase its power it often attempts to improve its economy, enlarge the area of land it controls, and expand its resources. With additional land, the country has more tax revenues, more citizens to draft into the armed forces, and a greater distance between its capital city and enemy borders. Increased resources provide the nation with more raw materials to manufacture the necessities of war—iron for guns, oil for tanks, and agricultural products to feed armies.

Imperialism is an extreme

form of nationalism where one nation seeks to dominate other nations by territorial expansion. This "empire building" is bound to create conflict as smaller, less powerful nations are absorbed by an imperialistic force. The Romans were one of the first groups to expand their empire beyond the boundaries that their nation occupied. The British and French did the same when they took over vast territories in Africa. The Second World War was waged in part because Germany wanted to obtain more territory and resources to fuel its rapidly growing industries. Germany initiated the African campaign to acquire oil from the Middle East for its war machine.

More recently, the Persian Gulf War was pri-marily about economics and resources. The oil-exporting country of Kuwait had loaned Iraq billions of dollars during the latter's war with Iran. After four years, Iraq had not won a hectare of Iranian territory and it was almost bankrupt. Saddam Hussein, the Iraqi leader, assumed that if he took over Kuwait, Iraq would not have to pay back its debts. But this was not the principal motivation for the invasion. Huge oil deposits lay underneath Iraq's less powerful and smaller neighbour. If Hussein could gain these resources, he would be able to bolster Iraq's economy and increase his military power.

As global populations rise and resources become increasingly scarce, conflicts over resources will become even more common.

CONSOLIDATING AND EXTENDING IDEAS

1 Define the term *war*. Working in a group, compare your definitions, then write one definition that all group members can accept.

2 Some people assert that war is an ineradicable part of human nature. Others contend that war can be avoided.

   Working in groups of four, pair up to address each side of the debate. Develop arguments and examples to support your chosen perspective. Present your ideas to the class.

3 a) Identify the difference between the terms *nation* and *state* or *country*.
   b) Explain what the doctrine of nationalism is.
   c) What view is expressed in the text regarding this doctrine?
   d) How can this doctrine result in conflict?
   e) What is your opinion regarding the doctrine of nationalism?
   f) How is imperialism an extension of nationalism?

4 a) Research one war from the list below:
   - War of 1812
   - Boer War
   - Korean War
   - Rwandan Civil War
   - Seven Day War
   - American Civil War
   - Spanish Civil War
   - Vietnam War
   - Falkland Islands War
   - Iran-Iraq War
   - another war

b) Create a timeline showing the chronology of major events.

c) List the primary reasons for the war.

d) Categorize the main cause of the war as political, cultural/religious, economic, or resource.

e) Describe the outcome.

5 Figure 23.5 illustrates the distribution of religion in the world. Do further research to determine the global distribution of religious groups other than Roman Catholics. Create a graded shading map similar to the one in Figure 23.3. Indicate where other Christians, Buddhists, Hindus, Muslims, and members of other religious faiths live.

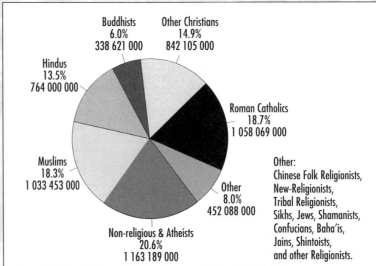

Buddhists
6.0%
338 621 000

Other Christians
14.9%
842 105 000

Hindus
13.5%
764 000 000

Roman Catholics
18.7%
1 058 069 000

Muslims
18.3%
1 033 453 000

Other
8.0%
452 088 000

Non-religious & Atheists
20.6%
1 163 189 000

Other:
Chinese Folk Religionists,
New-Religionists,
Tribal Religionists,
Sikhs, Jews, Shamanists,
Confucians, Baha'is,
Jains, Shintoists,
and other Religionists.

Figure 23.5 Distribution of religion in the world (as of mid-1994)

Source: Copyright ©World Eagle 1996. Reprinted with permission from World Eagle, 111 King St., Littleton, MA 01460 USA 1-800-854-8273. All rights reserved.

PATTERNS IN PHYSICAL GEOGRAPHY

Physical geography can play a significant role in war. Location, physiography, climate, weather, soils, and even natural vegetation have profoundly influenced all aspects of warfare. While today's modern technology does much to reduce the impact of the environment on military strategy, physical geography remains an important element for any military strategist.

LOCATION AND PHYSIOGRAPHY

Of all the elements of physical geography, location and physiography are probably the most critical factors in combat strategies. Military tactics often focus on the need to secure strate-

Figure 23.6 American soldiers tackling the jungle during the Vietnam War

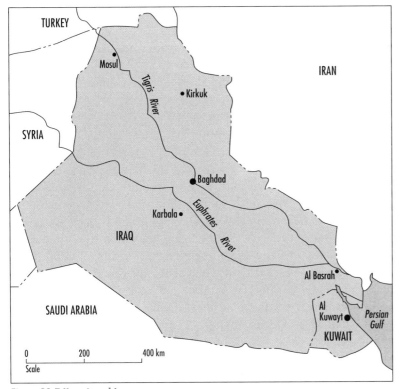

Figure 23.7 Kuwait and Iraq

gic locations as one force invades or another force defends its territory. Many battles make up a war. If one side fights and wins each battle successively closer to the enemy's base of operation, that side will undoubtedly win the war.

The US's invasions of isolated islands in the South Pacific during the Second World War illustrate the importance of location in a military operation. By systematically invading and holding Tarawa (1943), Tinian and Saipan (1944), Iwo Jima and Okinawa (1945), the American forces held strategic territories successively closer to Japan. With the defeat of each new island, American bombers could fly closer into Japan's heartland. The final assault, the flight of the *Enola Gay* with its atomic payload, took off from the island of Tinian. When Hiroshima and Nagasaki were bombed, Japan knew it had lost the war.

Location was a significant factor in the European theatre as well. Germany's location in the centre of Europe inspired the **blitzkrieg** strategy as it roared through Europe destroying all opposition. It was not until the Nazi forces declared war on the Soviet Union that Germany's fortunes turned. Faced with an eastern as well as a western front, German forces were split in two. The United States joined the Allied forces in 1941 and began attacking from the west at about the same time as the Soviet Union launched its assault from the east. With forces attacking from two sides, Germany's central location was no longer an advantage. It took almost four years, but eventually the Third Reich fell as the Soviets invaded from the east and the Allies attacked from the west.

Physiography includes relief and altitude. Relief describes how rough the ground is. Of course, mechanized units can manoeuvre more easily in areas with low relief. In areas with high relief, vehicles have problems going up and down hills. Altitude refers to the height above sea level. Today, because of aircraft, altitude has lost its importance in military operations. In the past, invasions over mountain ranges were difficult if not impossible. Switzerland's mountainous boundaries are one of the reasons it has remained a neutral country and has not been involved in a war for centuries. Crossing the Alps into Switzerland was so difficult that most would-be invaders ignored the tiny country.

Certain landforms are important for military or economic reasons. Countries often fight for land in order to gain a strategic benefit. A rise of land separating two countries may provide a defensive advantage for the country that controls this land. In 1967 Israel won the Golan Heights from Syria in the Six Day War. This land is not remarkable except for one detail—it commands a strategic position between Israel and one of its traditional enemies, Syria. Because it is higher in elevation than the surrounding land, Israeli defensive units can see what the Syrians are doing. Furthermore, an uphill infantry battle is difficult to wage against a heavily fortified piece of land such the Golan Heights.

Iraq invaded Kuwait in the Persian Gulf War partly because of Iraq's limited coastal access to the Persian Gulf. The coast Iraq does have is marshland unsuited to the port facilities needed for exporting Iraqi oil. Neighbouring Kuwait has a relatively long coastline and two offshore islands within sight of Iraq. One of the islands boasts a multimillion-dollar deep-water harbour suitable for oil supertankers. During the war itself, the Persian Gulf provided an excellent transportation route for American vessels. Aircraft carriers and destroyers could anchor off the coastline, and jets, missiles, and shells could be launched into enemy territory.

Physiography often determines the strategy of a war. If the ground is rough, infantry is probably the best force to deploy. In mountainous terrain soldiers on foot can manoeuvre more easily than artillery or armoured divisions. Unlike aircraft personnel, they can see guerrilla forces hiding in trenches or trees. This type of conflict can be costly in terms of human lives because troops end up fighting hand to hand. Coastal battles often require the use of transport ships and marine units, with air support and coastal bombardment from battleships off the coast. When the terrain is flat or gently rolling hills, mechanized cavalry followed by infantry is the best approach. Tanks and artillery can move in quickly to take strategic gun installations. Infantry can follow behind and capture enemy stragglers.

CLIMATE AND WEATHER

Weather has greatly influenced many wars. It affects visibility, troop movements, and communications. When the weather is foggy most military operations have to be postponed. Military units cannot attack if they are unable to see the enemy. On the other hand, fog is the ideal cover for small commando forces to move into strategic positions. In the past, communication was hindered during fog because visual contact could not be made using signal flags. Now, electronic communication has eliminated this problem.

At one time, heavy rain or snow often slowed down troop movements. Modern military equipment is more able to cope with inclement weather but air attacks may be postponed because of icy conditions.

Figure 23.8 A US aircraft carrier in the Gulf of Oman during the Persian Gulf War

Extreme cold or heat also limits the effectiveness of fighting units. However, it is an even playing field because both sides in a conflict experience the same difficulties.

Both Britain and Japan were saved from invasion by freak storms that destroyed invading armies. The Spanish Armada, a fleet of hundreds of Spanish galleons, was blown off course by heavy winds, saving the English people from conquest in 1588. Japan was similarly protected when a "divine wind," or kamikaze, blew Mongol ships off course in the thirteenth century.

In the War of 1812 between the United States and Britain, troops had to cease fighting for several months during winter until the spring thaw. The harsh North American climate with its below-freezing temperatures made military operations impossible. Water bodies froze and snow clogged travel routes. The only battle of the winter was the Battle of New Orleans, fought in late December and early January. Why do you think this battle was the exception?

During the Gulf War, fog rolling in off the waters of the Gulf made night bombing raids difficult because of reduced visibility.

Climate has been a significant factor in many wars. Climate includes temperature, precipitation, wind, and other elements of the atmosphere. Unlike weather, climate is predictable since it is defined as the prevailing weather conditions of an area. Generals need to know about the climate of a region before they begin military operations. Having this information enables them to select particular military

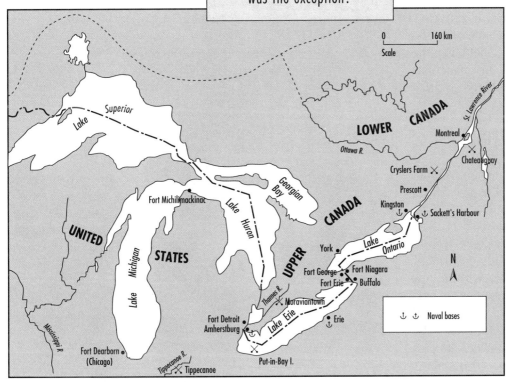

Figure 23.9 War of 1812

strategies that take advantage of the climate. If a region is known for its heavy snowfalls in winter, perhaps a military operation should be conducted in summer. If a campaign is unavoidable in winter, infantry should be outfitted with appropriate clothing and special equipment such as skis or snowshoes. In a desert region, provision must be made for extra supplies of water because dehydration could be a problem. The heat and constant rainfall of rainforest regions can complicate military operations. Troops must be inoculated against tropical diseases and use plenty of insect repellent.

In the Persian Gulf War, climate played an important role in the timing of the coalition offensive. Although Iraqi forces attacked Kuwait in August, the joint forces waited until winter before they counterattacked. Because the Gulf is a hot desert region where summer temperatures regularly exceed 40°C, the strategy was to launch an offensive in winter when temperatures were more hospitable. Of course, there were other reasons for the delayed attack, including the use of non-military sanctions such as the embargo on Iraqi oil products.

SOILS AND NATURAL VEGETATION

If soil is firm, military equipment can travel over it freely. When the ground is wet, it can easily turn into a sea of mud. Imagine 10 000 soldiers marching over someone's lawn after a rainfall, along with a few dozen tanks and a fleet of personnel carriers, jeeps, and field artillery. Picture the effect that mud could have on this advancing army. Many advancing armies utilize exist-

ing transportation routes, while most retreating armies blow up bridges, mine highways, and destroy railways. Often, overland invasions are the only option. The wetter the ground, the more difficult troop movement is. Bogs and swamps are particularly troublesome. They may be infiltrated more easily using helicopters instead of infantry or heavy vehicles.

Natural vegetation is significant in war because it determines the extent to which the enemy can be seen. If there are many trees the enemy can use guerrilla tactics. By hiding behind and in every tree, soldiers gain a strategic advantage. Military equipment is easily camouflaged under the cover of trees. In open grassland or desert regions, vegetation does not act as a hindrance to surveillance.

During the Second World War, soils and climate conspired with other factors to defeat Germany. The Nazi invasion of Russia was doomed from the start because Hitler failed to learn from Napoleon's tragic mistake 130 years earlier. The

Figure 23.10 The Nazi invasion of Russia during the Second World War

French emperor had sent his troops to conquer Russia, but the sheer size of the country and its harsh climate proved to be his downfall. Hitler was confident that his tanks and troops would crush the Soviets before winter. But he, too, would be thwarted by the vast expanse of the country in general and the hazards presented by the Pripet Marshes in particular. As the Soviet army retreated farther and farther east, the advancing German troops moved deeper into Soviet territory. But as fall gave way to winter, the German armies became bogged down in the muddy marshes. When the mud froze, the German war machine was literally frozen in position. Soldiers starved and equipment was lost. It was a blow from which the German army never fully recovered.

American forces in Vietnam had two enemies to fight: North Vietnamese troops and the jungles of the tropical rainforest. Dense vegetation camouflaged the North Vietnamese guerrillas. The harsh conditions of the environment challenged the US's military superiority. Jeeps, tanks, and field artillery were confined to the roads because of the jungles and boggy ground. As a result they became easy targets for enemy forces. In the end Vietnam became an unwinnable war, in part because of the physical geography.

The lack of vegetation during the Gulf War facilitated surveillance, especially with American spy satellites tracking hundreds of kilometres above the war zone. Remote-sensed images were so accurate that it was possible to read licence plates on vehicles hundreds of kilometres away. Recognizance experts were able to determine the location of munitions manufacturing plants and infrastructure installations vital to Iraq's military. Desert sand, however, hampered mechanical operations and surveillance. Tanks and other vehicles sometimes ground to a halt as the fine desert sands clogged air filters and reduced visibility.

CONSOLIDATING AND EXTENDING IDEAS

1 Describe how each of the following physical features can influence military strategy:
   - mountains
   - marshes or swamps
   - level plains
   - shorelines
   - islands
   - sand dunes

2 Explain how the following can influence military strategy:
   - climate and weather
   - natural vegetation and soil

3 Continue the research you started on a war of your choice (page 397, activity 4).
   a) Find out how physical geography influenced its causes and strategies.
   b) Determine the role technology played to overcome these elements of physical geography.
   c) In a ten- to fifteen-minute oral presentation to your class, describe the international

conflict you chose to research. Include the following:
- the major participants—countries, groups, individuals
- the causes—fundamental and immediate
- the course of the war
- the influences of physical geography on its causes and strategies
- the role of technology in overcoming physical geography
- the resolutions
- the present situation
- maps and other visuals

PATTERNS OF HUMAN GEOGRAPHY

The elements of human geography play just as important a role as those of physical geography in establishing the causes of war and determining military strategies. Human geography includes everything people and governments do that affect the land—political boundaries, the

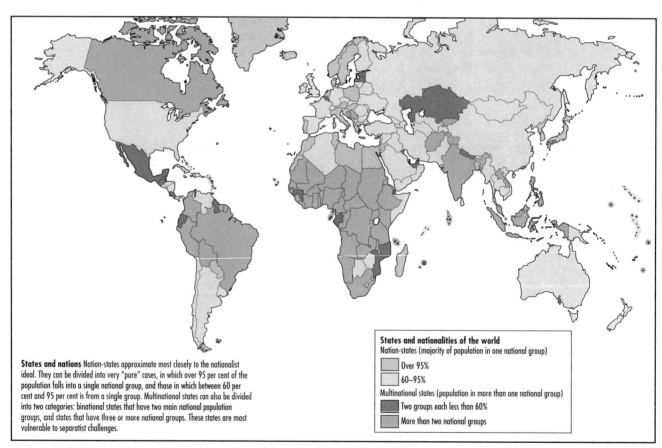

States and nations Nation-states approximate most closely to the nationalist ideal. They can be divided into very "pure" cases, in which over 95 per cent of the population falls into a single national group, and those in which between 60 per cent and 95 per cent is from a single group. Multinational states can also be divided into two categories: binational states that have two main national population groups, and states that have three or more national groups. These states are most vulnerable to separatist challenges.

States and nationalities of the world
Nation-states (majority of population in one national group)

- Over 95%
- 60–95%

Multinational states (population in more than one national group)

- Two groups each less than 60%
- More than two national groups

Figure 23.11 States and nations of the world
Source: *World Government*, rev. ed. (1994), by Peter Taylor. Copyright ©Andromeda Oxford Limited 1990. Used by permission of Oxford University Press, Inc.

size and shape of countries, population characteristics, industrialization, and urbanization.

POLITICAL BOUNDARIES

Political boundaries delineate the area of a country. They establish the contact zone between two countries or, in the case of coastlines, between a country and international waters. Whenever one country invades another, a political boundary is breached. Consequently, boundaries are important elements of human geography that have a profound effect on global conflict.

We all recognize political boundaries when we look at a map. But the concept has a much broader definition. Boundaries also extend in a vertical plane above and below the ground. Aircraft must have permission to fly over a country. During the Cold War, there were several incidents of suspected spy planes over military installations without permission to fly in foreign airspace. One of the most famous incidents occurred on 1 May 1960 when American pilot Gary Powers was shot down as he flew his U-2 spy plane over Soviet airspace. Today, sophisticated spy satellites are able to monitor military installations and activities from several kilometres in space.

Another reason for the Persian Gulf War was that Kuwait was believed to be extracting oil that lay beneath Iraqi territory. Though the oil wells themselves were on Kuwaiti land, the deposits they were pumping were buried on both sides of the border. Since oil is a liquid, it flows to the spot where it is being extracted. Iraq claimed that, in effect, Kuwait was taking oil that belonged to Iraq.

Boundaries are often drawn on maps before a region is explored, surveyed, or settled. These **antecedent boundaries** ensure that there will

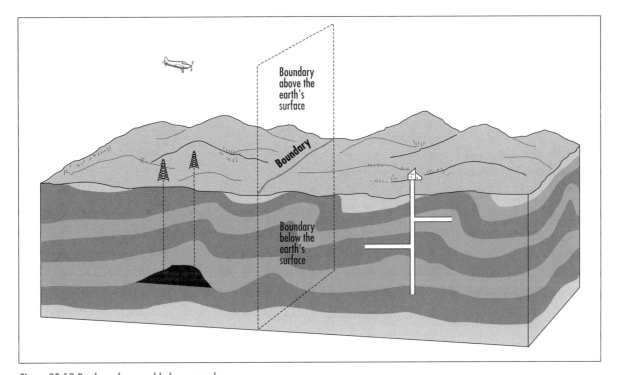

Figure 23.12 Borders above and below ground

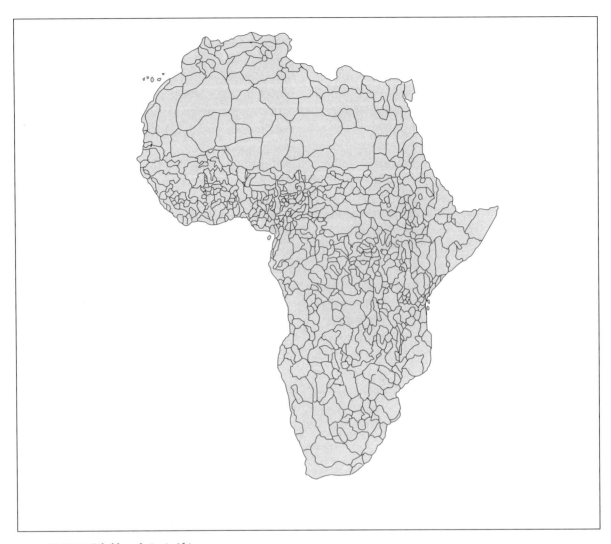

Figure 23.13 (a) Tribal boundaries in Africa
Source: G.P. Murdock, *Africa: Its Peoples and Their Culture History* (New York: McGraw-Hill).

be no other claims to the area when the region does become populated. Imperialist European countries moved into parts of Africa and North America in the eighteenth and nineteenth centuries. Antecedent borders were drawn on maps so that future demands by other imperialist powers would not follow years later when the land was settled. Some of the world's longest and most undisputed borders were established before people settled the region, for example,

the border between much of Canada and the United States. In Africa, antecedent borders have lead to conflict because indigenous peoples were grouped in countries along with their traditional enemies. The current problems in Burundi, Rwanda, and the Democratic Republic of Congo are evidence of this practice.

Boundaries that are drawn up after a region has been explored, settled, and surveyed are called **subsequent boundaries**. These are

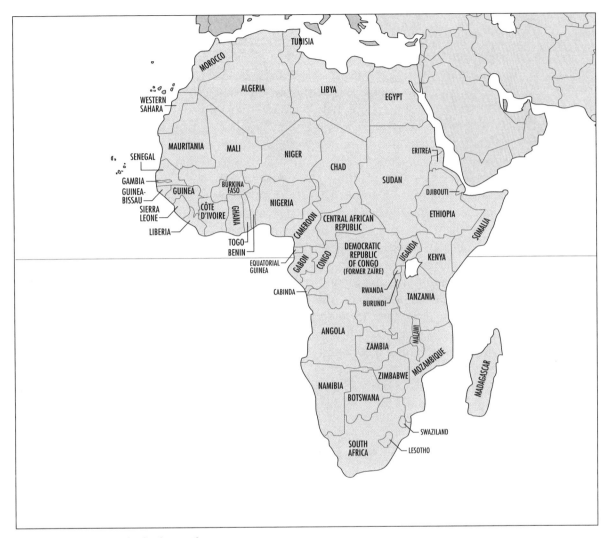

Figure 23.13 (b) Present-day borders in Africa

common in regions where large populations have been present for many centuries. Europe is notable in that its boundaries are subsequent. The advantage of these types of boundaries is that the lay of the land is familiar to everyone, so the borders take into consideration physical geography. Antecedent and subsequent boundaries can be further divided into natural, geometric, and cultural boundaries. In ancient times, countries preferred **natural boundaries**

because they were easier to defend than artificial ones. The borders of many European and South American countries run along rivers and mountain ranges.

Not only were natural borders easier to defend, but they were also clearly visible and did not require sophisticated surveying. Natural borders were usually sparsely populated; no one lives in the middle of a river, on mountain peaks, and or in hostile deserts. Therefore, there was a buffer between adjoining countries. Natural boundaries usually delineate regions in which people of similar cultural backgrounds have evolved. In Europe, a continent of many distinct peninsulas and valleys, there are many isolated regions in which diverse cultural groups have developed. Natural borders separated people for centuries, so distinctive cultures were able to grow. In this century, with advanced technology linking the diverse peoples of the continent, unification is occurring in most of Europe. The most striking example of technology linking one region to another is the 1995 opening of the Chunnel, a tunnel under the English Channel that links England and France.

There are problems with natural boundaries. First, they can change over time. The courses of rivers shift freely as they meander across flood plains. Does this mean that the border changes too? And where is the exact boundary? Is it in the centre of the river or along the river bank? Usually, borders are on the bank and the river is used by both countries for navigation. The Great Lakes form part of the border between the United States and Canada. While the border technically runs through the centre of the lakes, the lakes themselves are international waterways open to ships of all countries provided certain conditions are met. The owner or agent of each vessel must first request permission by filing a pre-clearance form, which ensures that the ship meets all financial and safety requirements.

When mountains serve as an international boundary, the exact border usually runs along the highest point of land between the two countries, known as the **divide**. It is possible that the height of the land could change as a result of erosion and tectonic uplift. Would this affect the exact location of the boundary?

Geometric boundaries are more common in Africa. Often, these antecedent boundaries did not take into consideration the indigenous peoples who occupied the region prior to imperialist claims. The indigenous nations did not have fixed borders. They were not usually necessary because population densities were so low, different groups seldom came in contact. Wars frequently developed when rival groups competed over land in buffer zones between their territories.

It is difficult to draw these **cultural boundaries** on maps, but Figure 23.13 (a) attempts to illustrate them. Much of the political tension that exists in Africa today is the result of the establishment of arbitrary borders following lines of latitude or longitude, or drainage basins. Cultural groups may have been separated, while conflicting interest groups were thrown together and had to occupy the same land. Of course, cultural boundaries also exist in other parts of the world. After the break-up of former Yugoslavia, one of the major causes of the ensuing conflict was disagreement over where new subsequent boundaries should be drawn and where cultural boundaries had evolved as various cultural groups occupied different areas over hundreds of years.

Geometric borders have some advantages over natural borders. Modern military technology is capable of breaching any border so natural borders have lost their importance. It could even be argued that artificial borders are easier to defend. Precisely surveyed and delineated on maps, there is no question where artificial boundaries occur. Because they are not dependent on natural features, artificial boundaries may be more accessible, and therefore more easily defended. In addition, artificial borders do not change as a result of gradational processes.

Coastlines are a special type of border. The exact line between land and water changes with tides and weather conditions. Coastal borders extend from the shoreline for several kilometres. In earlier times, this extended border was about 5.5 km—the distance a cannon-ball could be fired from the shore. Today, as countries seek to protect the marine resources of the continental shelf, territorial boundaries extend out to sea as far as 22, 370, or even as much as 650 km. Beyond this point the oceans are considered international waters. As with air space, foreign vessels need permission to enter a country's territorial waters. Landing must be made at a **port of entry**. It is possible for intruders to enter a country illegally, unnoticed in boats plying international waters.

THE SIZE AND SHAPE OF COUNTRIES

The size and shape of a country are elements of human geography because short borders are easier to defend than long ones. If a country is large, it has a long border to defend. China, the world's third largest country by land area, has a very long border with its traditional enemy, Russia. It takes considerable effort to secure the remote regions of northern and western China from eastern Russia. France is much smaller, thus it has a smaller border to defend.

The shape of a country may also affect the length of international borders. Elongated countries such as Chile are much more difficult to defend than compact countries such as nearby Colombia. Not only is the perimeter of the country long in ratio to its area, but its extremities are far away from the heart of the country. If the southern tip of Chile were attacked, it would take a long time for forces in the capital of Santiago to respond. By contrast, Bogotá, the capital of Colombia, is almost equidistant from the country's borders, therefore response time would be faster in any direction.

Canada's large size is a major consideration in planning for its defence, but it difficult to defend for a number of other reasons. While the concentration of its population in the southern ten per cent of its land mass might prevent invasion across the forty-ninth parallel, defending the north constitutes a serious problem. The region is a sparsely populated wilderness fragmented by many northern islands. Each island has a shoreline that is next to impossible to defend because it is very remote and because there is so much coastline. With the longest shoreline of any country, Canada could be invaded at many different locations.

Island nations are difficult to defend because they are often fragmented. In the Second World War, American forces were able to defeat Japan one island at a time until the enemy forces were isolated within the Japanese home islands. The United States itself is a fragmented nation because of its outlying states of Alaska and Hawaii. During the war, the US had great difficulty defending these two territories from Japan.

POPULATION CHARACTERISTICS

The citizens of a country are important to its military capability. Countries with large populations are able to produce larger armies than countries with small populations. With 1.2 billion people, China is capable of fielding the world's largest army. Yet even small nations can have considerable military might. Israel has a relatively small population, but all adults serve compulsory military service. Their training has ensured that Israel has one of the most successful military forces in the world.

During times of conflict some countries draft their citizens to serve in the military. The United States had a compulsory draft during the Vietnam War. All American males over the age of eighteen had to serve their country if they were selected in a lottery process. After two years they were allowed to return to civilian life.

Of course, the size of an army does not necessarily have anything to do with its effectiveness. A small, well-trained force is often much more effective than draftees who do not want to be involved in a conflict. Many countries have established military academies to train professional soldiers.

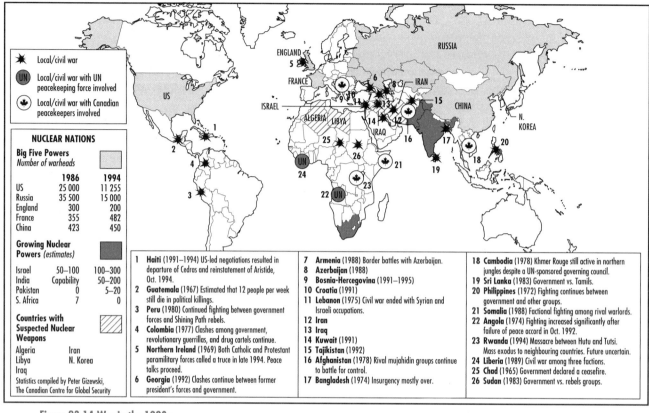

Figure 23.14 War in the 1990s

Source: *A Map History of the Modern World*, 2d ed., by Ian Hundey, M. Magarrey, and B. Catchpole (Toronto: Irwin,1995). Reprinted by permission of Stoddart Publishing Co. Limited.

INDUSTRIAL STRENGTH

The number of military personnel is no longer as important as it once was. Military might is now measured by a country's arsenal of military hardware. The arms trade is alive and well in the post-Cold War era. Multinational corporations make millions of dollars selling weapons to developing countries as well as to world powers. This has a negative impact on countries that are attempting to solve problems of overpopulation and poverty. With so much money going into the military, little is left for social programs. Military rulers in many developing countries are often more interested in quelling rebellious cultural groups than in improving living conditions. The threat to world peace is evident as the number

of conflicts in the developing world seems to be growing.

As developing nations expand their economies and develop technology, there is the possibility that they will acquire nuclear capability. The superpowers have agreed to more arms limitation treaties and the elimination of nuclear arsenals. The moratorium on the use of nuclear weapons does not necessarily extend to military dictators intent on expanding their territories.

A country's success in war is dependent on its ability either to buy weapons or manufacture them. One of the main reasons for the South's defeat in the American Civil War was its lack of industry. The South was an agricultural society that depended on trade, primarily with Britain. Once hostilities developed with the industrial North, the Union (northern) troops blockaded southern ports. The South was unable to import weapons and munitions needed to fight. The North, on the other hand, had steel mills, weapons factories, and the financial support of an industrial society. The South lost the war partly because it lacked the industrial infrastructure that modern warfare requires.

Prior to its invasion of Kuwait in the Gulf War, Iraq imported weapons from France and Russia. The country also used its oil wealth to establish weapons factories. After the invasion of Kuwait, many countries launched an **embargo** against Iraq to prevent new weapons supplies from reaching the country. Without external support, Saddam Hussein and his military leaders were forced to rely on their own military hardware. They underestimated the sophistication of the American arsenal. American-made smart bombs were able to knock out many of Iraq's military facilities and industrial plants. Thus, the superiority of the Allied forces' military machine made it almost impossible for Iraq to be victorious in this conflict.

CONSOLIDATING AND EXTENDING IDEAS

1 Define the different types of borders as listed below:
- antecedent boundaries
- subsequent boundaries
- geometric boundaries
- natural boundaries
- cultural boundaries.

2 a) Using an atlas, describe the borders of the following countries: Poland, Bolivia, Malaysia, Switzerland, Brazil, Democratic Republic of Congo (formerly Zaïre).
  b) Classify these borders as artificial or natural.

3 Study the two maps of Africa in Figures 23.13 (a) and (b).
  a) Choose one African country and estimate the number of tribal areas within it. What does this imply about the potential for political unrest?
  b) Which African countries have the most tribal regions? Which have the least?
  c) How can you explain the apparent negative correlation between the number of tribal regions and latitude?

4 a) Countries can be compact or elongated. Find two examples of each of these shapes in an atlas.
  b) For each example, explain why the country is easy or difficult to defend.
  c) Do research to determine if history supports your assessment.

5 How are population characteristics and industrial strength related to military strength?

STATLAB

FOCUS

The United States is the world's foremost military power. Using the following approach, determine whether Japan and the members of the European Union have enough military power to prevent the US from taking any action it sees fit.

ORGANIZE

Develop a list of possible criteria that could determine a country's military power. These might include population characteristics, industrial capacity, technological capabilities, research and development investment, and so on. Use Figure 23.15 as an example.

COUNTRY	A		B	C	D	E	F	G	H	TOTAL
United States	9167	1.0								
Belgium	30	0.003								
Denmark	43	0.005								
France	550	0.060								
Germany	350	0.038								
Greece	129	0.014								
Ireland	69	0.008								
Italy	294	0.032								
Netherlands	34	0.004								
Portugal	92	0.010								
Spain	499	0.054								
Sweden	412	0.045								
United Kingdom	242	0.026								
Japan	377	0.041								

A: Land area in 000 km^2 (column 1); power index (column 2)

Figure 23.15 Criteria for military power

LOCATE

Use the statistics in the Appendices and other reference materials to obtain recent data for each country.

SYNTHESIZE

Divide each value by the value given for the United States to obtain a "power index." For example, the land area of the US is 9 167 000 km^2 while the land area of the United Kingdom is 242 000 km^2.

Therefore, the power index would be 242 000 ÷ 9 167 000 = 0.026. (Most values will be less than 1.) Add the totals across for each country to get the "power total." The US will have a value equal to the number of criteria you use.

EVALUATE

Add together the power totals for Japan and for the members of the European Union to determine if their combined strength is capable of countering American domination.

CONCLUDE AND APPLY

What conclusions can you reach? What elements of military power do the statistics *not* include?

COMMUNICATE

Prepare an argument that either (i) supports the idea that America is the "peacekeeper of the world," or (ii) warns the nations of the world of America's military might.

GLOBAL CONFLICT: GLOBAL SYSTEMS AT WORK

When a country goes to war, the repercussions can be felt around the world. As with all world issues, wars do not occur in isolation. If an aggressive nation attacks a weaker neighbour, the effects are manifold. Refugees flee from the war-torn region into surrounding countries. Goods are often transported through these countries en route to one or both sides in the conflict. Frequently, environmental deterioration sets in, the result of smoke and other pollutants from the conflict. In addition, the fact that a war is being waged next door can have a psychological impact on neighbouring countries. When Iraq invaded Kuwait in 1991, Saudi Arabia was very anxious about Iraq's aspirations for the entire region. As a result, Saudi Arabia supported the alliance that fought Iraq.

In Chapter 22 we saw how the repercussions of war can extend beyond neighbouring countries. The United Nations will invariably get involved in an attempt to restore peace. Member states are frequently asked to contribute armed forces and military equipment to UN peacekeeping and peacemaking operations. The UN has been criticized for moving too slowly and entering conflicts too late—when military intervention is the only option. (See the article "Flashlights over Mogadishu" on page 383.)

ALLIANCES

Even if the UN stays out of a conflict, **alliances** between nations help smaller nations defend themselves against larger powers that may have military superiority. If one nation becomes too powerful, other countries can unite to maintain the balance of power. Small powers such as Kuwait and Canada defend themselves through alliances with other countries. Countries small in area may be less powerful than larger countries with military superiority, so they look to alliances with stronger nation states for security and protection.

In Europe, countries that have been traditional enemies for centuries have joined together for military as well as economic reasons. In 1949, Western Europe allied with the United States and Canada in the North Atlantic

Treaty Organization (NATO). The purpose of NATO was to provide collective security and defence against the perceived threat from the Soviet Union. In 1955 the Warsaw Pact allied Communist countries of Eastern Europe with the Soviet Union. This move was in response to the admission of West Germany into the NATO alliance. With the collapse of the Soviet Empire in 1991 and the shift towards democratic governments in many Eastern European nations, Europe is becoming even more unified. This is partly in response to the power of the United States, which now stands unchallenged by another superpower. A strong and united Europe is seen by many as the only chance to establish greater balance in world power.

Similar moves are being made in Asia. Many east Asian leaders are presently considering the idea of an East Asia Economic Group, put forward by Malaysian Prime Minister Mahathir. The proposed union would allow the state to continue controlling capitalist structures, encourage economic cooperation, and exclude Canada, the US, and Australia from membership. The United States is encouraging membership in the Asia Pacific Economic Council. Not

Canada has an alliance with the United States for the defence of its borders. During the Cold War, Canada's location directly between the Soviet Union and the United States made it a sitting target should hostilities have erupted between the two superpowers. NORAD and the DEW Line were two surveillance systems established to monitor possible enemy activity in northern Canada. If a Soviet missile had entered Canadian air space, American anti-missiles were poised to destroy it before it reached its target in the continental US. Today, satellites and so-called **Star Wars** technology have eliminated the need for ground-based early warning systems. American satellites observe military movement throughout the world. Powerful computers identify air traffic and determine if enemy planes or missiles are headed towards US territory.

only would economic cooperation be encouraged, but sustainable development and working conditions would be standardized across the region. Many countries in the rapidly growing eastern Pacific Rim are more concerned with development than with health standards for people and environments.

THE EFFECTS OF WAR ON THE ENVIRONMENT

War causes more damage to the environment than any other human activity. When lives are at stake, preserving the natural environment is not a consideration. To aid in enemy surveillance forests may be burned, or cleared with hazardous chemical exfoliants. Ecosystems are destroyed in the process. Plants die, followed by the animals that are dependent on them for food. Bombardment, troop movements, and the destruction of public property bring about devastating pollution. The soil is eroded and poisoned, water resources are contaminated, and the air is polluted with smoke and noxious gases. Chapter 18 described how ecocide during the Vietnam War destroyed 14 per cent of that country's forests and 8 per cent of its crop lands. The Persian Gulf War ravaged large areas of fragile desert terrain and once-fertile gardens in Kuwait and Saudi Arabia, and intensified water pollution in the Persian Gulf.

THE EFFECTS OF WAR ON PEOPLE

Frequently, it is innocent civilians who suffer the most in war. Children are caught between warring factions with little hope of survival. The environmental damage can have far-reaching effects on people. In some countries war has razed the land and destroyed its potential for agriculture. This was especially true in Cambodia. Ancient irrigation canals were destroyed when the Khmer Rouge took power. For centuries peasant farmers had relied on these canals to water the annual rice crop. With the canals gone, they could not grow what they needed for survival.

Land-mines pose a great threat to the human population. Approximately 1 million people have been killed by them since 1975. Every twenty minutes somebody is injured or killed by a land-mine. Eighty per cent of all victims are civilians. In many conflicts deadly explosives contained in a land-mine are laid to await their victims. When somebody steps on the trigger, the mechanism sets off an explosion. Even when hostilities have ceased these land-mines menace the population. It is estimated that 110 million land-mines are currently scattered in sixty-four countries. While efforts are being made to remove them, they are still being laid much faster than they can be cleared away. At the present rate, about 100 000 mines are removed each year; twenty to fifty times as many are laid annually. Even if land-mines were no longer used, it would take 4000 years to clear up just one-fifth of the land-mines in Afghanistan alone. Areas contaminated with land-mines remain inaccessible: people cannot return home, fields cannot be cultivated, economic productivity drops, and reconstruction is hindered.

FACTS ABOUT LAND-MINES

- They can be as small as a hockey puck and cost as little as two dollars.
- Cambodia has the most land-mines (estimated at 9 million), a result of the government's battle with rebel Khmer Rouge forces.
- MAG-Mines Advisory Group is a privately funded British organization that removes land-mines.
- Metal detectors can locate land-mines, but many of today's mines are made of plastic to avoid detection.
- China, Pakistan, and Russia manufacture the most land-mines, although there are many American mines in Southeast Asia.
- While land-mines usually maim adults, they kill children because of their smaller size.

The United Nations is involved in mine-clearing operations. While US $70 million is spent annually on removal and clearance programs, the solution lies in banning the manufacture of these weapons. In 1993 the UN General Assembly called for an export **moratorium** on land-mines. Two years later the European Parliament called for a complete ban on their manufacture. By January 1996, twenty-two countries had agreed to eliminate land-mines. However, there is still opposition to the moratorium. China, Pakistan, and India, in particular, support the manufacture and export of these lethal weapons. Another option, supported by the United States, is for the manufacture of self-destruct mines. These bombs would lose their effectiveness after a certain length of time. Whatever measures are taken, international cooperation is needed to end this destructive practice. In 1996 Canada invited forty countries to a conference in Ottawa to discuss the issue. While all nations condemned mines and most

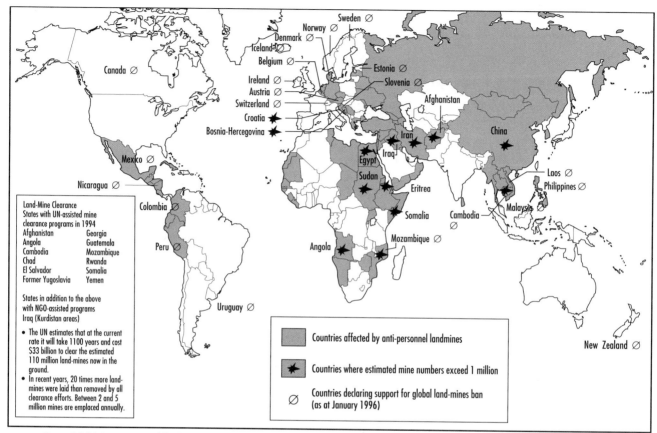

Figure 23.16 Anti-personnel land-mines

Source: *Armed Conflicts Report 1996*, annual publication of Project Ploughshares, Conrad Grebel College, Waterloo ON N2L 3G6; tel.; 519-888-6541; fax: 519-885-0806; e-mail: plough@watservl.uwaterloo.ca

signed an agreement banning their manufacture, Russia and the United States did not sign the declaration. Their rationale was that the issue is very complex and needs more study before their governments are prepared to take action.

The most devastating weapon of war is the nuclear bomb. Its predecessor, the atomic bomb, was first used against Japan in 1945 to force an end to the Second World War. Two of these bombs were dropped on Hiroshima and Nagasaki. The resultant desolation was witnessed through photographs and film and served to warn the world of the potential for annihilation. Today's nuclear weapons are even

more powerful. Not only can they cause mass destruction, they also leave behind nuclear radiation that would render a territory uninhabitable for thousands of years.

The nuclear Non-Proliferation Treaty was signed in 1968 and renewed indefinitely in 1995. Its intent was twofold: eventually to disarm countries that had nuclear bombs and to prevent other countries from developing the capability of making nuclear weapons. At the present time, the United States and Russia account for 97 per cent of this nuclear arsenal. Both are continuing to disarm. The US still has about 10 000 warheads while Russia has twice this number. Even as the two countries dismantle old warheads,

Figure 23.17 Demonstrators in Papeete, Tahiti, protesting French nuclear tests

new weapons are being developed, especially in the US. While these are not nuclear weapons, the US believes that it cannot stop developing new war matériel because its potential enemies will catch up and develop greater capabilities. France and China have not agreed to non-proliferation agreements. In 1995, France set off five nuclear bombs in the South Pacific despite international protests and condemnation. In January 1996 France set off another explosion. China detonated a nuclear bomb just three days after the Non-Proliferation Treaty meeting in 1995. The nuclear age is definitely not over yet—India is rumoured

For over twenty years Palestinian refugee camps existed in the Gaza Strip in southern Israel. Babies and young children grew up there to become rebellious youths and, finally, angry young adults who railed against living in such squalor. The hostilities came to a head in December 1987 and continued during a period of unrest known as the *intifada*.

to be preparing to explode a nuclear device. With the break-up of the Soviet Empire, the possibility exists that sophisticated weapons will go on sale to the highest bidder. This development alone makes world security tenuous at a time when the potential for world peace seems to be within reach.

Many people in war zones try to escape the terror by seeking refuge in neighbouring countries. Frequently, the sheer number of refugees is an overwhelming burden on the host country. Refugee camps are usually established along the border between the two countries. Overcrowding and poor sanitation lead to malnutrition and disease. Refugee camps, which are originally established as temporary facilities pending formal immigration, often end up being the permanent homes for the victims of war. Sometimes, refugee camps become the breeding ground for a generation of displaced and discontent young people.

THE THREAT OF FUTURE WAR

War remains a threat to world security. With one in five countries currently at war, we must recognize the global challenge to find means to end war and create peace. Approximately half of current wars are conflicts about nations seeking to establish their own states. The other half are about state control—conflicts between rival political groups struggling to gain control of the government. These are regional and civil wars that involve citizens of the same territory, not fighting against a common external enemy, but fighting each other for supremacy.

An extremely disturbing element of modern warfare is the employment of children as soldiers. As illustrated in Figure 23.18, children are increasingly being drawn into wars to augment adult forces. Of the thirty-nine countries engaged in armed conflict in 1995, 80 per cent deployed child soldiers, and 64 per cent deployed children under the age of fifteen. Over the past decade it has been estimated that 2 mil-

Increasingly, civilians are the major victims of war. In the first half of this century they represented about 50 per cent of war-related deaths. In more recent years the proportion of civilians in total deaths has been rising. Wars are now more life-threatening for non-combatants than for those fighting them. In the 1960s civilians accounted for 63 per cent of recorded war deaths; in the 1980s, 74 per cent; and in the 1990s the rate seems to be rising even higher. Today, more than 90 per cent of all casualties are non-combatants.

From "Priorities '96," in *World Military and Social Expenditures 1996* (Washington, DC: World Priorities), 717.

lion children have been killed, 4 to 5 million children have been disabled, 12 million have been rendered homeless refugees, over 1 million have become orphaned or separated from their parents, and some 10 million children have been psychologically traumatized by war. Too many children are learning the ways and means of armed conflict. We can only speculate about how they will apply this knowledge once they become adults.

Proliferation of weapons of mass destruction continues to be a huge menace. Chemical weapons such as blood agents, lung agents, and blister and nerve agents can be produced at relatively low cost, and delivery systems are readily available. Over 100 countries now have the industrial base to develop chemical weapons. Because these substances are dual-purpose (they also are used as fertilizers, insecticides, and detergents), chemical weapons arsenals cannot be detected. Between six and twenty countries are suspected of having some biological weaponry program for the development of biological agents (primarily viruses). These weapons of stealth are most useful during peacetime; like chemical weapons they are easily delivered to their targets. Biological weapons may become the tools of economic warfare as they can be genetically designed to destroy particular food crops.

The modern military industrial state is

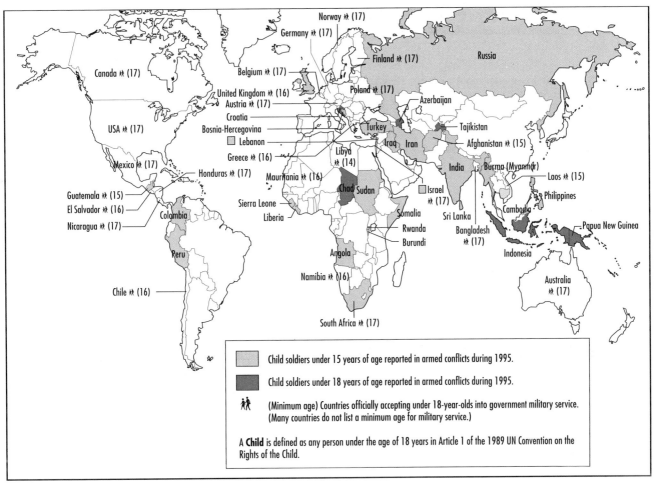

Figure 23.18 Child soldiers reported in government or insurgent military service
Source: *Armed Conflicts Report 1996,* annual publication of Project Ploughshares, Conrad Grebel College, Waterloo ON N2L 3G6; tel.; 519-888-6541; fax: 519-885-0806; e-mail: plough@watservl.uwaterloo.ca

defined by its nuclear capability. Six nations have demonstrated their nuclear power: the United States, the former USSR, India, Britain, France, and China. Other countries such as Israel, Pakistan, Libya, Algeria, Iraq, Iran, South Africa, and North Korea are believed by experts to have joined the "nuclear club." Some countries want nuclear capability for strategic status. They do not plan on using nuclear missiles for military purposes but as a deterrent for aggressor countries or as a buffer against political intimidation. Other nations want nuclear capability to destroy their enemies—nuclear weapons are immediate and decisive and a country cannot defend itself against them. The active black market in nuclear technology and ballistic missile delivery systems, and the free flow of information via sources such as the Internet increase the potential for any nation to become a nuclear power. The number of countries developing offensive weapons programs is doubling approximately every ten years.

CONSOLIDATING AND EXTENDING IDEAS

1 Outline how smaller countries seek to protect themselves from more powerful countries.

2 How does war affect people and environments?

3 Why is the world still in danger of war despite the end of the Cold War?

4 a) Research incidents of environmental and human destruction caused by war.
  b) Using this information, create a visual display illustrating current war zones.

5 Contact a national relief agency such as the Red Cross to find out what your class can do to raise money for young people in a war-torn region.

6 Study Figure 23.19.
  a) Which country is most affected by land-mines?
  b) Which country has the most land-mines?
  c) How long would it take to eliminate land-mines in each country at the rate of 100 000 mines per year?
  d) What solutions are there to the land-mine problem?

COUNTRY	NUMBER OF LAND-MINES (000 000)	MINES PER km^2
Afghanistan	10	15
Angola	15	12
Bosnia	3	59
Cambodia	10	55
Croatia	2	35
Egypt	23	23
Eritrea	1	11
Iran	16	10
Iraq	10	23
Kuwait	6	250
Mozambique	2	3
Somalia	1	2

Figure 23.19 Distribution of land-mines
Source: *Vital Signs 1996* (New York: W. W. Norton & Co., 1996), 133.

CHAPTER 24

GLOBAL CONFLICT: SOLVING THE ISSUE

There are no simple solutions to the issue of global conflict. In fact, many people would argue that this issue, above all others, is impossible to solve because conflict is part of human nature. Nevertheless, we must seek solutions to the problem in order to make the world more stable and secure.

The nature of this problem has changed over time. In the past, conflicts were over political and economic expansion. At present, most conflicts involve nationalist groups that are seeking political recognition. Thus, the solutions to today's conflicts are unlike those of the Cold War, when disarmament and nuclear non-proliferation were the key to the problem.

Possible answers to global conflict involve different levels of our global society. As individuals, we must examine our attitudes towards acceptance of increasing amounts of violence in our everyday lives. Globally, women must be

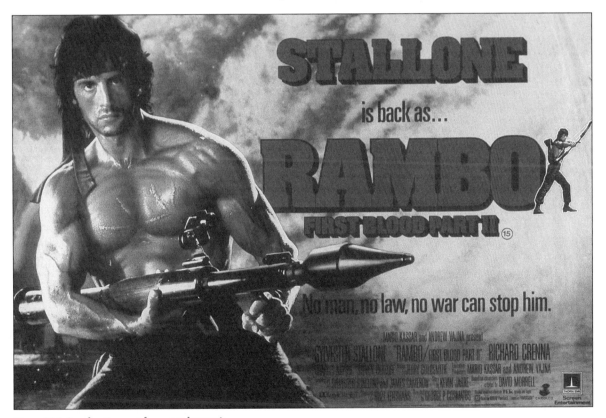

Figure 24.1 An advertisement for a popular movie.
As a society, are we becoming increasingly accepting of violence in our daily lives?

allowed to participate in political decision making. Their involvement may reduce the use of war as an element of foreign policy. Governments that recognize and respond to the needs of diverse cultural groups will be able to avoid armed conflicts within their borders. International organizations such as the United Nations are adapting to confront the changes in global conflict. New agencies are being established to address emerging global issues. Further solutions lie in continued disarmament and nuclear non-proliferation, and in controlling international weapons sales.

REDUCING VIOLENCE: CHANGING THE PARADIGM

We live in a society in which violence is evident everywhere. Newspapers and magazines provide vivid descriptions of brutal acts in our society; electronic media describe or show footage of grim crime scenes; the Vietnam War and, more recently, the conflict in Somalia offered nightly "entertainment" on our television sets. Although the constant barrage of violent acts that are presented in the media may desensitize people of all ages, does the public have a right to freedom of information? Scenes of cruelty, mutilation, and violent death on television programs and in movies are presented as entertainment. Perhaps people lose sight of the fact that it is human lives, and not merely objects, that are being destroyed.

One of the most extraordinary leaders of the twentieth century was Mahatma Gandhi. He led India to independence from Britain by practising a philosophy of **peaceful non-cooperation**. Gandhi preached love, religious tolerance, and non-violence, even while faced with armed opposition from British soldiers. Through his example an entire country refused to cooperate with its rulers and eventually gained independence using peaceful means. Ironically, Gandhi was assassinated by militant conspirators a year after independence had finally been achieved.

As technology advances, we are increasingly exposed to violent images. Whereas thirty years ago there were only a few television stations, cable and satellite dishes now offer the opportunity to view hundreds of programs. Many of these depict graphic violence, and with so much choice it is particularly difficult to monitor the programs people watch. The Internet has contributed to this problem by providing access to all kinds information, some of it pornographic, some of it promoting hatred.

The solution may lie within ourselves. As viewers, we can voice our objections about gratuitous violence we see in the media. Even if we watch violent movies, we must not allow ourselves to be lulled into accepting the behaviour depicted on the screen. Children, young people, and adults need to discuss with each other their concerns about violence in our society and in the media.

Another solution may be to allow people to take censorship into their own hands. New technology is allowing parents to regulate what their children watch on television. Combined with a program rating system, the V-chip can be set to block reception of programs a viewer considers inappropriate because of violence, language, and sexual content. Software programs are also being introduced that prevent children from gaining access to web sites on the Internet that contain violent or pornographic material.

Figure 24.2 Gandhi in Bengal, 1946

EMPOWERING WOMEN: CHANGING THE PARADIGM

In most aspects of their lives, women across the world continue to experience discrimination. In many countries, women do not share the same rights and freedoms as men, such as owning property. They also continue to face greater obstacles than men in education, the workforce, politics, and business.

Though there is no society yet in which women enjoy the same rights as men, thanks to years of struggle the status of women has improved significantly. Women's issues have been on the global agenda for over twenty years. Since 1975 was declared the UN International Women's Year, women's organizations have been drawing attention to the inequities women face and working to build more just societies. As a result, women's lives have improved. For example, in developed countries, more women graduate from university, more are represented in government, and more have access to the workforce where they can earn money to support themselves and their families. In developing countries, female literacy rates are improving, women have greater access to health care, and more women participate in economic activity.

However, the struggle for justice goes on. Over the years, this struggle has widened its scope and refined its strategies. For example, development projects aimed at women used to attempt to involve women in community development, even if that meant crafts and other traditionally female work. Now, development initiatives focus on understanding the power relationships that define the societies women live in and on making these societies more equitable. Empowering women no longer means just creating opportunity for women, but working at ensuring equal opportunity exists as part of the social fabric.

It is imperative that women be treated as equal to men in all societies. Education is the first step towards women's empowerment. If girls and women are better educated, they are able to make well-reasoned decisions about their lives. They need to be given the opportunity to develop self-esteem. Once this happens, and once women are able to participate in the

LOANS CITED AS ANSWER TO WORLD POVERTY

Among the royalty, potentates, government leaders, and financiers attending the first Microcredit Summit, the unassuming figure of Muhammad Yunus stands out. Yunus, a Bangladeshi economics professor, is inarguably the star of this conference that has brought together about 2000 world development advocates.

Although he is not an organizer of the event—called to spread the word about fighting poverty by extending tiny loans directly to the world's poor so they can invest in their own small, self-sustaining enterprises—no one begrudges Yunus's right to define the three-day conference.

"This summit is to pronounce good-bye to the era of financial apartheid," he says. "This summit declares that credit is more than just business, credit is a human right."

About twenty years ago, Yunus took $40 of his own money and lent it to a group of women in Bangladesh to help them buy materials for such enterprises as basket-weaving. The novel idea worked so well that two decades later his Grameen Bank has 2.1 million borrowers (mostly women) in 36 000 villages, with average loans of about $200. And around the world, aid agencies that have followed his example boast another 8 million borrowers, . . . [who] have been helped to buy a fruit stand, or a cow, or tools to make jewellery, enabling them to lift themselves out of poverty.

So successful has the approach been that summit organizers have set a goal of 100 million borrowers by the year 2005.

But Huguette Labelle, president of the Canadian International Development Agency, a featured speaker at Sunday's summit opening, sounded a warning. Calling the education of girls and microcredit the two most powerful tools for fighting world poverty, she said the danger with the 100 million goal is to set "a false market" before the structure and training are ready to accommodate the explosion in borrowers. She also noted that donor nations and institutions must not lose track of the fact that traditional forms of foreign aid—building roads, establishing clean water, education and health programs—remain preconditions for alleviating poverty.

Julian Beltrame, Southam News. *From Calgary Herald*, 3 February 1997: A7

political process of their countries, they will no longer be treated as subordinates.

Empowering women and their daughters economically will ultimately lead to greater political participation and increased influence. It may also lead to fewer wars, as research indicates that women prefer to use strategies of mediation, compromise, and consensus in the settlement of conflicts.

REFORMING THE UNITED NATIONS

The United Nations was created in 1947 to ensure world peace and security. Today, with new players and a different world, the organization's role has to be expanded. Its purpose is no longer solely to maintain peace, but to intervene before situations escalate to the point of armed conflict.

The UN is in need of reform to meet the changing realities of today's world. Created after the Second World War, the UN started with just fifty-one members. By 1995, 184 countries had joined the organization. The fact that virtually every country in the world is represented makes this organization a truly global forum. The General Assembly is the main body of the organization. Other bodies include the Security Council, the Secretariat, the Economic and Social Council, the International Court of Justice, and the Trusteeship Council. The Security Council has become the most important branch of the organization because it deals with peacekeeping operations. Unlike decisions made by the General Assembly, the Security Council's decisions have to be followed by member nations. The US, Britain, France, China, and Russia are permanent members of the Security Council. In addition to these five member countries, ten other countries are elected on a rotating basis. The five permanent members can veto any decision made by the Council. During the Cold War, this veto hindered the UN's ability to keep world peace. Either the US, the USSR, or China disallowed most peacekeeping actions.

Many experts believe that it is time for the Security Council to expand its membership. Japan and Germany are each seeking permanent seats on the Council. Although they were defeated in the Second World War over fifty years ago, both countries contend that they should now be helping to maintain world peace. There is also a need for greater representation of developing countries on the Security Council. The interests of these countries have to be considered since they represent most of the world's people.

ADDRESSING NATIONALIST GOALS OF CULTURAL GROUPS

In its most recent annual publication, *Armed Conflicts Report: 1996*, Project Ploughshares Institute of Peace and Conflict Studies suggests that it is more accurate to refer to the majority of current wars as "local" rather than "regional." These wars are local in the sense that they are not disputes between states (even though they can still affect surrounding states, as explained in Chapter 23); rather, they are within a single state, often within a sub-region of a single country. The points of conflict are also local issues:

land ownership, access to resources, ethnic rivalry. In the Middle East, where more than 40 per cent of states in the region are at war, and in Africa, where more than 28 per cent of the states are sites of armed conflict, the primary causes of war involve precisely these local issues.

Governments that recognize and respond to the needs of diverse nationalist groups will be able to avoid armed conflicts within their borders. A combination of factors can contribute towards the peaceful resolution of political conflict within countries: negotiation and compromise, demilitarization, and security in the form of equity and social justice for all groups.

ARMS REDUCTION AND CONTROL: MILITARY EXPENDITURES

Further solutions to global conflict lie in continued disarmament and nuclear non-proliferation, and in controlling international weapons sales. The top six conventional weapons exporters are permanent members of the United Nations. The United States, the former Soviet Union, France, Germany, China, and the United Kingdom accounted for 89 per cent of the total conventional arms exported from 1988 through 1992. Russia and the United States are the world leaders in arms sales. Russia's sales exceed US $21 billion a year, while American sales are more than $14 billion. The European Union (primarily France) exports another $4.2 billion worth of arms annually.

While each of these six member countries publicly endorses disarmament and nuclear non-proliferation treaties, the world's stockpile of nuclear weapons still represents over 700 times the explosive power used in the twentieth century's three major wars, which killed 44 000 000 people ("Priorities '96," in *World Military and Social Expenditures 1996* [Washington, DC: World Priorities], 5). Despite a decline in world military expenditures in the past five years, in 1995 global spending on weaponry, military technology, and waging wars amounted to more than US $1.4 million per minute.

The truth of the matter is that arms manufacture, export, and import are big business. The economies in these six countries have been closely tied to the manufacture of war machinery for decades, so much so that this industry forms a major part of total manufacturing output. Proponents of arms sales in these countries argue that they need to sell the weapons to keep their economies stable. If weapons sales were drastically reduced, they predict that the global economy could be plunged into a deep recession.

Countries that currently produce and sell arms need to reach an agreement to impose a ban on the manufacture of armaments. Arms manufacturers must adjust to the changing world and retool to manufacture consumer goods instead of weapons. These changes have to take place in the countries of the developed world (particularly the US). It is these countries that perpetuate the research, design, and manufacture of advanced nuclear and conventional weapons technology.

Countries of the developing world that spend much of their annual budgets on weapons to expand their territory, quell rebel outbreaks, and control an oppressed populace need to develop other means to address and solve conflicts. Between 1960 and 1994, arms

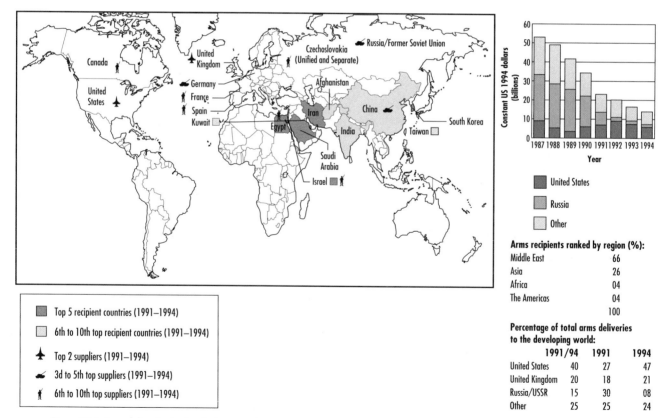

Figure 24.3 Arms deliveries to developing countries

Source: *Armed Conflicts Report 1996*, annual publication of Project Ploughshares, Conrad Grebel College, Waterloo ON N2L 3G6; tel.; 519-888-6541; fax: 519-885-0806; e-mail: plough@watservl.uwaterloo.ca

imports by developing countries amounted to US $775 billion—69 per cent of the total trade in arms. Because of increased pressure from aid donors, arms imports by these countries have declined. Any decline in spending by developing countries can result in the channelling of more funds into social programs, public works, and land reform.

The World Game Institute (an international futurist group) in Philadelphia claims that military spending could be redirected to solve the world's most pressing problems. The Institute has calculated that $19 billion is needed each year to eradicate hunger through food aid and agricultural programs. A further $15 billion needs to be directed towards primary health care, while $21 billion is required to finance housing projects in developing countries. Providing safe drinking water would cost $50 billion, while illiteracy could be eliminated at a cost of $5 billion a year (all figures are in US dollars). The money that is currently being spent on military hardware could find far better use if it were applied to these pressing issues.

CONSOLIDATING AND EXTENDING IDEAS

1 Refer to the statistics in Appendix C on pages 450 to 457.
   a) Prepare a graded shading map for defence expenditures, 1991.
   b) What patterns are evident?

2 For each of the following issues, write two brief paragraphs that detail opposing viewpoints:
   a) To what extent does violence in the media affect violence in society?
   b) Should the participation of women in the military include active soldiering (waging war)?
   c) To what extent should the United Nations involve itself in civil wars or ethnic struggles *within* nations? *between* nations?
   d) Should arms production and sales be banned?

3 a) What suggestions does this chapter make for world peace?
   b) Do you think the solutions are overly idealistic or can they be achieved?
   c) What could you do as an individual to promote world peace? Be specific and include a variety of ideas.

4 Refer to Figure 24.4
   a) Calculate the average difference in dollars per capita spent on health versus defence for the twelve African countries represented in the graph.
   b) Using traditional and technological resources in your school's library, gather the most recent statistics available on these twelve countries for each of the following indicators:
      • crude birth rate
      • crude death rate
      • infant mortality rate
      • life expectancy
      • population per physician
      • percentage of safe drinking water
   c) Chart your findings.

d) Evaluate your findings and discuss your opinion on the results of this research.

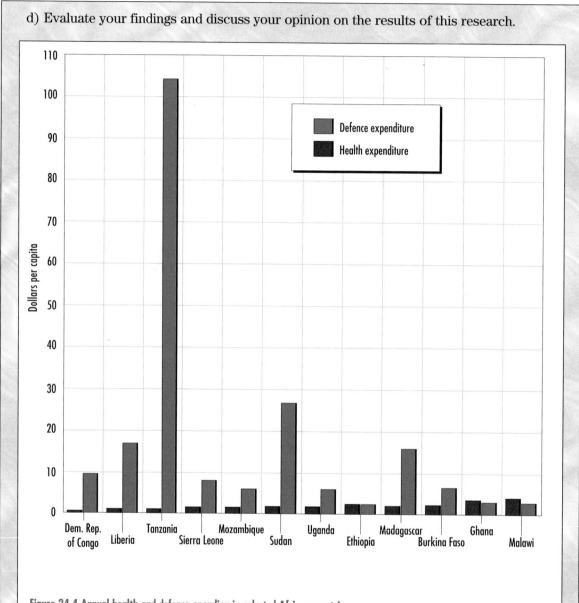

Figure 24.4 Annual health and defence spending in selected African countries

5 a) What suggestions does the World Game Institute offer for military spending?
   b) Are these ideas reasonable? Explain.

COUNTRY PROFILE: CUBA

Cuba is one of the few remaining Communist countries in the world. Situated less than 100 km away from the coast of Florida, and at one time strongly allied with the former Soviet Union, it has been a thorn in the side of the United States for almost forty years. Now that the Soviet Empire has dissolved, the island country of Cuba finds itself isolated politically as well as physically from the rest of North America.

Fidel Castro, charismatic freedom-fighter or terrorist (depending on your point of view), seized power from the fascist dictatorship of Fulgencio Batista in 1959. Initially, the US supported Castro, but never forgave him when he seized all American properties and investments in Cuba, nationalized industry, and set up a Communist regime. By 1961, the US had broken off relations with the island state and Castro had forged an alliance with the Soviets. That same year the US tried to overthrow the Cuban leader by sending an army of just under 2000 Cuban exiles and CIA agents to the Bay of Pigs on the southwest coast of Cuba. The Cuban army easily routed the invaders, and Castro's prestige and popularity in Cuba rose as a result. Cuba's links with the Soviet Union strengthened and the Soviets installed missile-launching bases on the island. In October 1962, the US threatened

CUBA—FACTS AND FIGURES

GNP per capita	n/a
Population	11 000 000
Birth rate*	17
Death rate*	7
Infant mortality**	12
Female literacy	94%
Defence expenditures	$1 272 000 000

*per 1000 population
**per 1000 births

Figure 24.5

Source: Statistics (1990–95) from *World Resources*, 1996–97.

Figure 24.6 School children in Havana

to blockade Cuban ports and search Soviet ships to prevent missiles from reaching the launching platforms. After an incredibly tense six days, the Soviet Union backed down and agreed to dismantle the launching pads. In return the Americans pledged not to attack Cuba again. The prospect of nuclear war had never seemed closer.

Today, foreign relations between Cuba and its superpower neighbour have not improved. In 1995, American politician Jessie Helms introduced the Helms-Burton bill, which has since become law. This legislation allows US courts to punish foreign companies carrying on business in Cuba on property that the government seized from the US after Castro came to power. This means that an American citizen could sue a Canadian company that trades with Cuba. Several countries throughout the world have criticized the US for this manoeuvre.

What is the situation in present-day Cuba? The country has been economically sluggish since the collapse of the Soviet Union. Foreign aid has dwindled and there is no longer a market in Eastern Europe for Cuban cigars, sugar, and other subtropical agricultural products. Through political pressure (the Helms-Burton Bill), the US is trying to prevent the country from expanding its markets elsewhere. Havana, Cuba's capital, is described as a beautiful old colonial city gone to seed. Basic infrastructure is decaying—streets are full of pot-holes and buildings are literally falling down. However, social programs have made Cuba the envy of many developing countries in Latin America. Castro has encouraged the growth of learning in his country. Education and health care are free. Illiteracy and many diseases common in developing countries have been almost eliminated. Cuba has approximately four times as many scientists, engineers, and technicians engaged in research per capita as Mexico (0.19 per cent of the population in Cuba compared to 0.05 per cent in Mexico). By contrast, Canada's ratio is 0.3 per cent of the total population.

Some Canadian businesses, especially mining enterprises, are now active in the Cuban economy despite US censure. And, Cuba has become a favourite tourist destination for Europeans and winter-weary Canadians. As Castro is now over seventy, the future direction of Cuba—a topic of wide speculation—is anybody's guess.

FACING THE FUTURE

WHAT WILL THE FUTURE LOOK LIKE?

Anticipating the twenty-first century gives us the opportunity to make educated guesses about issues and events that are likely to arise and dominate global interaction for decades to come.

POPULATION ISSUES

We can predict, with a fair degree of confidence, that the population by the year 2020 will near the 8 billion mark. The largest percentage of this population will be in Asia, outdistancing by far the populations of Africa, South America, Europe, the former Soviet Union, North America, and Oceania. The proportion of elderly people in these populations will grow, particularly in Indonesia, Thailand, Mexico, Turkey, Brazil, the Philippines, India, Egypt, China, Bangladesh, and Pakistan. This will increase the strain on the working populations to provide essentials to elderly and young people.

The spread of HIV, the human immunodeficiency virus, will be catastrophic throughout central and East Africa. In Uganda, Zimbabwe, and Congo, most of the population will die of AIDS and AIDS-related diseases.

The earth can likely provide sufficient food for even 8 billion people if sustainable agricultural practices and improved distribution systems are developed. The rise in population will put serious pressure on world resources, but advances in technology will allow us to manage this demand.

ECONOMIC ISSUES

In the next decade, the world economy is unlikely to be stronger than it has been during the 1990s. Recessions and slow growth will continue to force governments to deal with the demands of domestic pressure groups. Even though the concept of a nation as an economic unit is fading, economic nationalism, as evidenced by protectionism, will persist.

China will emerge as a new economic superpower as it leans towards "red capitalism." The countries of Western Europe will remain economically integrated and politically cooperative, and will maintain their alliance with the United States. The indicators of quality of life (see Chapter 3, pages 19 to 20) will continue to reveal the contrast between quality of life in the Northern Hemisphere as compared to that in the Southern Hemisphere. Countries of the South will struggle to attain similar standards that populations in the North enjoy—longer life expectancy, sufficient healthy food, better access to clean water, and so on. North-South relations will remain an uneven mixture. In the North, foreign policy will continue to focus on humanitarian concerns; a determination to influence global decision making in order to safeguard economic and political interests; and a readiness to use diplomatic and military intervention to protect clients and allies. The South will be trying to set the agenda to make global economics more responsive to its needs.

GEOPOLITICAL ISSUES

It is highly probable that world politics will be relatively unchanged. Russia will still be more engrossed in domestic developments than in foreign policy, as it continues to recover from the break-up of the USSR. Eastern Europe will go through periodic unrest, which will be answered by various forms of Russian intervention and by UN involvement.

As regional and civil conflicts persist, a variety of small wars will break out, more frequently over state control than over state formation. These wars will be resolved through collective action by regional states and major powers, often under the auspices of the United Nations. It is expected that the current models of international dialogue will improve and become more responsive to issues that arise between nations and between economic and political groups.

CONSOLIDATING AND EXTENDING IDEAS

1 Create future wheels (see page 130) to illustrate your vision of each of the following in 2020:
   a) resource issues—food, hunger, and famine
   b) environmental issues—earth, water, and air

2 Build a three-dimensional "sculpture" to describe/illustrate what "global citizenship" means to you.

POPULATION AND CULTURE INDICATORS*

REGION	POPULATION (000 000) 1950	POPULATION (000 000) 1995	POPULATION (000 000) 2025	LAND AREA (000 km²) 1993	POPU-LATION DENSITY PER km² 1993	BIRTH RATE 1990–95	DEATH RATE 1990–95	POPULATION GROWTH RATE (% p.a.) 1980–85	POPULATION GROWTH RATE (% p.a.) 1990–95	POPULATION GROWTH RATE (% p.a.) 2000–2005	INFANT MORT-ALITY RATE 1990–95	LIFE EXPECT-ANCY 1990–95	FOOD ENERGY AVAIL-ABLE AS A % OF NEED (per cap.) 1990–92	LITERACY RATE FEMALE/MALE (%) 1993	WOMEN AS A % OF LABOUR FORCE 1992	FERT-ILITY RATE 1990–95	POPU-LATION <15 (%) 1995
WORLD	2 516	5 716	8 294	30 984	42.7	25	9	1.8	1.6	1.4	64	65		69/86		3.1	32
AFRICA	222	728	1 496	29 634	23.7	42	14	2.9	2.8	2.6	93	53		40/62		5.8	44
Algeria	8.8	27.9	45.5	2 382	11.4	29	6	3.1	3.0	2.0	55	66	123	41/68	10	3.9	39
Angola	4.1	11.1	26.6	1 247	8.2	51	19	2.6	3.7	3.1	124	47	80	29/56	38	7.2	47
Benin	2.1	5.4	12.3	111	45.9	49	18	2.8	3.1	2.8	86	46	104	19/42	47	7.1	47
Botswana	0.4	1.5	3.0	567	2.4	37	7	3.4	3.1	2.7	43	61	97	55/68	35	4.9	43
Burkina Faso	3.7	10.3	21.7	274	35.7	47	18	2.5	2.8	2.5	130	48	94	7/26	46	6.5	45
Burundi	2.5	6.4	13.5	26	233.7	46	16	2.8	3.0	2.6	102	48	84	19/45	47	6.8	46
Cameroon	4.5	13.2	29.2	465	27.0	41	12	2.8	2.8	2.8	63	56	95	44/70	33	5.7	44
Central African Rep.	1.3	3.3	6.4	623	5.2	42	17	2.3	2.5	2.3	102	47	82	41/60	45	5.7	43
Chad	2.7	6.4	12.9	1 259	4.8	44	18	2.3	2.7	2.5	122	48	73	29/57	21	5.9	43
Congo	0.8	2.6	5.7	342	7.1	45	15	2.8	3.0	2.6	84	52	103	59/78	39	6.3	46
Côte d'Ivoire	2.8	14.3	36.8	318	42.1	50	15	3.9	3.5	3.2	92	52	111	24/44	34	7.4	49
Dem. Rep. of Congo	12.2	43.9	104.6	2 268	18.2	48	15	3.2	3.2	3.0	93	52	96	61/83	35	6.7	48
Egypt	20.3	62.9	97.3	995	56.3	29	8	2.6	2.2	1.7	67	62	132	34/60	10	3.9	38
Equatorial Guinea	0.2	0.4	0.8	28	13.5	44	18	7.2	2.6	2.4	117	48	n/a	61/86	40	5.9	43
Eritrea	1.1	3.5	7.0	101	34.6	43	15	2.5	2.7	2.5	105	50	n/a	n/a	n/a	5.8	44
Ethiopia	19.6	55.1	126.9	1 101	49.6	49	18	2.5	3.0	2.9	119	47	73	21/41	37	7.0	46
Gabon	0.5	1.3	2.7	258	5.0	37	16	4.0	2.8	2.5	94	54	104	45/68	37	5.3	39
Gambia	0.3	1.1	2.1	10	93.2	44	19	3.0	2.8	2.4	132	45	n/a	20/48	40	5.6	41
Ghana	4.9	17.5	38.0	228	72.3	42	12	3.6	3.0	2.8	81	56	93	46/71	40	6.0	45
Guinea	2.6	6.7	15.1	246	25.6	51	20	2.2	3.0	2.9	134	45	97	18/45	39	7.0	47
Guinea-Bissau	0.5	1.1	2.0	28	36.6	43	21	1.9	2.1	2.1	140	44	97	36/63	40	5.8	42

* See page 441 for definitions

REGION	POPULATION (000 000)			LAND AREA (000 km²)	POPU-LATION DENSITY PER km²	BIRTH RATE	DEATH RATE	POPULATION GROWTH RATE (% p.a.)			INFANT MORT-ALITY RATE	LIFE EXPECT-ANCY	FOOD ENERGY AVAIL-ABLE AS A % OF NEED (per cap.)	LITERACY RATE FEMALE/MALE (%)	WOMEN AS A % OF LABOUR FORCE	FERT-ILITY RATE	POPU-LATION <15 (%)
	1950	1995	2025	1993	1993	1990–95	1990–95	1980–85	1990–95	2000–2005	1990–95	1990–95	1990–92	1993	1992	1990–95	1995
Kenya	6.3	28.3	63.4	570	45.8	44	12	3.6	3.6	3.0	69	59	89	62/82	39	6.3	48
Lesotho	0.7	2.1	4.2	30	62.0	37	10	2.8	2.7	2.6	79	61	93	57/78	43	5.2	42
Liberia	0.8	3.0	7.2	97	29.4	47	14	3.2	3.3	3.1	126	55	98	18/49	30	6.8	46
Libya	1.0	5.4	12.9	1 760	2.9	42	8	4.4	3.5	3.2	68	63	140	54/84	10	6.4	45
Madagascar	4.2	14.8	34.4	582	22.8	44	12	3.1	3.3	3.1	93	56	95	73/88	39	6.1	46
Malawi	2.9	11.1	22.3	94	113.7	51	20	3.2	3.5	2.0	143	44	88	37/69	41	7.2	47
Mali	3.5	10.8	24.6	1 220	8.3	51	19	2.9	3.2	2.9	159	46	96	17/32	16	7.1	47
Mauritania	0.8	2.2	4.4	1 025	2.2	40	14	2.6	2.5	2.5	101	48	106	24/47	23	5.4	45
Mauritius	0.5	1.1	1.5	2	546.3	21	7	1.0	1.1	1.1	18	70	128	75/85	27	2.4	28
Morocco	9.0	27.0	40.7	446	60.4	29	8	2.4	2.1	1.6	68	63	n/a	26/52	21	3.8	37
Mozambique	6.2	16.0	35.1	784	20.4	45	19	2.3	2.4	2.8	148	47	77	18/52	47	6.5	45
Namibia	0.5	1.5	3.0	823	1.9	37	11	2.7	2.7	2.5	60	59	n/a	n/a	24	5.3	42
Niger	2.4	9.2	22.4	1 267	6.7	53	19	3.4	3.4	3.2	124	47	95	5/18	46	7.4	48
Nigeria	32.9	111.7	238.4	911	131.0	45	15	2.9	3.0	2.7	84	53	93	39/61	34	6.5	46
Rwanda	2.1	8.0	15.8	25	315.7	44	17	3.2	2.6	2.5	110	46	82	44/65	47	6.6	46
Senegal	2.5	8.3	16.9	193	41.3	43	16	2.8	2.5	2.6	68	49	98	19/39	39	6.1	45
Sierra Leone	2.0	4.5	8.9	72	62.7	49	25	2.0	2.4	2.3	166	43	83	14/40	32	6.5	44
Somalia	3.1	9.3	21.3	627	15.2	50	19	3.2	1.3	3.0	122	47	81	14/36	38	7.0	48
South Africa	13.7	41.5	71.0	1 221	33.4	31	9	2.5	2.2	2.1	53	63	128	79/80	36	4.1	37
Sudan	9.2	28.1	58.4	2 376	11.5	40	13	2.8	2.7	2.6	78	52	87	28/53	22	5.7	44
Swaziland	0.3	0.9	1.6	17	47.3	39	11	3.0	2.8	2.6	75	58	n/a	71/74	38	4.9	43
Tanzania	7.9	29.7	62.9	886	32.5	43	14	3.2	3.0	2.6	85	51	95	49/75	47	5.9	46
Togo	1.3	4.1	9.4	54	71.4	45	13	2.9	3.2	2.9	85	55	99	30/61	36	6.6	46
Tunisia	3.5	8.9	13.3	155	55.2	26	6	2.6	1.9	1.5	43	68	131	56/73	25	3.2	35

REGION	POPULATION (000 000)			LAND AREA (000 km²)	POPU-LATION DENSITY PER km²	BIRTH RATE	DEATH RATE	POPULATION GROWTH RATE (% p.a.)			INFANT MORT-ALITY RATE	LIFE EXPECT-ANCY	FOOD ENERGY AVAIL-ABLE AS A % OF NEED (per cap.)	LITERACY RATE FEMALE/MALE (%)	WOMEN AS A % OF LABOUR FORCE	FERT-ILITY RATE	POPU-LATION <15 (%)
	1950	1995	2025	1993	1993	1990–95	1990–95	1980–85	1990–95	2000–2005	1990–95	1990–95	1990–92	1993	1992	1990–95	1995
Uganda	4.8	21.3	48.1	200	96.4	52	19	2.8	3.4	2.7	115	42	93	44/70	14	7.3	49
Zambia	2.4	9.5	19.1	743	12.0	45	15	3.6	3.0	2.4	104	44	87	65/82	30	6.0	47
Zimbabwe	2.7	11.3	19.6	387	28.2	39	12	3.3	2.6	2.0	67	56	94	77/88	34	5.0	44
ASIA	1 377.3	3 458.0	4 960.0	26 790	122.9	25	8	1.9	1.6	1.4	65	65		60/79		3.0	32
Afghanistan	9.0	20.1	45.3	652	31.5	50	22	-2.0	5.8	2.7	163	44	72	11/42	9	6.9	41
Armenia	1.4	3.6	4.7	28	125.6	21	7	1.0	1.4	1.0	21	71	n/a	97/99	n/a	2.6	30
Azerbaijan	2.9	7.6	10.1	86	85.2	23	6	1.6	1.2	1.0	28	70	n/a	96/99	n/a	2.5	32
Bangladesh	41.8	120.4	196.1	130	938.8	36	12	2.2	2.2	2.0	108	53	88	23/47	8	4.4	39
Bhutan	0.7	1.6	3.1	47	35.1	40	15	2.1	1.2	2.3	124	49	128	23/51	32	5.9	41
Cambodia	4.4	10.3	19.7	177	51.0	44	14	3.0	3.0	2.3	116	51	96	22/48	37	5.3	45
China	554.8	1 221.5	1 526.1	9 326	129.2	19	7	1.4	1.1	0.8	44	71	112	68/87	43	2.0	26
Georgia	3.7	5.5	6.1	70	78.5	16	9	0.8	0.1	0.4	19	72	n/a	98/99	n/a	2.1	24
India	357.6	935.7	1 392.1	2 973	301.6	29	10	2.2	1.9	1.6	82	60	101	34/62	25	3.8	35
Indonesia	79.5	197.6	275.6	1 812	107.4	25	8	2.1	1.6	1.3	58	63	121	75/88	31	2.9	33
Iran	16.9	67.3	123.5	1 636	38.6	36	7	4.4	2.7	2.9	36	67	125	56/74	19	5.0	44
Iraq	5.2	20.4	42.7	437	45.5	38	7	3.3	2.5	2.8	58	66	128	38/66	22	5.7	44
Israel	1.3	5.6	7.8	20	266.2	21	7	1.8	3.8	1.3	9	76	125	93/97	34	2.9	29
Japan	83.6	125.1	121.6	377	331.9	10	8	0.7	0.3	0.1	4	79	125	99/99	38	1.5	16
Jordan	1.2	5.4	12.1	89	49.9	39	6	5.4	3.9	3.0	36	68	110	73/91	11	5.6	43
Kazakhstan	6.8	17.1	21.7	2 670	6.4	20	8	1.1	0.5	0.8	30	69	n/a	96/99	n/a	2.5	30
Korea, North	9.7	23.9	33.4	120	191.5	24	5	1.7	1.9	1.3	24	71	121	n/a	46	2.4	29
Korea, South	20.4	45.0	54.4	99	450.8	16	6	1.4	1.0	0.8	11	71	120	95/99	34	1.8	24
Kuwait	0.2	1.5	2.8	18	102.4	24	2	4.5	-6.5	2.5	18	75	n/a	72/78	16	3.1	41

REGION	POPULATION (000 000)			LAND AREA (000 km²) 1993	POPULATION DENSITY PER km² 1993	BIRTH RATE 1990–95	DEATH RATE 1990–95	POPULATION GROWTH RATE (% p.a.)			INFANT MORTALITY RATE 1990–95	LIFE EXPECTANCY 1990–95	FOOD ENERGY AVAILABLE AS A % OF NEED (per cap.) 1990–92	LITERACY RATE FEMALE/MALE (%) 1993	WOMEN AS A % OF LABOUR FORCE 1992	FERTILITY RATE 1990–95	POPULATION <15 (%) 1995
	1950	1995	2025					1980–85	1990–95	2000–2005							
Kirgyz Rep.	1.7	4.7	7.1	192	23.9	29	7	2.0	1.7	1.5	35	68	n/a	96/99	n/a	3.7	38
Laos	1.8	4.9	9.7	231	20.0	45	15	2.3	3.0	2.6	97	51	111	39/65	44	6.7	45
Lebanon	1.4	3.0	4.4	10	283.6	27	7	0	3.3	1.5	34	69	127	88/94	28	3.1	34
Malaysia	6.1	20.1	31.6	329	58.6	29	5	2.6	2.4	1.7	13	71	120	74/87	35	3.6	38
Mongolia	0.8	2.4	3.8	1 567	1.5	28	7	2.8	2.0	1.9	60	64	97	73/87	46	3.6	38
Myanmar	17.8	46.5	75.6	658	67.8	33	11	2.1	2.1	1.9	84	58	114	75/88	37	4.2	37
Nepal	8.2	21.9	40.7	137	154.1	39	13	2.6	2.6	2.4	99	54	100	11/37	33	5.4	42
Oman	0.4	2.2	6.1	212	8.0	44	5	4.8	4.2	3.7	30	70	n/a	n/a	9	7.2	48
Pakistan	39.5	140.5	284.8	771	166.1	41	9	3.7	2.8	2.7	91	59	99	21/46	13	6.2	44
Philippines	21.0	67.6	104.5	298	223.2	30	6	2.5	2.1	1.8	44	65	104	93/94	31	3.9	38
Saudi Arabia	3.2	17.9	42.7	2 150	7.7	35	5	5.5	2.2	3.1	29	69	121	44/69	8	6.4	42
Singapore	1.0	2.8	3.4	1	4 586.9	16	6	1.2	1.0	0.6	6	75	136	83/95	32	1.7	23
Sri Lanka	7.7	18.4	25.0	65	276.9	21	6	1.7	1.3	1.1	18	72	101	85/93	27	2.5	31
Syria	3.5	14.7	33.5	184	74.8	41	6	3.5	3.4	3.2	39	67	126	49/82	18	5.9	47
Tajikistan, Rep.	1.6	6.1	11.8	143	40.1	37	6	2.8	2.9	2.5	48	70	n/a	96/99	n/a	4.9	43
Thailand	20.0	58.8	73.6	511	111.3	19	6	1.8	1.1	0.9	37	69	103	91/96	44	2.1	28
Turkey	20.8	61.9	90.9	770	77.4	27	7	2.5	2.0	1.5	65	67	127	69/90	34	3.4	34
Turkmenistan	1.2	4.1	6.7	488	8.1	32	8	2.4	2.0	1.9	57	65	n/a	97/99	n/a	4.0	40
United Arab Emirates	0.1	1.9	3.0	84	20.4	23	3	6.1	2.6	1.8	19	71	n/a	76/77	7	4.2	31
Uzbekistan	6.4	22.8	37.7	425	51.5	32	6	2.6	2.2	2.0	41	69	n/a	96/98	n/a	3.9	40
Vietnam	30.0	74.5	118.2	325	217.8	31	8	2.2	2.2	1.9	42	64	103	87/95	47	3.9	38
Yemen	4.3	14.5	34.7	528	24.6	49	16	3.1	5.0	3.1	119	53	n/a	n/a	14	7.6	47

REGION	POPULATION (000 000) 1950	POPULATION (000 000) 1995	POPULATION (000 000) 2025	LAND AREA (000 km²) 1993	POPULATION DENSITY PER km² 1993	BIRTH RATE 1990-95	DEATH RATE 1990-95	POPULATION GROWTH RATE (% p.a.) 1980-85	POPULATION GROWTH RATE (% p.a.) 1990-95	POPULATION GROWTH RATE (% p.a.) 2000-2005	INFANT MORTALITY RATE 1990-95	LIFE EXPECTANCY 1990-95	FOOD ENERGY AVAILABLE AS A % OF NEED (per cap.) 1990-92	LITERACY RATE FEMALE/MALE (%) 1993	WOMEN AS A % OF LABOUR FORCE 1992	FERTILITY RATE 1990-95	POPULATION <15 (%) 1995
NORTH & CENTRAL AMERICA	**202.6**	**454.2**	**615.5**	**21 377**	**19.0**	**20**	**8**	**1.3**	**1.4**	**1.1**	**19**	**73**		**n/a**		**2.5**	**27**
Belize	0.1	0.2	0.4	23	8.9	35	5	2.6	2.6	2.3	33	n/a	n/a	n/a	n/a	4.2	42
Canada	13.7	29.5	38.3	9 221	3.0	15	8	1.1	1.2	0.9	7	77	122	99/99	40	1.9	21
Costa Rica	0.9	3.4	5.6	51	64.0	26	4	2.9	2.4	1.8	14	76	121	89/94	22	3.1	35
Cuba	5.9	11.0	12.7	110	99.3	17	7	0.8	0.8	0.5	12	76	135	94/95	32	1.8	23
Dominican Rep.	2.4	7.8	11.2	48	157.5	27	6	2.4	1.9	1.4	42	68	102	80/80	16	3.1	35
El Salvador	1.9	5.8	9.7	21	266.3	34	7	0.9	2.2	2.0	46	66	102	67/71	25	4.0	41
Guatemala	3.0	10.6	21.7	108	92.5	39	8	2.8	2.9	2.7	48	65	103	44/60	17	5.4	44
Haiti	3.2	7.2	13.1	28	250.1	35	12	1.8	2.0	2.1	86	57	89	38/44	41	4.8	40
Honduras	1.4	5.7	10.7	112	50.3	37	6	3.2	3.0	2.5	43	66	98	69/70	20	4.9	44
Jamaica	1.4	2.4	3.3	11	230.4	22	6	1.6	0.7	1.0	14	74	114	87/79	46	2.4	31
Mexico	27.3	93.7	136.6	1 909	47.2	28	4	2.4	2.1	1.5	36	70	131	85/90	27	3.2	36
Nicaragua	1.1	4.4	9.1	119	34.6	41	7	2.8	3.7	2.8	52	67	99	65/63	26	5.0	46
Panama	0.9	2.6	3.8	76	33.7	25	5	2.2	1.9	1.4	25	73	98	88/89	28	2.9	33
Trinidad and Tobago	0.6	1.3	1.8	5	249.3	21	6	1.4	1.1	1.1	18	71	114	90/96	30	2.4	32
United States	152.3	263.3	331.2	9167	28.1	16	9	0.9	1.0	0.8	9	76	138	99/99	41	2.1	22
SOUTH AMERICA	**111.6**	**319.8**	**462.7**	**17 529**	**17.7**	**25**	**7**	**2.2**	**1.7**	**1.4**	**48**	**67**		**n/a**		**3.0**	**33**
Argentina	17.2	34.6	46.1	2 737	12.2	20	8	1.5	1.2	1.1	24	71	131	96/96	28	2.8	29
Bolivia	2.8	7.4	13.1	1 084	7.1	36	10	1.9	2.4	2.2	75	61	84	71/88	26	4.8	41
Brazil	53.4	161.8	230.3	8 457	18.5	25	8	2.2	1.7	1.4	58	66	114	81/82	28	2.9	32
Chile	6.1	14.3	19.8	749	18.4	22	6	1.6	1.6	1.2	16	72	102	94/94	29	2.5	30

REGION	POPULATION (000 000)			LAND AREA (000 km²)	POPU-LATION DENSITY PER km²	BIRTH RATE	DEATH RATE	POPULATION GROWTH RATE (% p.a.)			INFANT MORT-ALITY RATE	LIFE EXPECT-ANCY	FOOD ENERGY AVAIL-ABLE AS A % OF NEED (per cap.)	LITERACY RATE FEMALE/MALE (%)	WOMEN AS A % OF LABOUR FORCE	FERT-ILITY RATE	POPU-LATION <15 (%)
	1950	1995	2025	1993	1993	1990–95	1990–95	1980–85	1990–95	2000–2005	1990–95	1990–95	1990–92	1993	1992	1990–95	1995
Columbia	12.0	35.1	49.4	1 039	32.7	24	6	2.1	1.7	1.3	37	69	106	89/90	22	2.7	33
Ecuador	3.3	11.5	17.8	277	40.9	28	6	2.7	2.2	1.7	50	67	105	86/90	19	3.5	36
Guyana	0.4	0.8	1.1	197	4.1	25	7	0.8	0.9	1.1	48	65	n/a	96/98	25	2.6	32
Paraguay	1.4	5.0	9.0	397	11.7	33	6	3.2	2.8	2.3	38	67	116	89/93	21	4.3	40
Peru	7.6	23.8	36.7	1 280	17.9	27	7	2.3	1.9	1.7	64	65	87	80/92	24	3.4	35
Suriname	0.2	0.4	0.6	156	2.9	25	6	1.2	1.1	1.1	28	70	n/a	89/94	30	2.7	35
Uruguay	2.2	3.2	3.7	175	18.0	17	10	0.6	0.6	0.6	20	73	101	97/96	31	2.3	24
Venezuela	5.0	21.8	34.8	882	23.4	27	5	2.5	2.3	1.8	33	70	99	89/91	28	3.3	36
EUROPE	**548.7**	**727.0**	**718.2**	**4 727**	**108.6**	**12**	**11**	**0.4**	**0.2**	**0**	**12**	**75**		**97/99**		**1.6**	**19**
Albania	1.2	3.4	4.7	27	121.8	24	6	2.1	0.9	1.1	30	73	107	n/a	41	2.9	31
Austria	6.9	8.0	8.3	83	94.3	12	11	0	0.7	0.2	7	76	133	99/99	40	1.5	18
Belarus	7.8	10.1	9.9	208	49.6	12	12	0.7	-0.1	-0.1	16	72	n/a	96/99	n/a	2.0	22
Belgium	8.6	10.1	10.4	30	331.1	12	11	0	0.3	0.1	6	76	149	99/99	34	1.6	18
Bosnia and Hercegovina	2.7	3.5	4.5	51	68.6	13	7	1.0	-4.4	0.2	15	n/a	n/a	n/a	n/a	1.6	22
Bulgaria	7.3	8.8	7.7	111	80.7	10	13	0.2	-0.5	-0.4	14	72	n/a	97/99	46	1.5	18
Croatia	3.9	4.5	4.2	56	80.3	11	12	0.4	-0.1	-0.1	9	71	n/a	95/99	n/a	1.7	19
Czech Rep.	8.9	10.3	10.6	77	125.8	13	13	0	0	0.1	9	73	145	n/a	n/a	1.8	19
Denmark	4.3	5.2	5.1	43	121.9	13	12	0	0.2	0	7	76	135	99/99	45	1.7	17
Estonia	1.1	1.5	1.4	43	36.5	11	13	0.8	-0.6	-0.3	16	72	n/a	100/100	n/a	1.6	21
Finland	4.0	5.1	5.4	305	16.5	13	10	0.5	0.5	0.3	5	76	113	97/99	47	1.9	19
France	41.8	58.0	61.2	550	104.3	13	10	0.5	0.4	0.2	7	77	143	99/99	40	1.7	20
Germany	68.4	81.6	76.4	350	230.8	10	12	-0.2	0.6	-0.1	6	76	n/a	99/99	39	1.3	16

REGION	POPULATION (000 000)			LAND AREA (000 km²)	POPU-LATION DENSITY PER km²	BIRTH RATE	DEATH RATE	POPULATION GROWTH RATE (% p.a.)			INFANT MORT-ALITY RATE	LIFE EXPECT-ANCY	FOOD ENERGY AVAIL-ABLE AS A % OF NEED (per cap.)	LITERACY RATE FEMALE/MALE (%)	WOMEN AS A % OF LABOUR FORCE	FERT-ILITY RATE	POPU-LATION <15 (%)
	1950	1995	2025	1993	1993	1990–95	1990–95	1980–85	1990–95	2000–2005	1990–95	1990–95	1990–92	1993	1992	1990–95	1995
Greece	7.6	10.5	9.9	129	79.2	10	10	0.6	0.4	0	10	78	151	93/98	27	1.4	17
Hungary	9.3	10.1	9.4	92	113.6	12	15	-0.2	-0.5	-0.3	15	70	137	98/98	45	1.7	18
Iceland	0.1	0.3	0.3	100	2.6	18	7	1.1	1.1	0.9	5	78	n/a	n/a	43	2.2	25
Ireland	3.0	3.6	3.9	69	50.5	15	9	0.9	0.3	0.4	7	75	157	n/a	29	2.1	24
Italy	47.1	57.2	52.3	294	196.6	10	10	0.1	0.1	-0.2	8	77	139	96/98	32	1.3	15
Latvia	2.0	2.6	2.3	62	43.0	12	13	0.6	-0.9	0.5	14	71	n/a	99/100	n/a	1.6	21
Lithuania	2.6	3.7	3.8	46	82.6	14	11	0.9	-0.1	0.1	13	73	n/a	98/99	n/a	1.8	22
Macedonia	1.2	2.2	2.6	25	88.0	16	7	1.4	1.1	0.7	27	72	n/a	n/a	n/a	2.0	24
Moldova, Rep.	2.5	4.4	5.1	33	129.3	16	11	1.0	0.3	0.5	25	68	n/a	94/99	n/a	2.1	26
Netherlands	10.1	15.5	16.3	34	450.2	13	9	0.5	0.7	0.3	7	77	114	n/a	31	1.6	18
Norway	3.3	4.3	4.7	307	14.0	14	11	0.3	0.5	0.3	8	77	120	n/a	41	1.9	20
Poland	27.8	38.4	41.5	304	126.5	13	11	0.9	0.1	0.3	15	72	131	97/99	46	1.9	22
Portugal	8.4	9.8	9.7	92	107.3	12	11	0.3	-0.1	0	10	75	136	82/89	37	1.6	19
Romania	16.3	22.8	21.7	230	101.5	11	11	0.5	-0.3	-0.2	23	70	116	95/98	47	1.5	20
Russian Federation	103.3	147.0	138.5	16 996	8.8	11	12	0.7	-0.1	-0.2	22	70	n/a	97/99	n/a	1.5	21
Slovak Rep.	3.5	5.4	6.0	48	112.5	14	11	0.7	0.4	0.4	12	71	n/a	n/a	n/a	1.9	23
Slovenia Rep.	1.5	1.9	1.8	20	95.0	11	11	0.5	0.3	-0.1	8	73	n/a	n/a	n/a	1.5	18
Spain	28.0	39.6	37.6	499	78.4	10	9	0.5	0.2	0	7	78	141	93/97	24	1.2	17
Sweden	7.0	8.8	9.8	412	21.1	14	11	0.1	0.5	0.3	5	78	111	99/99	45	2.1	19
Switzerland	4.7	7.2	7.8	40	172.5	13	9	0.7	1.1	0.5	6	78	130	99/99	36	1.6	18
Ukraine	37.0	51.4	48.7	604	86.5	12	13	0.4	-0.1	-0.2	16	71	n/a	97/99	n/a	1.6	20
United Kingdom	50.6	58.3	61.5	242	239.3	14	11	0.1	0.3	0.2	7	76	130	99/99	39	1.8	20
Yugoslavia	7.1	10.8	11.5	255	94.0	14	10	0.7	1.3	0.4	20	72	140	89/97	n/a	2.0	22

REGION	POPULATION (000 000)			LAND AREA (000 km²)	POPU-LATION DENSITY PER km²	BIRTH RATE	DEATH RATE	POPULATION GROWTH RATE (% p.a.)			INFANT MORT-ALITY RATE	LIFE EXPECT-ANCY	FOOD ENERGY AVAIL-ABLE AS A % OF NEED (per cap.)	LITERACY RATE FEMALE/MALE (%)	WOMEN AS A % OF LABOUR FORCE	FERT-ILITY RATE	POPU-LATION <15 (%)
	1950	1995	2025	1993	1993	1990–95	1990–95	1980–85	1990–95	2000–2005	1990–95	1990–95	1990–92	1993	1992	1990–95	1995
OCEANIA	**12.6**	**28.5**	**41.0**	**8 453**	**3.3**	**19**	**8**	**1.5**	**1.5**	**1.3**	**27**	**73**		**n/a**		**2.5**	**26**
Australia	8.2	18.1	24.7	7 644	2.3	15	7	1.4	1.4	1.1	7	77	124	99/99	38	1.9	22
Fiji	0.3	0.7	1.1	18	40.9	24	5	2.0	1.5	1.5	23	72	n/a	86/92	21	3.0	35
New Zealand	1.9	3.6	4.4	268	13.0	17	8	0.8	1.2	0.8	9	76	131	n/a	n/a	2.1	23
Papua New Guinea	1.6	4.3	7.5	453	9.2	33	11	2.2	2.3	2.1	68	56	114	57/78°	35	5.1	40
Solomon Islands	0.1	0.4	0.8	28	12.6	38	4	3.5	3.3	3.1	27	70	n/a	n/a	n/a	5.4	44

DEFINITIONS

Population: The number of people living in a country. Projections are based on UN Population Division.

Land Area: All the land contained within a country excluding inland water bodies, claims to disputed territory, and continental shelves.

Population Density: The total population expressed as a ratio of the total land area.

Birth Rate: The total number of live births per 1000 population.

Death Rate: The total number of deaths per 1000 population.

Population Growth Rate: The death rate subtracted from the birth rate.

Infant Mortality Rate: The number of infants per 1000 births who die before the age of one.

Life Expectancy: The average lifespan of a population.

Food Energy Available as a % of Need: Available food expressed as a ratio of food needed.

Literacy Rate: The percentage of (i) women and (ii) men who can read and write.

Women as a % of the Labour Force: The proportion of the labour force that employs women.

Fertility Rate: The average number of live births per year for each woman of childbearing years.

Population <15: The percentage of the population under 15 years of age.

ECONOMIC AND RESOURCE INDICATORS*

REGION	GNP PER CAP. (US$) 1993	GDP PER CAP. (US$) 1993	AVERAGE ANNUAL GROWTH RATE OF GDP PER CAP. 1983–93	% OF GDP FROM AGRICULTURE 1993	% OF GDP FROM INDUSTRY 1991	% OF GDP FROM SERVICES 1991	FOREIGN AID AS A % OF GNP 1991–93	DEFENCE EXPENDITURES (000 000 US$) 1991	CROP LAND (ha per cap.) 1993	CEREAL PRODUCTION (000 t) 1990–92	WOOD PRODUCTION (000 m³) 1989–91	COMMERCIAL ENERGY PRODUCTION (pJ*) 1991	TOTAL VALUE OF MINERAL RESERVES (000 000 US$) 1991
WORLD									0.26	1 928 044	3 462 348	334 890	4 190 350
AFRICA									0.27	90 225	513 545	21 335	589 615
Algeria	1 780	1 862	1.0	13	43	43	0	971	0.29	2 818	2 163	4 392	2 094
Angola	n/a	n/a	n/a	n/a	n/a	n/a	n/a	n/a	0.34	372	6 440	1 047	468
Benin	430	418	2.6	36	13	51	13.5	13	0.37	558	5 046	12	n/a
Botswana	2 790	2 722	8.8	6	47	47	3.3	79	0.30	38	1 389	0	4 534
Burkina Faso	300	288	3.2	44	20	37	15.6	62	0.36	2 102	8 951	0	0
Burundi	180	157	3.8	52	21	27	22.8	n/a	0.23	300	4 215	0	n/a
Cameroon	820	885	-2.2	29	25	47	5.6	94	0.56	915	14 225	332	15 096
Central African Rep.	400	391	0.8	50	14	36	13.6	n/a	0.64	86	3 491	0	n/a
Chad	210	199	4.2	44	22	35	19.3	n/a	0.54	776	4 037	0	n/a
Congo	950	976	0.5	11	35	53	5.0	n/a	0.07	26	3 670	331	74
Côte d'Ivoire	630	698	-0.4	36	24	39	8.4	n/a	0.28	1 239	12 635	19	n/a
Dem. Rep. of Congo	264	274	-0.5	n/a	n/a	n/a	n/a	n/a	0.19	1 347	38 933	83	73 903
Egypt	660	697	2.9	18	22	60	10.0	3 582	0.05	13 844	2 248	2 320	2 808
Equatorial Guinea	420	413	3.4	47	26	27	41.0	n/a	0.61	n/a	607	0	n/a
Eritrea	n/a	n/a	n/a	13	21	66	n/a	n/a	0.38	n/a	n/a	n/a	n/a
Ethiopia	100	n/a	n/a	60	10	29	n/a	1 217	0.27	6 587	42 536	5	0
Gabon	4 960	5 383	1.3	8	45	47	2.1	n/a	0.37	24	4 130	733	5 946
Gambia	350	346	3.2	28	15	58	28.7	n/a	0.17	108	930	0	n/a
Ghana	430	370	4.7	48	16	36	10.2	69	0.26	1 083	17 343	22	17 737
Guinea	500	503	3.7	24	34	45	14.1	n/a	0.12	922	3 870	1	166 320
Guinea-Bissau	240	235	5.0	45	19	36	44.0	n/a	0.33	172	567	0	n/a
Kenya	270	219	4.0	29	18	54	11.7	n/a	0.17	2 853	35 599	21	6
Lesotho	650	350	6.0	10	47	43	12.0	n/a	0.16	155	613	0	n/a

*1 pJ = 1 x 10¹⁵ pJ

* See page 449 for definitions

REGION	GNP PER CAP. (US$) 1993	GDP PER CAP. (US$) 1993	AVERAGE ANNUAL GROWTH RATE OF GDP PER CAP. 1983–93	% OF GDP FROM AGRICULTURE 1993	% OF GDP FROM INDUSTRY 1991	% OF GDP FROM SERVICES 1991	FOREIGN AID AS A % OF GNP 1991–93	DEFENCE EXPENDITURES (000 000 US$) 1991	CROP LAND (ha per cap.) 1993	CEREAL PRODUCTION (000 t) 1990–92	WOOD PRODUCTION (000 m³) 1989–91	COMMERCIAL ENERGY PRODUCTION (pJ*) 1991	TOTAL VALUE OF MINERAL RESERVES (000 000 US$) 1991
Liberia	560	596	-1.4	n/a	n/a	n/a	n/a	n/a	0.13	106	6 056	1	16 750
Libya	6 125	5 645	-5.0	n/a	n/a	n/a	n/a	n/a	0.43	290	643	3 417	0
Madagascar	220	242	1.4	34	14	52	14.0	39	0.22	2 561	8 099	1	454
Malawi	200	188	2.7	39	18	43	26.4	n/a	0.16	1 259	8 210	3	n/a
Mali	270	263	3.3	42	15	42	16.3	n/a	0.25	2 114	5 592	1	n/a
Mauritania	500	438	2.1	28	30	42	23.1	27	0.10	95	12	0	5 120
Mauritius	3 030	3 006	6.5	10	33	57	1.5	n/a	0.10	2	25	0	n/a
Morocco	1 040	1 027	3.6	14	32	53	3.6	730	0.38	5 966	2 221	22	2 044
Mozambique	90	97	4.5	33	12	55	96.4	230	0.21	506	16 037	1	328
Namibia	1 820	1 716	3.3	10	27	63	6.3	33	0.45	94	n/a	0	3 139
Niger	270	260	0.3	39	18	44	15.2	n/a	0.42	2 073	4 958	5	36
Nigeria	300	300	4.6	34	43	24	0.8	814	0.31	13 113	107 761	4 145	145
Rwanda	210	198	1.1	41	21	38	18.6	n/a	0.15	313	5 936	1	n/a
Senegal	750	730	2.4	20	19	61	10.4	68	0.30	926	4 787	0	n/a
Sierra Leone	150	164	1.4	38	16	46	21.2	n/a	0.13	503	3 082	0	3 580
Somalia	131	115	2.7	65	9	26	n/a	n/a	0.11	346	7 129	0	n/a
South Africa	2 980	2 961	1.0	5	39	56	n/a	2 063	0.33	8 868	19 709	3 964	229 647
Sudan	493	621	n/a	34	17	50	n/a	n/a	0.49	3 999	22 798	3	108
Swaziland	1 190	1 179	3.9	12	39	50	5.4	n/a	0.24	100	2 223	1	n/a
Tanzania	90	85	4.9	56	14	30	41.5	n/a	0.12	3 723	34 295	2	0
Togo	340	322	0.8	49	18	33	11.9	n/a	0.63	469	1 002	0	841
Tunisia	1 720	1 691	3.7	18	31	51	2.5	323	0.58	2 131	3 249	232	n/a
Uganda	180	179	3.8	53	12	35	17.4	70	0.34	1 580	15 149	3	22
Zambia	380	412	1.3	34	36	30	28.8	61	0.59	1 012	13 195	37	30 430
Zimbabwe	520	525	2.9	15	36	48	9.4	312	0.27	1 715	7 893	177	7 985

*1 pJ = 1 x 10^15 pJ

REGION	GNP PER CAP. (US$) 1993	GDP PER CAP. (US$) 1993	AVERAGE ANNUAL GROWTH RATE OF GDP PER CAP. 1983–93	% OF GDP FROM AGRICULTURE 1993	% OF GDP FROM INDUSTRY 1991	% OF GDP FROM SERVICES 1991	FOREIGN AID AS A % OF GNP 1991–93	DEFENCE EXPENDITURES (000 000 US$) 1991	CROP LAND (ha per cap.) 1993	CEREAL PRODUCTION (000 t) 1990–92	WOOD PRODUCTION (000 m³) 1989–91	COMMERCIAL ENERGY PRODUCTION (pJ*) 1991	TOTAL VALUE OF MINERAL RESERVES (000 000 US$) 1991
ASIA	n/a	n/a							0.14	875 970	1 071 682	94 351	416 993
Afghanistan	n/a	n/a	n/a	n/a	n/a	n/a	n/a	n/a	0.46	2 616	6 480	99	n/a
Armenia	660	587	-6.7	48	30	22	0.6	n/a	n/a	274	n/a	n/a	n/a
Azerbaijan	730	676	-5.2	22	52	26	n/a	n/a	0.27	1 328	n/a	n/a	n/a
Bangladesh	220	208	3.9	30	18	52	6.9	234	0.08	28 203	30 944	192	n/a
Bhutan	n/a	n/a	6.4	41	29	30	n/a	n/a	0.08	106	1 532	6	n/a
Cambodia	n/a	206	5.6	47	14	38	n/a	n/a	0.25	2 371	6 048	0	n/a
China	n/a	361	8.9	19	48	33	0.5	12 025	0.08	399 927	281 371	29 720	176 310
Georgia	580	550	-10.9	58	22	20	0.2	n/a	0.18	535	n/a	n/a	n/a
India	300	279	5.0	31	27	41	0.8	7 990	0.19	196 173	274 510	7 327	123 116
Indonesia	740	773	5.9	19	39	42	1.6	1 739	0.16	52 871	167 822	5 291	30 223
Iran	2 159	1 821	1.6	21	36	43	n/a	4 270	0.28	14 912	6 745	8 335	9 829
Iraq	2 363	2 923	-14.9	n/a	n/a	n/a	n/a	n/a	0.28	2 482	153	606	0
Israel	13 920	13 362	4.7	n/a	n/a	n/a	2.5	3 239	0.08	256	108	1	n/a
Japan	31 490	33 857	4.1	2	41	57	-0.3	16 464	0.04	13 985	29 582	3 099	7 245
Jordan	1 190	1 265	0.1	8	26	66	12.5	594	0.08	128	9	0	n/a
Kazakhstan	1 560	1 459	-2.1	29	42	30	0.1	n/a	2.05	23 218	n/a	n/a	n/a
Korea, North	n/a	n/a	n/a	n/a	n/a	n/a	n/a	5 328	0.09	10 086	4 693	2 464	11 134
Korea, South	7 660	7 497	9.0	7	43	50	0	6 359	0.05	8 058	6 592	916	1 929
Kuwait	19 360	12 711	-2.1	0	55	45	1.3	7 959	0	2	n/a	436	n/a

*1 pJ = 1 x 10¹⁵ pJ

REGION	GNP PER CAP. (US$) 1993	GDP PER CAP. (US$) 1993	AVERAGE ANNUAL GROWTH RATE OF GDP PER CAP. 1983–93	% OF GDP FROM AGRICULTURE 1993	% OF GDP FROM INDUSTRY 1991	% OF GDP FROM SERVICES 1991	FOREIGN AID AS A % OF GNP 1991–93	DEFENCE EXPENDITURES (000 000 US$) 1991	CROP LAND (ha per cap.) 1993	CEREAL PRODUCTION (000 t) 1990–92	WOOD PRODUCTION (000 m³) 1989–91	COMMERCIAL ENERGY PRODUCTION (pJ*) 1991	TOTAL VALUE OF MINERAL RESERVES (000 000 US$) 1991
ASIA									0.14	875 970	1 071 682	94 351	416 993
Kirgyz Rep.	850	853	0.6	43	35	22	n/a	n/a	0.31	1 432	n/a	n/a	n/a
Laos	280	290	-4.7	51	18	31	15.3	n/a	0.17	1 470	4 194	3	0
Lebanon	n/a	1 955	n/a	n/a	n/a	n/a	n/a	20	0.11	80	473	2	n/a
Malaysia	3 140	3 384	6.7	n/a	n/a	n/a	0.4	1 670	0.25	1 928	49 938	1 950	12 132
Mongolia	390	471	n/a	21	46	33	n/a	268	0.60	537	2 390	88	7 190
Myanmar	n/a	1 238	-0.1	63	9	28	n/a	298	0.23	14 137	22 850	84	909
Nepal	190	180	4.9	43	21	36	10.3	35	0.11	5 359	18 244	3	n/a
Oman	4 850	5 879	5.8	3	53	44	0.5	1 182	0.03	4	n/a	1 581	1 801
Pakistan	430	422	4.8	25	25	50	2.3	3 014	0.16	21 391	26 183	707	444
Philippines	850	834	2.6	22	33	45	2.8	843	0.14	14 086	38 466	258	24 459
Saudi Arabia	7 953	7 410	2.5	6	50	43	n/a	35 438	0.22	4 495	n/a	19 367	118
Singapore	19 850	19 769	6.9	0	37	63	0	1 518	0	0	n/a	0	n/a
Sri Lanka	600	585	3.9	25	26	50	7.4	340	0.11	2 432	9 037	11	772
Syria	1 219	1 413	0.8	30	23	48	n/a	3 095	0.42	3 584	56	1 192	n/a
Tajikistan, Rep.	470	437	-3.0	33	35	32	n/a	n/a	0.15	300	n/a	n/a	n/a
Thailand	2 110	2 150	8.9	10	39	51	0.6	1 761	0.36	22 438	37 739	523	3 241
Turkey	2 970	2 922	4.2	15	30	55	0.5	2 014	0.46	30 129	15 681	771	5 987
Turkmenistan	1 416	1 376	1.9	32	31	37	n/a	n/a	0.38	571	n/a	n/a	n/a
United Arab Emirates	21 430	19 592	1.6	2	57	40	0.8	4 249	0.02	8	n/a	5 968	n/a
Uzbekistan	970	934	1.4	23	36	41	n/a	n/a	0.21	1 985	n/a	n/a	n/a
Vietnam	170	180	n/a	29	28	42	n/a	n/a	0.09	20 874	28 970	332	132
Yemen	n/a	958	n/a	21	24	55	n/a	910	0.11	n/a	324	391	n/a

*1 pJ = 1 x 10^{15} pJ

REGION	GNP PER CAP. (US$) 1993	GDP PER CAP. (US$) 1993	AVERAGE ANNUAL GROWTH RATE OF GDP PER CAP. 1983–93	% OF GDP FROM AGRICULTURE 1993	% OF GDP FROM INDUSTRY 1991	% OF GDP FROM SERVICES 1991	FOREIGN AID AS A % OF GNP 1991–93	DEFENCE EXPENDITURES (000 000 US$) 1991	CROP LAND (ha per cap.) 1993	CEREAL PRODUCTION (000 t) 1990–92	WOOD PRODUCTION (000 m³) 1989–91	COMMERCIAL ENERGY PRODUCTION (pJ*) 1991	TOTAL VALUE OF MINERAL RESERVES (000 000 US$) 1991
NORTH & CENTRAL AMERICA									0.61	398 318	749 939	88 467	859 660
Belize	2 450	2 568	7.3	19	28	53	5.5	9	0.28	24	188	0	n/a
Canada	19 970	18 982	2.4	3	32	65	-0.4	7 358	1.58	52 855	179 004	11 851	328 644
Costa Rica	2 150	2 317	5.0	15	26	59	2.2	48	0.16	265	4 123	13	1 732
Cuba	n/a	n/a	n/a	n/a	n/a	n/a	n/a	1 272	0.31	491	3 134	35	53 394
Dominican Rep.	1 230	1 261	3.4	15	23	62	0.5	22	0.19	522	982	3	1 548
El Salvador	1 320	1 382	2.8	9	25	66	5.7	201	0.13	849	4 566	21	n/a
Guatemala	1 100	1 128	2.8	25	19	55	2.1	158	0.19	1 385	7 825	15	172
Haiti	477	211	-1.5	39	16	46	n/a	21	0.13	312	5 841	1	222
Honduras	600	627	3.1	20	30	50	10.7	82	0.38	690	6 189	3	62
Jamaica	1 440	1 587	1.1	8	41	51	3.7	23	0.09	3	204	0	59 800
Mexico	3 610	3 815	2.4	8	28	63	0.1	917	0.27	24 662	23 514	8 053	43 012
Nicaragua	340	437	-4.1	30	20	50	47.8	225	0.31	446	4 077	12	n/a
Panama	2 600	2 587	1.4	10	18	72	1.9	73	0.26	306	1 872	7	0
Trinidad and Tobago	3 830	3 654	-1.8	3	43	55	0.1	n/a	0.10	17	75	533	n/a
United States	24 740	24 279	2.5	n/a	n/a	n/a	-0.2	227 055	0.73	315 486	508 200	67 936	371 074
SOUTH AMERICA									0.33	75 293	343 918	14 541	677 080
Argentina	7 220	7 567	1.4	6	31	63	0.2	1 161	0.81	21 874	10 819	2 145	2 305
Bolivia	760	762	2.4	n/a	n/a	n/a	11.5	122	0.34	853	1 595	164	1 899
Brazil	2 930	3 242	2.2	11	37	52	0	1 081	0.31	37 816	262 439	2 434	324 640
Chile	3 170	3 302	6.8	n/a	n/a	n/a	0.4	735	0.31	2 915	18 309	211	208 632
Columbia	1 400	1 516	4.0	16	35	50	0.3	1 403	0.16	3 979	19 384	1 807	2 729
Ecuador	1 200	1 303	2.7	12	38	50	2.0	401	0.28	1 462	9 233	684	3
Guyana	350	400	0.1	30	38	32	42.2	n/a	0.61	225	190	0	47 530

*1 pJ = 1 x 10^15 pJ

REGION	GNP PER CAP. (US$) 1993	GDP PER CAP. (US$) 1993	AVERAGE ANNUAL GROWTH RATE OF GDP PER CAP. 1983–93	% OF GDP FROM AGRICULTURE 1993	% OF GDP FROM INDUSTRY 1991	% OF GDP FROM SERVICES 1991	FOREIGN AID AS A % OF GNP 1991–93	DEFENCE EXPENDITURES (000 000 US$) 1991	CROP LAND (ha per cap.) 1993	CEREAL PRODUCTION (000 t) 1990–92	WOOD PRODUCTION (000 m³) 1989–91	COMMERCIAL ENERGY PRODUCTION (pJ*) 1991	TOTAL VALUE OF MINERAL RESERVES (000 000 US$) 1991
Paraguay	1 510	1 452	3.6	26	21	53	2.0	n/a	0.48	1 018	8 430	107	n/a
Peru	1 490	1 794	-0.5	11	43	46	1.7	605	0.15	1 691	8 061	320	31 289
Suriname	1 180	1 015	0.9	22	24	54	13.0	16	0.16	208	147	14	14 375
Uruguay	3 830	4 174	3.0	9	27	64	0.8	143	0.41	1 224	3 729	22	n/a
Venezuela	2 840	2 869	3.1	5	42	53	0.1	1 525	0.19	2 003	1 328	6 633	42 744
EUROPE									0.20	281 421	366 822	44 335	226 503
Albania	340	211	n/a	40	14	37	n/a	103	0.25	700	2 307	83	3 365
Austria	23 510	23 159	2.7	2	35	62	-0.3	813	0.20	4 882	16 865	244	1 178
Belarus	2 870	2 704	n/a	17	52	29	0.6	n/a	0.65	6 387	n/a	n/a	n/a
Belgium	21 650	20 957	2.8	2	30	68	-0.4	1 505	n/a	2 213	5 182	488	6 296
Bosnia and Herzegovina	n/a	n/a	n/a	n/a	n/a	n/a	n/a	n/a	n/a	n/a	n/a	n/a	n/a
Bulgaria	1 140	1 169	2.5	13	28	49	1.7	1 790	0.46	7 977	3 975	360	n/a
Croatia	n/a	2 591	n/a	11	30	58	n/a	n/a	0.37	n/a	n/a	n/a	n/a
Czech Rep.	2 710	3 070	1.3	6	40	54	0.7	2 800	n/a	11 599	17 521	1 780	1 057
Denmark	26 730	26 333	1.8	4	27	69	-1.0	1 272	0.51	8 645	2 223	451	n/a
Estonia	3080	3 281	n/a	8	29	63	1.0	n/a	0.65	830	n/a	n/a	n/a
Finland	19 300	16 566	1.7	5	31	64	-0.6	1 084	0.51	3 566	41 589	317	6 993
France	22 490	21 779	2.5	3	29	69	-0.6	18 044	0.35	58 595	44 946	4 372	24 993
Germany	23 560	23 679	n/a	1	38	61	-0.4	16 450	0.16	24 672	59 236	8 002	2 016
Greece	7 390	7 060	2.4	18	32	50	0.1	1 977	0.40	5 396	2 424	342	17 131
Hungary	3 350	3 732	-1.1	6	28	66	1.1	1 230	0.50	12 931	6 265	591	3 646
Iceland	24 950	23 075	2.4	12	28	60	0	n/a	0.03	n/a	n/a	25	n/a
Ireland	13 000	13 495	4.0	8	10	82	-0.2	278	0.29	2 091	1 619	139	4 571

*1 pJ = 1 x 10¹⁵ pJ

REGION	GNP PER CAP. (US$) 1993	GDP PER CAP. (US$) 1993	AVERAGE ANNUAL GROWTH RATE OF GDP PER CAP. 1983–93	% OF GDP FROM AGRI-CULTURE 1993	% OF GDP FROM INDUSTRY 1991	% OF GDP FROM SERVICES 1991	FOREIGN AID AS A % OF GNP 1991–93	DEFENCE EXPEND-ITURES (000 000 US$) 1991	CROP LAND (ha per cap.) 1993	CEREAL PRODUCTION (000 t) 1990–92	WOOD PRODUCTION (000 m³) 1989–91	COMMERCIAL ENERGY PRODUCTION (pJ*) 1991	TOTAL VALUE OF MINERAL RESERVES (000 000 US$) 1991
Italy	19 840	17 356	2.6	3	32	65	-0.3	9 146	0.22	18 744	8 426	1 165	2 015
Latvia	2 010	1 762	n/a	15	32	53	0.5	n/a	0.67	1 340	n/a	n/a	n/a
Lithuania	1 320	1 168	n/a	21	41	38	0.7	n/a	0.89	2 807	n/a	n/a	n/a
Macedonia	820	821	n/a	n/a	n/a	n/a	n/a	n/a	n/a	n/a	n/a	n/a	n/a
Moldova, Rep.	1 060	974	n/a	35	48	18	-0	n/a	0.54	2 512	n/a	n/a	n/a
Netherlands	20 950	20 237	2.8	4	n/a	62	-0.8	3 947	0.06	1 302	1 386	3 041	n/a
Norway	25 970	24 060	2.1	3	35	62	-1.1	1 864	0.21	1 312	11 437	5 438	13 426
Poland	2 260	2 241	0.1	6	39	55	2.1	2 200	0.40	25 265	18 788	3 872	26 434
Portugal	9 130	8 705	5.6	6	38	56	-0.3	638	0.32	1 473	10 929	38	9 200
Romania	1 140	1 141	2.9	21	40	40	1.0	1 150	0.47	16 253	15 789	1 448	2 728
Russian Federation	2 340	2 214	n/a	9	51	39	0.4	n/a	0.96	100 220	n/a	n/a	n/a
Slovak Rep.	1 960	2 085	n/a	7	44	49	0.8	n/a	n/a	n/a	n/a	n/a	n/a
Slovenia Rep.	6 490	6 182	n/a	6	36	58	n/a	n/a	n/a	n/a	n/a	n/a	n/a
Spain	13 590	12 122	4.2	5	35	61	-0.3	3 484	0.54	17 526	17 477	1 314	11 737
Sweden	24 740	21 320	1.7	2	31	67	-0.9	2 788	0.35	5 098	53 691	1 072	61 515
Switzerland	35 760	32 919	2.5	n/a	n/a	n/a	-0.4	1 853	0.06	1 265	4 970	371	n/a
Ukraine	2 210	2 116	n/a	35	47	18	0.3	n/a	0.70	39 994	n/a	n/a	n/a
United Kingdom	18 060	16 255	2.7	2	33	65	-0.3	22 420	0.12	22 466	6 485	8 773	1 309
Yugoslavia	n/a	n/a	-1.3	n/a	n/a	n/a	n/a	3 490	0.80	14 917	13 291	626	27 050
OCEANIA									**1.86**	**23 085**	**41 043**	**6 867**	**421 350**
Australia	17 500	16 444	2.8	3	29	67	-0.3	4 210	2.64	22 214	19 630	6 402	386 121
Fiji	2 130	2 210	2.8	18	20	62	3.7	23	0.34	35	307	1	n/a
New Zealand	12 600	12 530	1.1	7	26	67	-0.2	423	1.09	832	12 243	460	5 075
Papua New Guinea	1 130	1 239	3.7	26	43	31	9.2	37	0.10	3	8 202	2	15 960
Solomon Islands	262	718	5.4	n/a	n/a	n/a	18.6	n/a	0.16	0	449	0	1 110

*1 pJ = 1 x 10^{15} pJ

DEFINITIONS

GNP per cap.: The gross domestic product (GDP) per capita plus income that residents receive from abroad, divided by the population.

GDP per cap.: The total value of goods and services produced by the domestic economy including net exports of goods and services, divided by the total population.

Foreign Aid as a % of GNP: The net amount of grants and concessional loans given to a country as a percentage of GNP. Negative numbers indicate the amount of foreign aid given by a country; positive numbers indicate aid received.

Defence Expenditures: The amount of money spent on regular military forces and military aid to other countries.

Cereal Production: The production of all grains for food and seed.

Commercial Energy Production (pJ): The total of all energy produced including coal, oil, natural gas, and electricity, expressed in petajoules (pJ). One petajoule (1×10^{15}) is the equivalent of 163 400 "UN Standard" barrels of oil, or 34 140 "UN Standard" tonnes of coal.

Total Value of Mineral Reserves: The total value of known reserves of fifteen important minerals. Reserves are defined as mineral deposits whose quantity and grade have been determined using ore samples.

TECHNOLOGY INDICATORS*

REGION	URBAN POPULATION AS A % OF TOTAL POPULATION 1995	AVERAGE ANNUAL FERTILIZER USE (kg/ha) 1993	AVERAGE NO. OF TRACTORS (000) 1991–93	COMMERCIAL ENERGY CONSUMPTION PER CAP. (pJ*) 1993	HYDRO-ELECTRIC INSTALLED CAPACITY (000 mW) 1993	CO₂ EMISSIONS PER CAP. (t) 1992	PASSENGER CARS (000) 1992	PERSONS PER VEHICLE 1992	SCIENTISTS, ENGINEERS, AND TECHNICIANS ENGAGED IN RESEARCH 1991	TELEVISION SETS PER 1000 PEOPLE 1991	TELEPHONES PER 100 PEOPLE 1991	METHANE EMISSIONS (000 000 t) 1991	CFC EMISSIONS (000 t) 1991
WORLD	45	83	25 879.4	325 296	612.5	4.1	43 825	9				270	400
AFRICA	35	21	515.9	8 805	20.7			47				21	12
Algeria	56	17	91.3	1 183	0.3	3.0	725	20	n/a	73	4.3	1.7	n/a
Angola	32	2	10.3	26	0.3	0.4	122	60	n/a	6	n/a	0.3	n/a
Benin	42	9	0.1	7	0	0.1	22	125	1 036	5	0.3	0	n/a
Botswana	31	2	6.0	n/a	0	1.7	21	21	n/a	12	3.9	0.1	n/a
Burkina Faso	20	6	0.1	8	0	0	11	356	n/a	5	0.2	0.3	n/a
Burundi	6	3	0.2	3	0	0	14	290	338	1	n/a	0	n/a
Cameroon	45	3	0.5	36	0.7	0.2	63	67	n/a	22	0.6	0.3	n/a
Central African Rep.	51	1	0.2	3	0	0	12	184	579	3	0.3	0.1	n/a
Chad	37	1	0.2	1	0	0	11	379	n/a	1	0.2	0.2	n/a
Congo	43	12	0.7	24	0.1	1.7	26	54	2 335	5	0.7	0	n/a
Côte d'Ivoire	44	15	3.7	109	0.9	0.5	155	47	n/a	59	n/a	0.1	n/a
Dem. Rep. of Congo	29	1	2.4	73	2.8	0.1	92	203	n/a	1	n/a	0.4	n/a
Egypt	45	357	60.3	1 226	2.8	1.5	1 054	67	28 425	98	4.2	1.0	n/a
Equatorial Guinea	31	0	0.1	2	0	0.3	n/a	40	n/a	9	n/a	0	n/a
Eritrea	17	0	0.9	n/a	n/a	n/a	n/a	n/a	n/a	n/a	n/a	0.1	n/a
Ethiopia	13	6	3.9	45	0.4	0	39	875	217	2	0.3	1.2	n/a
Gabon	50	1	1.5	32	0.3	4.5	17	30	n/a	36	n/a	0.3	n/a
Gambia	26	4	0	3	0	0.2	5	104	n/a	n/a	1.2	0	n/a
Ghana	36	1	4.1	67	1.1	0.2	82	124	1 893	15	0.5	0.1	n/a
Guinea	30	2	0.3	15	0	0.2	11	220	n/a	5	n/a	0.5	n/a
Guinea-Bissau	22	1	0	3	0	0.2	4	173	n/a	5	1.5	0.1	n/a
Kenya	28	27	14.0	90	0.6	0.2	n/a	81	n/a	9	1.5	0.5	n/a
Lesotho	23	19	1.8	n/a	0	n/a	n/a	n/a	n/a	3	1.2	0	n/a
Liberia	51	0	0.3	5	0	0.1	n/a	222	n/a	18	n/a	0	n/a

*1 pJ = 1 × 10¹⁵ pJ

* See page 457 for definitions

REGION	URBAN POPULATION AS A % OF TOTAL POPULATION 1995	AVERAGE ANNUAL FERTILIZER USE (kg/ha) 1993	AVERAGE NO. OF TRACTORS (000) 1991–93	COMMERCIAL ENERGY CONSUMPTION PER CAP. (pJ*) 1993	HYDRO-ELECTRIC INSTALLED CAPACITY (000 mW) 1993	CO_2 EMISSIONS PER CAP. (t) 1992	PASSENGER CARS (000) 1992	PERSONS PER VEHICLE 1992	SCIENTISTS, ENGINEERS, AND TECHNICIANS ENGAGED IN RESEARCH 1991	TELEVISION SETS PER 1000 PEOPLE 1991	TELEPHONES PER 100 PEOPLE 1991	METHANE EMISSIONS (000 000 t) 1991	CFC EMISSIONS (000 t) 1991
AFRICA	35	21	515.9	8 805	20.7			47				21	12
Libya	86	49	34.0	127	0	8.1	n/a	6	2 600	91	n/a	0.5	n/a
Madagascar	27	3	2.9	1	0.1	0.1	n/a	156	1 225	20	n/a	0.9	n/a
Malawi	14	51	1.4	1	0.1	0.1	n/a	294	n/a	n/a	0.6	0	n/a
Mali	27	10	0.8	7	0	0	n/a	307	n/a	0	n/a	0.4	n/a
Mauritania	54	22	0.3	39	0	1.4	n/a	139	n/a	23	n/a	0.1	n/a
Mauritius	41	245	0.4	21	0	1.3	n/a	18	365	215	7.2	0	n/a
Morocco	48	29	41.7	297	0.7	1.0	n/a	26	n/a	70	1.9	0.4	n/a
Mozambique	34	1	5.8	14	2.1	0.1	n/a	129	n/a	2	0.4	0.1	n/a
Namibia	31	0	3.1	n/a	n/a	n/a	n/a	n/a	n/a	16	n/a	0.1	n/a
Niger	23	0	0.2	15	0.2	0.2	35	222	n/a	4	n/a	0.2	n/a
Nigeria	39	16	11.9	705	2.0	0.8	391	80	7 380	29	n/a	4.5	n/a
Rwanda	6	2	0	7	0	0	7	291	138	n/a	0.2	0	n/a
Senegal	42	11	0.5	38	0	0.4	63	55	4 610	35	n/a	0.2	n/a
Sierra Leone	36	6	0.5	6	0	0.1	33	89	n/a	10	n/a	0.2	n/a
Somalia	26	0	2.1	n/a	0	0	5	396	n/a	14	n/a	0.5	n/a
South Africa	51	64	130.7	3 578	0.6	7.3	3 600	7	n/a	101	15.2	2.4	n/a
Sudan	25	5	10.5	48	0.2	0.2	185	150	n/a	61	0.3	1.1	n/a
Swaziland	31	61	4.4	n/a	0	0.3	27	21	n/a	16	3.3	0	n/a
Tanzania	24	14	6.6	30	0.3	0.1	n/a	318	n/a	1	0.6	0.8	n/a
Togo	31	4	0.4	9	0	0.2	3	87	n/a	6	0.3	0	n/a
Tunisia	59	22	26.8	218	0	1.7	179	16	n/a	75	5.1	0.1	n/a
Uganda	13	0	4.7	16	0.2	0	13	702	n/a	8	0.3	0.3	n/a
Zambia	43	16	6.0	51	2.3	0.3	75	49	n/a	25	1.3	0.2	n/a
Zimbabwe	32	55	16.1	208	0.7	1.8	283	39	n/a	27	3.2	0.2	n/a

*1 pJ = 1 x 10^{15} pJ

REGION	URBAN POPULATION AS A % OF TOTAL POPULATION	AVERAGE ANNUAL FERTILIZER USE (kg/ha)	AVERAGE NO. OF TRACTORS (000)	COMMERCIAL ENERGY CONSUMPTION PER CAP. (pJ*)	HYDRO-ELECTRIC INSTALLED CAPACITY (000 mW)	CO_2 EMISSIONS PER CAP. (t)	PASSENGER CARS (000)	PERSONS PER VEHICLE	SCIENTISTS, ENGINEERS, AND TECHNICIANS ENGAGED IN RESEARCH	TELEVISION SETS PER 1000 PEOPLE	TELEPHONES PER 100 PEOPLE	METHANE EMISSIONS (000 t)	CFC EMISSIONS (000 t)
	1995	1993	1991–93	1993	1993	1992	1992	1992	1991	1991	1991	1991	1991
ASIA	**34**	**118**	**5 565.4**	**95 679**	**157.8**			**32**				**140.0**	**100**
Afghanistan	20	5	0.8	22	0.3	0.1	31	242	n/a	8	n/a	0.2	n/a
Armenia	69	39	14.6	49	0.8	1.2	n/a	n/a	n/a	n/a	n/a	0	n/a
Azerbaijan	56	27	35.7	546	1.7	8.8	n/a	n/a	n/a	n/a	n/a	0.5	n/a
Bangladesh	18	98	5.3	313	0.2	0.2	41	896	n/a	4	n/a	3.9	n/a
Bhutan	6	1	0	2	0.4	0.1	n/a	n/a	n/a	n/a	n/a	n/a	n/a
Cambodia	21	6	1.4	7	0	0	n/a	n/a	n/a	8	n/a	0.1	n/a
China	30	261	770.3	29 679	59.7	2.3	n/a	191	n/a	27	1.1	47.0	8
Georgia	58	54	22.5	159	1.7	2.5	n/a	n/a	n/a	n/a	n/a	0.1	n/a
India	27	73	1 131.4	9 338	19.8	0.9	2 790	185	199 983	27	0.7	33.0	3
Indonesia	35	85	234.7	2 658	2.2	1.0	1 294	63	32 038	55	0.6	10.0	1
Iran	59	52	117.3	3 264	2.0	3.8	2 000	27	5 048	66	3.9	3.3	2
Iraq	75	52	32.3	933	0.9	3.3	672	19	n/a	68	n/a	0.5	2
Israel	91	225	25.5	505	0	8.1	n/a	n/a	n/a	n/a	n/a	0	n/a
Japan	78	407	2 003.3	17 505	21.0	8.8	34 924	2	742 247	610	42.1	3.9	64
Jordan	71	34	5.8	147	0	2.6	n/a	20	447	77	n/a	0.1	n/a
Kazakhstan	60	14	203.3	3 381	3.5	17.5	n/a	n/a	n/a	n/a	n/a	2.5	n/a
Korea, North	61	315	74.7	2 925	5.0	11.2	n/a	n/a	n/a	14	n/a	1.6	n/a
Korea, South	81	474	64.6	4 504	2.5	6.6	n/a	10	92 265	207	36.7	1.4	4
Kuwait	97	200	0.1	471	0	8.1	n/a	3	2 072	281	18.9	0.2	n/a
Kirgyz Rep.	39	20	25.1	150	2.8	3.4	n/a			n/a	n/a	0.2	n/a

*1 pJ = 1 x 10^15 pJ

REGION	URBAN POPULATION AS A % OF TOTAL POPULATION	AVERAGE ANNUAL FERTILIZER USE (kg/ha)	AVERAGE NO. OF TRACTORS (000)	COMMERCIAL ENERGY CONSUMPTION PER CAP. (pJ*)	HYDRO-ELECTRIC INSTALLED CAPACITY (000 mW)	CO_2 EMISSIONS PER CAP. (t)	PASSENGER CARS (000)	PERSONS PER VEHICLE	SCIENTISTS, ENGINEERS, AND TECHNICIANS ENGAGED IN RESEARCH	TELEVISION SETS PER 1000 PEOPLE	TELEPHONES PER 100 PEOPLE	METHANE EMISSIONS (000 000 t)	CFC EMISSIONS (000 t)
	1995	1993	1991–93	1993	1993	1992	1992	1992	1991	1991	1991	1991	1991
Laos	22	4	0.9	5	0.2	0.1	n/a	255	n/a	5	0.2	0.3	n/a
Lebanon	87	118	3.0	121	0.3	3.9	n/a	n/a	186	327	n/a	0.3	n/a
Malaysia	54	212	12.4	996	1.4	3.7	n/a	8	6 707	144	11.3	1.0	n/a
Mongolia	61	4	11.7	105	0	4.0	n/a	n/a	n/a	38	n/a	0.3	n/a
Myanmar	26	19	10.7	71	0.3	0.1	18	611	n/a	2	n/a	2.3	n/a
Nepal	14	31	4.6	19	0.2	0.1	n/a	n/a	409	2	n/a	0.6	n/a
Oman	13	143	0.1	162	0	6.1	n/a	7	n/a	762	5.0	0.2	n/a
Pakistan	35	101	279.5	1 135	4.7	0.6	715	132	15 927	16	n/a	3.3	4
Philippines	54	61	11.3	787	2.1	0.7	455	104	6 685	41	1.7	1.9	n/a
Saudi Arabia	80	122	2.1	2 933	0	13.9	1 468	3	n/a	277	9.9	2.6	n/a
Singapore	100	n/a	0	745	0	18.0	287	7	4 887	372	38.7	0	n/a
Sri Lanka	22	111	32.7	78	1.2	0.3	174	53	3 483	32	1.0	0.6	n/a
Syria	52	65	69.2	565	0.9	3.2	126	50	n/a	59	5.6	0.6	n/a
Tajikistan	32	81	35.0	258	4.1	0.7	n/a	n/a	n/a	n/a	n/a	0.2	n/a
Thailand	20	54	74.9	1 628	2.5	2.0	1 222	20	8 324	109	2.3	5.5	n/a
Turkey	69	80	724.4	1 979	9.8	2.5	1 650	22	18 643	174	15.1	1.1	n/a
Turkmenistan	45	97	62.0	555	0	11.0	n/a	n/a	n/a	n/a	n/a	1.0	n/a
United Arab Emirates	84	710	0.2	1 039	0	42.3	n/a	4	n/a	109	41.2	0.5	n/a
Uzbekistan	41	150	178.3	1 903	1.9	5.8	n/a	n/a	n/a	n/a	n/a	1.3	n/a
Vietnam	21	136	36.7	306	1.9	0.3	n/a	n/a	n/a	38	n/a	4.4	n/a
Yemen	34	n/a	n/a	n/a	0	0.8	28	30	n/a	17	1.5	0.1	n/a
NORTH & CENTRAL AMERICA	**68**	**95**	**5 831.2**	**97 154**	**151.9**			**2**				**35.0**	**100**
Belize	47	114	1.1	4	0	1.3	1.7	40	n/a	165	n/a	0	n/a
Canada	77	60	778.1	9 198	62.7	15.0	12 622	2	88 210	626	57.0	3.6	8
Costa Rica	50	208	6.8	63	0.9	1.2	169	12	1 528	136	14.9	0.1	n/a

*1 pJ = 1 x 10^{15} pJ

REGION	URBAN POPULATION AS A % OF TOTAL POPULATION	AVERAGE ANNUAL FERTILIZER USE (kg/ha)	AVERAGE NO. OF TRACTORS (000)	COMMERCIAL ENERGY CONSUMPTION PER CAP. (pJ*)	HYDRO-ELECTRIC INSTALLED CAPACITY (000 mW)	CO_2 EMISSIONS PER CAP. (t)	PASSENGER CARS (000)	PERSONS PER VEHICLE	SCIENTISTS, ENGINEERS, AND TECHNICIANS ENGAGED IN RESEARCH	TELEVISION SETS PER 1000 PEOPLE	TELEPHONES PER 100 PEOPLE	METHANE EMISSIONS (000 000 t)	CFC EMISSIONS (000 t)
	1995	1993	1991–93	1993	1993	1992	1992	1992	1991	1991	1991	1991	1991
Cuba	76	52	78.2	369	0	2.6	241	n/a	20 882	203	5.8	0.4	n/a
Dominican Rep.	65	61	2.3	148	0.4	1.4	152	30	n/a	82	n/a	0.2	n/a
El Salvador	45	106	3.4	72	0.4	0.7	52	33	1 775	87	4.8	0	n/a
Guatemala	41	87	4.3	72	0.4	0.6	95	38	1 783	45	2.1	0.2	n/a
Haiti	32	5	0.2	9	0	0.1	26	120	n/a	5	0.9	0.1	n/a
Honduras	44	32	3.9	43	0.5	0.6	27	39	n/a	70	1.9	0.1	n/a
Jamaica	54	107	3.1	104	0	3.3	69	21	33	124	n/a	0	n/a
Mexico	75	71	172.0	4 941	8.2	3.8	n/a	8	46 146	127	11.8	3.1	3
Nicaragua	63	21	2.7	52	0.1	0.6	n/a	54	1 027	61	n/a	0.1	n/a
Panama	53	48	5.0	61	0.6	1.7	133	13	n/a	165	11.1	0.1	n/a
Trinidad & Tobago	72	51	2.6	267	0	16.3	244	6	529	301	18.4	0.3	n/a
United States	76	108	4 800.0	81 751	77.4	19.1	143 550	2	949 200	814	50.9	27.0	90
SOUTH AMERICA	78	59	1 223.6	10 095	90.1			12				21.0	10
Argentina	88	11	280.0	2 019	7.2	3.5	4 284	6	17 329	219	14.5	3.6	2
Bolivia	61	6	5.3	86	0.4	0.9	261	23	n/a	98	2.9	0.5	n/a
Brazil	78	85	733.3	3 800	48.2	1.4	10 598	11	52 863	204	9.4	9.9	4
Chile	84	58	39.7	539	2.4	2.6	711	13	7 570	201	8.3	0.4	n/a
Columbia	73	94	36.3	829	7.8	1.8	842	22	2 107	108	8.8	2.1	1
Ecuador	58	31	8.9	245	1.5	1.7	166	44	n/a	82	5.0	0.5	n/a
Guyana	36	24	3.6	15	0	1.0	24	24	267	31	2.0	0	n/a
Paraguay	53	14	16.3	51	6.5	0.6	165	40	807	48	3.0	0.4	n/a
Peru	72	44	16.5	314	2.5	1.0	373	35	4 858	95	3.5	0.5	n/a
Suriname	50	49	1.3	24	0.3	4.6	37	9	n/a	133	11.5	0	n/a
Uruguay	90	72	32.9	77	2.3	1.6	380	10	2 093	227	19.6	0.7	n/a
Venezuela	93	165	48.8	2 083	11.0	5.8	1 582	10	7 260	156	9.1	2.0	1

*1 pJ = 1 x 10^{15} pJ

REGION	URBAN POPULATION AS A % OF TOTAL POPULATION 1995	AVERAGE ANNUAL FERTILIZER USE (kg/ha) 1993	AVERAGE NO. OF TRACTORS (000) 1991–93	COMMERCIAL ENERGY CONSUMPTION PER CAP. (pJ*) 1993	HYDRO-ELECTRIC INSTALLED CAPACITY (000 mW) 1993	CO_2 EMISSIONS PER CAP. (t) 1992	PASSENGER CARS (000) 1992	PERSONS PER VEHICLE 1992	SCIENTISTS, ENGINEERS, AND TECHNICIANS ENGAGED IN RESEARCH 1991	TELEVISION SETS PER 1000 PEOPLE 1991	TELEPHONES PER 100 PEOPLE 1991	METHANE EMISSIONS (000 000 t) 1991	CFC EMISSIONS (000 t) 1991
EUROPE	**74**	**115**	**9 791.2**	**108 523**	**179.4**			**3**				**53.0**	**120**
Albania	37	17	9.2	43	1.4	1.2	n/a	n/a	n/a	83	n/a	0.1	n/a
Austria	61	175	348.6	966	11.7	7.3	2 991	2	14 426	475	58.9	0.3	2
Belarus	71	136	125.2	1 249	n/a	10.0	n/a	n/a	44 100	n/a	n/a	0.5	n/a
Belgium	97	403	116.7	1 976	0.1	10.2	3 875	2	36 770	447	54.6	0.2	3
Bosnia and Hercegovina	49	11	205.9	29	1.2	3.4	n/a	n/a	n/a	n/a	n/a	0	n/a
Bulgaria	71	54	48.4	965	2.2	6.1	1 310	6	62 247	249	n/a	8.8	1
Croatia	64	172	4.2	263	2.1	3.3	n/a	n/a	n/a	n/a	n/a	0	n/a
Czech Rep.	65	81	78.0	1 659	1.1	13.0	3 242	4	108 351	410	27.3	0.4	3
Denmark	85	191	157.1	762	0	10.4	1 604	3	25 448	528	97.2	0.3	2
Estonia	73	57	19.6	214	2.5	13.2	n/a	n/a	n/a	n/a	n/a	0	n/a
Finland	63	132	233.3	1 014	2.6	8.2	1 940	2	21 195	488	53.0	0.2	1
France	73	237	1 460.0	9 153	24.8	6.3	23 550	2	283 099	400	48.2	1.8	17
Germany	87	221	1 374.0	13 724	4.4	11.0	35 502	2	483 145	650	67.1	3.4	23
Greece	65	148	215.8	989	2.6	7.3	1 738	4	1 022	195	45.8	0.3	3
Hungary	65	40	41.4	990	0	5.7	1 945	5	34 544	409	17.8	0.3	2
Iceland	92	n/a	11.1	54	0.9	6.9	120	2	n/a	319	49.6	0	n/a
Ireland	58	769	167.3	428	0.2	8.9	803	3	7 642	271	27.9	0.5	1
Italy	67	148	1 439.0	6 749	17.8	7.0	27 300	2	113 120	423	55.5	1.5	17
Latvia	73	56	52.0	187	1.5	5.5	n/a	n/a	n/a	n/a	n/a	0.1	n/a
Lithuania	72	27	47.4	368	0.1	5.9	n/a	n/a	n/a	n/a	n/a	0.2	n/a

*1 pJ = 1 x 10^{15} pJ

REGION	URBAN POPULATION AS A % OF TOTAL POPULATION 1995	AVERAGE ANNUAL FERTILIZER USE (kg/ha) 1993	AVERAGE NO. OF TRACTORS (000) 1991–93	COMMERCIAL ENERGY CONSUMPTION PER CAP. (pJ*) 1993	HYDRO-ELECTRIC INSTALLED CAPACITY (000 mW) 1993	CO_2 EMISSIONS PER CAP. (t) 1992	PASSENGER CARS (000) 1992	PERSONS PER VEHICLE 1992	SCIENTISTS, ENGINEERS, AND TECHNICIANS ENGAGED IN RESEARCH 1991	TELEVISION SETS PER 1000 PEOPLE 1991	TELEPHONES PER 100 PEOPLE 1991	METHANE EMISSIONS (000 000 t) 1991	CFC EMISSIONS (000 t) 1991
Macedonia	60	18	50.0	139	n/a	2.0	n/a	n/a	n/a	n/a	n/a	0	n/a
Moldova, Rep.	52	52	53.8	234	n/a	3.3	n/a	n/a	n/a	n/a	n/a	0.1	n/a
Netherlands	89	560	182.0	3 306	0	9.2	5 371	2	64 420	485	46.2	1.4	4
Norway	73	229	156.0	904	27.0	14.0	1 612	2	20 700	423	50.2	2.4	1
Poland	65	87	1 168.8	4 056	0.9	8.9	5 261	5	32 500	292	13.7	1.8	4
Portugal	36	75	131.2	603	3.4	4.8	2 552	4	8 575	176	26.3	0.2	3
Romania	55	39	142.1	1 762	6.2	5.2	n/a	14	102 601	194	n/a	0.8	1
Russian Federation	76	29	1 283.3	30 042	42.9	14.1	n/a	n/a	n/a	n/a	n/a	17.0	n/a
Slovak Rep.	59	50	33.7	672	1.2	n/a	n/a	n/a	n/a	n/a	n/a	0.3	n/a
Slovenia Rep.	64	249	69.8	194	0.8	2.8	n/a	n/a	n/a	n/a	n/a	0	n/a
Spain	76	93	765.8	3 359	14.7	5.7	11 996	3	29 086	389	32.3	1.4	11
Sweden	83	120	165.8	1 660	16.6	6.6	3 601	2	51 811	471	68.1	0.2	2
Switzerland	61	321	114.0	985	11.8	6.4	2 994	2	25 620	406	90.5	0.1	2
Ukraine	70	39	430.5	8 058	4.7	11.7	n/a	n/a	348 600	n/a	n/a	3.6	n/a
United Kingdom	89	338	500.0	9 518	1.1	9.8	20 807	2	n/a	434	43.4	3.8	17
Yugoslavia	57	99	n/a	381	4.1	3.6	3 511	5	53 550	197	16.1	0.2	3
OCEANIA	**70**	**41**	**400.5**	**4 595**	**12.7**			**2**				**5.8**	**6**
Australia	85	32	315.3	3 917	7.2	15.2	7 442	2	55 103	484	44.8	4.8	5
Fiji	41	56	7.1	11	0	1.0	40	12	126	14	10.1	0	n/a
New Zealand	86	154	75.7	565	5.1	7.6	1 498	2	n/a	372	43.0	1.0	1
Papua New Guinea	16	31	1.1	33	0.2	0.6	17	n/a	n/a	2	1.7	0	n/a
Solomon Is.	17	0	n/a	2	0	0.5	n/a	n/a	n/a	n/a	2.2	0	n/a

*1 pJ = 1 x 10^{15} pJ

Source: (Appendices A to C): Statistics from *World Resources 1996–97* (New York: Oxford University Press, 1996).

DEFINITIONS

Hydroelectric Installed Capacity: Combined
generating capacity of hydroelectric plants.

GLOSSARY

agglomeration. An economic strategy in which a country's industry is concentrated in a few favoured locations in a country; the opposite of dispersion.

agribusiness. A large agricultural operation that is run like an industry.

aid. Money, goods, or services provided to the people of a country in an effort to help them.

alley cropping. The planting of crops between rows of trees, usually acacia.

alliances. International political agreements (often military) that define the level of co-operation between states.

alluvium. A deposit of fertile soil laid down by rivers, especially at times of flood.

anomaly. A component of a set of data that does not fit into an established pattern; an oddity.

anoxic. The condition of oxygen deficiency or its absence.

antecedent boundaries. Political borders established before a region was populated. *See also* subsequent boundaries.

aquaculture. The growing and harvesting of aquatic plants and animals.

aqueduct. A structure, often built like a bridge, that carries water over a valley or low ground.

aquifer. An underground deposit of sand and gravel that holds water, underlain with an impervious layer of rock.

archipelago. A chain of islands.

artesian well. A well from which water flows upwards without pumping.

aristocracy. A form of government by a select group of individuals who control the affairs of state.

assimilate. To become part of a dominant society.

badlands. A barren region marked by pinnacles, gullies, and sharp-edged ridges caused by erosion.

beriberi. A disease causing inflammation of the nerves due to a deficiency of Vitamin B.

bilateral aid. Aid given by one country to another.

biodiversity. The existence of a wide variety of species of plants, animals, and micro-organisms in a natural community or habitat. The maintenance of biodiversity is important for the stability of ecosystems.

biomass. The total mass of living organisms in a particular ecosystem; or organic waste.

birth rate (or crude birth rate). The number of live births per 1000 population in a given year; calculated by dividing total births by the mid-year population, then multiplying the result by 1000. *See also* death rate.

blitzkrieg. An intense military strategy to bring about a swift victory, particularly as used by Germany in the Second World War.

carnivores. Animals that eat meat.

carrying capacity. The maximum number of people a given environment can support indefinitely.

cartel. A monopoly exercised by an association of commodity producers; e.g., OPEC.

cash crops. Agricultural products that are grown solely for sale, rather than for consumption by the growers.

catchment basin. A depression or storage tank used for collecting seasonal rain.

central planning system. An economic system controlled by a central government.

chlorine sink. A naturally occurring condition in which chlorine (or any other element) is stored and rendered inactive.

chlorofluorocarbons (CFCs). Mainly synthetic chemicals containing chlorine, fluorine, and carbon that destroy the ozone; often found in refrigerants and aerosol cans.

chronic hunger. A condition in which essential nutrients are excluded from the diet over an extended period.

civil war. An internal military conflict between rival factions within the same country. *See also* revolutionary war.

Cold War. The period of tension that began after the Second World War during which the United States and the Soviet Union were competing militarily even though they were not combatants.

collectivism. A political system that works to the advantage of the state or the collective. *See also* individualism.

colonialism. The political domination of a nation's territory by a foreign power.

communism. A socio-economic and political system advocating the collective ownership of the means of production, central economic planning, and rule by a single political party.

competitive advantage. The ability of one organism to supersede another because of its genetic or organizational superiority in a particular ecosystem.

conservation. The philosophy and policy of preserving the earth's natural resources.

consumer products. Goods produced for mass consumption.

consumer society. A society in which the consumption of goods and services is an important social and economic activity.

cooperative. An organization in which a group of people share land, equipment, or other capital items for their mutual economic benefit.

correlation. The relationship that exists between two sets of data.

cost-benefit analysis. The process of considering the social and financial advantages and disadvantages before making a decision.

cottage industry. A business activity carried on in the home.

country. An area on the earth's surface that has definite, widely acknowledged boundaries and sovereignty over the people who live within these boundaries. *See also* nation and state.

cultural boundaries. Loosely delineated borders between nations.

cultural issue. An issue that relates to people, and their values, attitudes, and institutions.

death rate (or crude death rate). The number of deaths per 1000 population in a given year; calculated by dividing total deaths by the mid-year population, then multiplying the result by 1000. *See also* birth rate.

demand-induced scarcity. A situation that arises when the rate of population increase exceeds the growth of essential resources. *See also* supply-induced scarcity and structural scarcity.

democratic. Favouring social equality with a system of government by the whole population, usually through elected representatives. *See also* direct democracy.

democratic socialism. A left-wing democratic political system in which the state provides extensive universal social programs to the people.

demographic transition model. A hypothetical model that illustrates the stages of demographics.

demographics. Statistical data and analyses of trends relating to human populations and their size, distribution, composition, and development.

dependant. A person who relies on another, especially for financial support.

dependency load. The percentage of a population that is under the age of 15 and over the age of 65.

desertification. The degradation of land into desert, caused by a combination of naturally occurring droughts and unsustainable agricultural practices.

desiccation. The drying out of any naturally occurring or synthetic substance.

developed world. Those countries where the standard of living is high.

developing world. Those countries where the standard of living is low compared to what we are used to in Canada.

deviation. The variance multiplied by itself (squared).

dictatorship. Rule by one person without any opposition. *See also* fascism.

direct democracy. A system of government based on people voting directly on the affairs of state instead of having representatives vote for them. *See also* democratic

dispersion. An economic strategy in which industry is spread throughout a country; the opposite of agglomeration.

diversified economies. Economic regions or countries where there is a balance between primary, secondary, tertiary, and quaternary industries.

domestic system. Organization within the household or native country.

domestication. The intentional planting of crops and the taming and raising of animals for human use.

drawing down. The diminishing or reduction of something.

drought. A period of lower than normal or absent precipitation, which results in a shortage of water.

ebola. A severe, infectious tropical disease that causes haemorrhaging and is generally fatal.

ecocide. The killing of ecosystems.

economic imperialism. The taking over of a country's economy by another country or a corporation.

economic issue. An issue that relates to money and finance.

élite. A group of people who have more political, social, and/or economic power and distinction than others.

embargo. An official suspension of trade or other activity with an enemy country.

emerging industrial economies. Economic regions (countries) where secondary industries are taking over from primary industries as the main economic activity.

emigration. The act of leaving one's country to settle in another. *See also* immigration.

entrepôt. A commercial centre for import and export, and for collection and distribution.

environmental determinists. People who believe that we are affected primarily by natural factors; opposite to environmental possibilists.

environmental issue. An issue that affects the earth's environment.

environmental possibilists. People who believe that we can overcome the limitations of our environment through the use of our intelligence; opposite of environmental determinists.

erosion. The gradual wearing away of land or rock by natural forces.

ethnic cleansing. The mass expulsion or extermination of people from opposing cultural, national, or religious groups within a certain area.

evapotranspiration. The release of water vapour from the earth and surface by the evaporation of surface water and the transpiration of plants.

extirpated. Referring to life forms no longer living in a region because they have been removed or destroyed.

famine. An extreme scarcity of food that occurs when the resource base shrinks because of a natural phenomenon such as drought.

fascism. Extreme right-wing government in which the rights of the people are considered less important than the glorification of the state and the nation. *See also* dictatorship.

fluidized bed combustion. The industrial process whereby sulphur and nitrogen are removed from coal.

free-enterprise system. An economic system, based on laissez-faire economics.

genetic engineering. The manipulation of an organism's genetic material; in agriculture, the manipulation of plant genes to improve crop strains.

genocide. The mass extermination of humans, especially of a particular race, nation, or religion.

geometric boundaries. Borders between states that follow lines of latitude or longitude, or other artificial lines of demarcation.

geometric growth. A trend in which the rate of increase is constantly growing.

geopolitics. The study of politics and its relationship to geographical features.

global warming. The potential increase in the temperature of the earth's surface caused by the greenhouse effect. *See also* greenhouse gases.

graded-shading map. A map in which different shades of one colour are used to represent statistical categories; the darkest shades representing the highest values and the lightest shades representing the lowest values.

grass-roots development. Development aid invested in small projects that benefit people living in local villages.

Green Revolution. The development of high-yield crops in conjunction with improved agricultural technologies.

greenhouse effect. The trapping of the sun's warmth in the lower atmosphere caused by high levels of carbon dioxide and other gases that are more transparent to incoming solar radiation. *See also* greenhouse gases.

greenhouse gases. Any of various gases, especially carbon dioxide, that contribute to the greenhouse effect.

gross domestic product. The total value of goods produced and services provided within a country in one year.

growing season. The period of time in which there is enough sunlight and heat for crops to be grown.

growth pole theory. An economic development theory that combines the advantage of agglomeration and dispersion. The strategy involves developing new industry in an urban location (or growth pole) in the hope that prosperity will spread to the surrounding area.

harvested. Taken out of the environment for human consumption.

herbivores. Animals that eat plants.

hinterland. The area surrounding a city that provides the natural resources it needs to support its industries.

Holocaust. The genocide practised against the Jewish nation by the German state during the Second World War.

horticulture. The intensive production of fruits, vegetables, and ornamental plants.

humanitarian view. An attitude towards aid that encourages the reduction or elimination of suffering and promotes human welfare. *See also* pragmatic view, nationalist view, and internationalist view.

humus. Organic matter, found in topsoil, that absorbs moisture.

ideology. A set of beliefs and ideas that form the basis of social, political, and economic systems.

illiterate. Unable to read or write.

immigration. The act of coming as a permanent resident to a country other than one's home land. *See also* emigration.

imperialism. The policy of one country attempting to dominate other, less powerful ones through territorial expansion or the control of markets.

import substitution. The production of goods by a country in order to avoid costly imports.

indigenous people. People native to a specific region over thousands of years.

individualism. A political system that works to the advantage of the individual. *See also* collectivism.

industrial revolution. A historical period when technical and economic development accelerated, mechanization took over from muscle power, and an urban working class emerged.

infant mortality rate. The number of infants per 1000 births who die before the age of one.

information age. The development, beginning in the 1980s, of computer technology that allowed for increased storage, accessibility, and organization of knowledge.

intercropping. The planting of one crop between the rows of another.

International Monetary Fund (IMF). An international agency that promotes monetary cooperation and the growth of trade, and makes loans available to developing countries experiencing debt problems.

internationalist view. An attitude towards aid that encourages intervention and cooperation among countries. *See also* pragmatic view, nationalist view, and humanitarian view.

Iron Curtain. The term coined by Winston Churchill after the Second World War to describe the division between Communist countries in the East and capitalist countries in the West.

irrigation. The supply of water to crop land by means of channels, streams, and sprinklers.

issue. A topic that has engendered considerable disagreement and has reached the stage of open debate.

kwashiorkor. A nutritional disease caused by a protein deficiency of diet, especially in young children in the tropics.

labour-intensive industries. Industries that require a lot of human effort and relatively few natural resources.

laissez-faire economics. An economic theory that favours the development of economic systems free from government interference.

landfill. A place where waste is disposed of by burying it under a shallow layer of ground.

leaching. The process of excess water dissolving and moving minerals down from the upper layers of the soil horizon, making the land more difficult to cultivate.

left wing. A political orientation that favours the individual or people over the state. *See also* right wing.

line of best fit. A straight line on a scattergraph, drawn between plotted ordered pairs, which shows the type of correlation.

linear correlation coefficient. A statistical calculation expressed as a three-digit decimal between 1 and -1 where the relationship between one set of data is measured in relation to another set of data.

literacy. The ability to read and write.

malnutrition. A condition in which there is a deficiency of one or more proteins, minerals, or vitamins in a diet.

marasmus. A nutritional disease characterized by a wasting away of the body and caused by a diet deficient in calories and protein; especially common in bottle-fed babies in the developing world.

marginal lands. Lands that are difficult to cultivate and that yield little profit.

materialism. The belief that material possessions are more important than social concerns.

mean. The average, calculated by dividing the sum of samples in a statistical population by the total number of samples.

melanoma. A skin cancer caused by overexposure to UV rays.

microclimatology. The study of climate conditions for a particular region that may be an anomaly in the larger region in which it is located; e.g., a location in a city may be warmer and less windy than elsewhere because it has southern exposure and is protected by buildings.

migrate. To move, permanently or semi-permanently, from one place to another, especially to a different country.

migratory. Moving from place to place; opposite of sedentary.

monarchy. A form of government in which an individual, usually a king or queen, controls affairs of state through divine right.

monoculture. The exclusive planting of only one form of vegetation in farms or in forests; examples include corn fields and white pine forests in Canada.

moratorium. A temporary prohibition or suspension of a particular activity.

multiculturalism. A government policy that encourages the development and nurturing of different cultural practices and values.

multilateral aid. Aid given by several countries to another.

multiplier effect. The economic consequences of an action, whether in terms of jobs created or extra income. When a given amount is injected into an economy, the income of the region increases by some multiplier of that amount.

nation. A group of people having common beliefs, heritage, and ancestry, not necessarily synonymous with the term *country*.

nationalism. A sense of devotion and pride for one's nation; the belief that every nation has the right to territory and independence.

nationalist view. An attitude towards aid that supports the needs of the home nation. *See also* pragmatic view, humanitarian view, and internationalist view.

natural boundaries. Borders between states that follow physical features of the land, such as rivers or mountain divides.

needs assessment. A technique used in decision making that outlines the problem and a plan of action to reach the stated goal.

negative correlation. The relationship that exists when the value of one set of data decreases as the value of another set of data increases. *See also* correlation and positive correlation.

nomadic herders. Pastoralists who constantly move from one place to another in search of grazing land and water; usually found in arid and semi-arid regions.

non-governmental organizations (NGOs). Religious groups, service agencies, and other non-profit organizations that provide aid; e.g., the Red Cross.

non-renewable resources. Resources that can not be restored or regenerated once they have been harvested. *See also* renewable resources, sustainable resources, and ubiquitous resources.

oasis. A fertile spot in a desert, where water is found.

oedema. A condition in which swelling occurs due to malnutrition and kwashiorkor.

old-growth forests. Forests that have never been harvested.

oligarchy. A form of government by a few influential individuals who control the affairs of state. *See also* aristocracy.

open-pit mines. Excavations that allow people to access mineral deposits directly, without the need of tunnels or shafts. Often, they are gigantic holes in the ground into which trucks drive to obtain the ore.

optimum population. A theoretically perfect situation in which a region has exactly the correct number of people for the resources available at a given level of technology. *See also* underpopulation and overpopulation.

ordered pairs. Two coordinates that name a precise place in a two-dimensional graph.

orographic precipitation. Rainfall, snow, etc., that falls because air is forced to rise over high ground, which causes water vapour to condense and then precipitation to occur.

overpopulation. The situation in which a region has too many people for a given resource base at a given level of technology. *See also* underpopulation and optimum population.

ozone. A layer of a form of oxygen that stretches from 15 to 50 km above the surface of the earth and shields it from ultraviolet radiation.

ozone depletion. The belief that increased chlorofluorocarbons are reducing stratospheric ozone and thereby allowing more ultraviolet radiation to reach the earth's surface.

pandemic. An epidemic that reaches such a large scale that the disease is found everywhere.

paradigm. The psychological rules and conditions we use to understand those things we perceive.

parliamentary system. A political system where elected representatives control affairs of state.

pastoralism. The practice of breeding and rearing animals for their meat, milk, and fur.

peaceful non-cooperation. The philosophy, popularized by Mahatma Gandhi, of opposing authority without using violence.

peacekeeping. A UN initiative that preserves and enforces peace between hostile nations; often undertaken to improve the chances of a peace agreement being negotiated. *See also* peacemaking.

peacemaking. A UN initiative that sides with one combatant over another through economic or military action so that peace can be achieved. *See also* peacekeeping.

pellagra. A chronic disease of the skin, gastrointestinal tract, and nervous system, caused by a lack of the vitamin niacin and the amino acid tryptophan. It is fatal if left untreated.

petrochemicals. Chemicals derived from oil and natural gas, used in the manufacture of plastics, paints, synthetic fibres, etc.

pH (percentage hydrogen) levels. Refers to the level of acidity of a substance measured on a scale of 0–14. The lower the number the higher the acidity; the higher the number the higher the alkalinity.

photolysis. A chemical reaction produced by exposure to light or ultraviolet radiation—a key reaction in photosynthesis.

photosynthesis. The process by which the energy of sunlight is used by plants to synthesize carbohydrates from carbon dioxide (in the air) and water.

phytoplankton. Minute, often microscopic, plants that live near the surface of water and are of great importance as they constitute the basis of food for all other forms of aquatic life.

political coup. The seizure of power from a government by a faction opposed to it; originally from the French *coup d'état*.

political issue. An issue that involves power struggles within and between governments.

population momentum. The tendency for high population growth to continue long after birth rates have dropped.

population projections. Predictions about how many people a place may have in the future.

population pyramid. Back-to-back bar graphs set on a vertical axis that show the age/sex breakdown of a country's population.

port of entry. The place where people or vessels enter a country's territorial boundaries.

positive correlation. The relationship that exists when the value of one set of data increases as the value of another set of data increases. *See also* correlation and negative correlation.

power grid. The system by which electrical energy is transported from power stations to users within a region.

pragmatic view. An attitude towards aid that claims to support logic and reason and discourages interference. *See also* humanitarian view, nationalist view, and internationalist view.

primary industry. An industry, such as lumbering, mining, and farming, that involves the extraction of raw materials from the environment.

protectionism. The economic theory or practice of protecting home industries.

proletariat. The collectivity of workers, especially those without capital who depend on selling their labour.

pull factors. Influencing factors that cause people to move to a country or region. *See also* push factors.

push factors. Influencing factors that cause people to leave a country or region. *See also* pull factors.

quality of life. Concerns the standard of living and the non-material things that people desire to improve the way they live.

quantify. To be measured statistically; defined as a quantity.

quaternary industry. An industry that involves the development of high-technology or information services such as software development and medical research.

rate of natural increase/decrease. The birth rate minus the death rate divided by the total population. A positive result means an increase; a negative one, a decrease.

raw materials. Materials that will be treated, prepared, or manufactured into consumer products; e.g., lumber or pulp.

renewable resources. Resources that can be restored or regenerated once they have been harvested. *See also* non-renewable resources, sustainable resources, and ubiquitous resources.

resource base. The total resources available to a population at a given level of technology.

resource-based economies. Economic regions (countries) where the extraction and exportation of one or two natural resources is the main economic activity; e.g., oil-rich Saudi Arabia.

resource issue. An issue that relates to people's use of natural resources to supply their basic needs or improve their living conditions.

revolutionary war. An internal military conflict that arises when people successfully overthrow the government in power. *See also* civil wars.

rickets. A disease of children that results in softening of the bones, caused by a deficiency of vitamin D, which leads to a poor absorption of calcium.

right wing. A political orientation that favours the state over the individual or the people. *See also* left wing.

rural. An area that is well removed from large urban centres.

sample population. A statistical cross-section of people that is representative of a given group.

scattergraph. A graphic technique used to show the type of correlation between two sets of data. If the line of best fit goes from lower left to upper right, it is a positive correlation; if the line of best fit goes from upper left to lower right, it is a negative correlation.

scurvy. A disease caused by a deficiency of vitamin C, characterized by swollen bleeding gums and the opening of previously healed wounds.

secondary industry. Industry that involves the manufacture of products from raw materials; examples include auto-making, pulp and paper, and computer manufacturing.

sedentary. Staying in one place; opposite of migratory.

selective cutting. A method of forest harvesting in which only the mature trees of a certain species are cut singly or in small groups.

sexual revolution. A period beginning in the 1970s when women asserted their rights to the same economic, social, and political freedoms as men.

sharecroppers. Subsistence farmers who pay to rent their land from the landowner with a portion of their harvest.

shifting cultivation. A subsistence farming technique in which a patch of land is cleared, crops are grown, and then the patch is deserted until the soil regains its fertility.

slash-and-burn agriculture. A subsistence farming technique in which vegetation is cut down, left to dry, and then burned down before crops are grown.

smart bombs. Bombs capable of finding their targets through a sophisticated guidance system.

socio-economic. Relating to the interaction of social and economic factors.

sodicity. A natural process that occurs in dry regions when salt rises to the surface of the soil, leaving a residue and rendering the soil unproductive.

sovereignty. A country's right to govern the people who live within its boundaries.

spatial distribution. The way in which something, such as a population, is spread out within a given space.

stakeholders. People or institutions that are affected by an issue.

standard of living. The quantity and quality of goods and services that people are able to purchase or otherwise attain to accommodate their needs and wants.

Star Wars. A defence system developed by the US that uses satellites and other extraterrestrial weaponry to track and defend against foreign military attacks.

starvation. A state of suffering caused by not having enough food to sustain life.

state. A set of sovereign institutions in a delineated territory that claims to act in the common interest of the people it represents.

static population. A population that is neither growing nor declining.

stratospheric ozone. Ozone that is found in the stratosphere (a layer of the atmosphere 10 to 50 km above the earth's surface).

structural scarcity. The situation that arises when a government is unable to distribute essential needs to the people. *See also* demand-induced scarcity and supply-induced scarcity.

structural-adjustment policies. A system of austerity programs imposed by development agencies on a developing country to allow it to reduce its foreign debt. The resultant spending cuts in social programs often cause hardship for citizens..

subsequent boundaries. Political borders that evolve together with the population they encompass. *See also* antecedent boundaries.

subsistence economies. Economic regions (countries) characterized by a lack of food and other basic resources. The predominant economic activities of the people are farming or pastoralism so that they can obtain food.

supply-induced scarcity. The situation that arises when the environment has been so degraded that it can no longer supply people's basic needs. *See also* demand-induced scarcity and structural scarcity.

sustainable resources. Resources that are harvested in such a way that they are able to regenerate. *See also* non-renewable resources, renewable resources, and ubiquitous resources.

sustainable development. Economic development that attempts to manage the environment and resources in a way that will allow future generations to use them.

tar sands. An unconventional oilfield where the oil is mixed with sand.

tariffs. Taxes placed on imports to reduce competition with domestically produced goods; a form of protectionism.

terraces. Steps cut into the hillsides to allow for cultivation on slopes that would normally be too steep.

terrene. Relating to the earth.

tertiary industry. A service industry such as medicine, education, or accounting.

theocracy. A form of government, said to be by God, based on religious laws or controlled by religious leaders.

tied aid. Aid given by one country to another in exchange for economic or political concessions.

total deviation. The sum of deviations for each member of a statistical population.

totalitarian. A dictatorship that regulates every aspect of state and private behaviour.

trade balance. The state of a country's trading position; essentially, the balance of the value of exports and imports.

trade deficit. Where the value of exports is less than the value of imports.

traditional. Pertaining to the customary way a nation carries out a process over a long period of time.

transhipment. Products moving from one mode of transportation to another.

transnational. Corporations that operate in many countries.

ubiquitous resources. Resources that are always present such as soil, air, and water. *See also* non-renewable resources, renewable resources, and ubiquitous resources.

ultraviolet radiation. Very short electromagnetic waves coming from the sun that can prove harmful to plant and animal life if not filtered by stratospheric ozone.

unconventional oilfield. An oilfield where extraction is complicated by geology; e.g., the tar sands of northern Saskatchewan and Alberta.

underpopulation. The situation in which a region has too few people to develop fully a given resource base at a given level of technology. *See also* overpopulation and optimum population.

urbanization. The migration of rural populations into towns or urban areas (cities).

UV rays. *See* ultraviolet radiation.

value-added. The increased worth given to a product because of the labour or enhancement that has gone into its production.

variance. The difference between the mean and each sample in a statistical population expressed as an integer.

wadi. In arid areas, a valley or stream course that is usually dry but may sometimes have a stream after a burst of heavy rain.

weighting factor. A numeric value assigned to a criterion in the decision-making process, based on a person's value system.

xenophobia. A fear and distrust of outsiders or strangers.

xerophyte. A plant that has adapted to growing in arid conditions.

zero population growth. When the birth rate and the death rate are equal, so the population does not grow naturally.

INDEX